Science – A Challenge to Philosophy?

NORDEUROPÄISCHE BEITRÄGE
aus den Human- und Gesellschaftswissenschaften

SCANDINAVIAN UNIVERSITY STUDIES
in the Humanities and Social Sciences

Herausgegeben von / Edited by Hartmut Schröder

Bd./Vol. 27

PETER LANG

Frankfurt am Main · Berlin · Bern · Bruxelles · New York · Oxford · Wien

Heikki J. Koskinen/Sami Pihlström/Risto Vilkko
(eds.)

Science – A Challenge to Philosophy?

PETER LANG
Europäischer Verlag der Wissenschaften

Bibliographic Information published by Die Deutsche Bibliothek
Die Deutsche Bibliothek lists this publication in the Deutsche Nationalbibliografie; detailed bibliographic data is available in the internet at <http://dnb.ddb.de>.

ISSN 0942-0657
ISBN 3-631-54260-7
US-ISBN 0-8204-7757-5

www.peterlang.de

Contents

II Perspectives on Naturalism

III Science, Values, and Literature

Preface

This volume is based on papers presented on the occasion of the XV Internordic Philosophical Symposium, "Science – a Challenge to Philosophy?", which was jointly organized by the Nordic Institute of Philosophy and the Philosophical Society of Finland in Helsinki in May, 2004. It covers a number of important and timely philosophical issues: naturalism – its strengths, weaknesses, and limits – in metaphysics, epistemology, and the philosophy of science; the relation between philosophical and scientific methodology; the ethics of science and, more generally, the place of science in society; the relation between the natural sciences and the humanities; as well as the ways in which scientific progress may challenge traditional philosophical concepts and problems (e.g., metaphysical categories, epistemic norms, ethical values, or religious beliefs) and, on the other hand, the ways in which philosophical arguments and/or artistic innovations – or philosophical interpretations of them – may challenge some of the claims made in the name of science.

The core of the symposium program consisted of eight invited plenary talks by distinguished philosophers from Canada, Germany, the United States, Sweden, Norway, Denmark, Iceland, and Finland. In other words, one speaker was invited from each of the five Nordic countries, i.e., countries regularly involved in organizing the biannual Internordic Philosophical Symposia, and three from outside Scandinavia. Most of the invited speakers kindly submitted their papers for publication in this volume: Cheryl Misak (University of Toronto), Kenneth R. Westphal (University of East Anglia), Ólafur Páll Jónsson (University of Iceland), Gunnar Svensson (University of Stockholm), and Mogens Pahuus (University of Aalborg). Among the parallel session papers of the symposium, there were contributions from several non-Scandinavian countries: Canada, Estonia, Germany, Italy, Spain, United Kingdom, and Turkey. All the Nordic countries, and Finland in particular, were, naturally, well represented. Altogether forty-one papers were presented in the meeting; the best of the contributed papers have been selected for publication in this anthology. In any event, both the symposium and this proceedings volume can be said to be truly international, not simply inter-Nordic.

Why did we choose the general title we did, with the question mark? First, it might be noted that the Internordic Philosophical Symposia have in recent years been fairly metaphilosophical. They have carried titles such as "Philosophy with a Human Face" (Reykjavík, 1998) and "Philosophy at the Turn of the Millennium" (Bergen, 2000). Moreover, as the XIV Symposium, which took place in Stockholm in 2002, examined the relation between philosophy and religion, the organizers felt that it would be a good idea to talk about science in the following Symposium. We did not, however, want to organize a conference exclusively dedicated to technical issues in the philosophy of science. Rather, through the choice of our topic we are suggesting that science, and especially the relation between science and philosophy, can and should also be approached from a more general philosophical and metaphilosophical point of view. The notion of naturalism, popular in (at least analytic and post-analytic) philosophy especially since W. V. Quine's famous 1969 paper, "Epistemology Naturalized", might also have been mentioned in the title, given its importance in many of the contributions to this volume, but it is used in such a variety of different ways and is often so carelessly defined that we decided it would be better to leave it out from the title and let the papers themselves throw light on how this concept should (or should not) be used and what kind of challenges naturalism poses for us today.

Of course, not all the papers included in this volume are "metaphilosophical" in nature. We have also included a number of historically oriented papers, as well as papers on both general and more specialized philosophy of science. In many cases, there is no clear demarcation between the philosophy of science, on the one hand, and such traditional fields of philosophical reflection as metaphysics or epistemology. Topics such as intentionality and normativity – central in the humanities – as well as the place of science in society and the relation between science and values are taken up in several contributions. There are also articles specifically on naturalism (including Quine's philosophy, in particular) and on the philosophy of mind. Obviously, the rise of naturalism – as well as some philosophers' worry that naturalists are going too far in their censure of traditional foundationalist and aprioricist philosophy – is a phenomenon that cannot be neglected in philosophy today. The place of science in the philosophical reflections of the twentieth century was so central that hardly any philosopher can simply overlook the conceptual, methodological, and ethical problems raised by the advancement of scientific knowledge, especially when applied to the "human world". The papers in this volume, as different as they are, contribute to this general task of evaluating the philosophical importance of scientific progress, both in its positive and negative aspects, particularly in relation to the classical philosophical task of understanding the place of human beings in the world.

We have organized this book in three parts. Part I, "Mind, Metaphysics, and the Philosophy of Science", contains ten papers on general issues in metaphysics, epistemology, the philosophy of mind, and the philosophy of science, in many cases with a historical orientation. Historical and systematic approaches are often combined. Part II, "Perspectives on Naturalism", contains eleven papers on various aspects of naturalism in semantics, epistemology, and the philosophy of science; almost all of them touch Quine's work in one way or another. Part III, "Science,

Values, and Literature", adopts a broader perspective, with six contributions on moral realism, on the relations between philosophy, science, and literature, and on the place, task, and justification of values in scientific practices.

There are several institutions and individuals we want to acknowledge for their kind assistance in making the symposium and, finally, this book possible. First of all, we would like to thank all the plenary speakers who kindly accepted the invitation to give a talk in the symposium, as well as all the contributing speakers, many of whom contributed to this volume. We also wish to thank the three other members of the organizing committee, that is, Professor Leila Haaparanta, Professor Lars Hertzberg (who has been actively involved in the Nordic Institute of Philosophy much longer than any of us), and Dr. Arto Siitonen. Sincere thanks are also due to the students who gave us hand with all kinds of practical matters: Sanna Hirvonen and Anna Myllymäki. We would also like to thank the Philosophical Society of Finland for acting as the local organizer of the symposium, particularly the President of the Society and the Rector of the University of Helsinki, Professor Ilkka Niiniluoto, who hosted a memorable reception. The symposium was financially supported by the Academy of Finland and the Finnish Cultural Foundation, for which we are deeply grateful. In the initial stages of the planning of the meeting, we also received some useful practical advise from the organizers of the previous Internordic Philosophical Symposium: Professor Hans Ruin and his colleagues at Södertörns högskola, Stockholm. We also thank Professor Hartmut Schröder for the decision to publish this proceedings volume in the Peter Lang special series, *Scandinavian University Studies in the Humanities and Social Sciences*, and Ute Winkelkötter (Peter Lang, Hamburg office) for her help in this process and for her efficient handling of technical matters. In the final stages of the preparation of this volume, Ms. Auli Kaipainen at the Department of Philosophy, University of Helsinki, has given us invaluable technical assistance.

With this volume, we want to contribute to the current lively discussions on the relations between science, philosophy, and society, and to provide the public with an enjoyable selection of papers drawn from a stimulating symposium.

Helsinki, September 2005

Heikki J. Koskinen, Sami Pihlström, and Risto Vilkko
(The Editors)

I

Mind, Metaphysics, and Philosophy of Science

Common Experience, Scientific Intelligence, and the Literary Spirit. Reflections on Charles S. Peirce's Conception of Philosophy

Mats Bergman

Charles S. Peirce was both a practising scientist and a philosopher. Although his achievements in philosophy have tended to overshadow his work in sciences such as experimental psychology and metrology, many commentators have emphasised the relevance of Peirce's scientific labours for his thought. Indeed, Kenneth L. Ketner (1987) has even maintained that Peirce was *primarily* a scientist and mathematician, and only secondarily a philosopher.

Although Ketner's claim could be disputed on the ground that Peirce's contributions to philosophy far exceed his scientific accomplishments, it would be difficult to deny that Peirce's conception of philosophy is intimately connected to his active engagement in the life of science. Certainly, Ketner's viewpoint seems to receive ample support from Peirce, who describes himself as a "scion of natural science" (MS 326:4 [late]).[1] Furthermore, Peirce urges philosophers to adopt the methods of the successful scientific disciplines, and in a letter to William James, he bluntly asserts that "philosophy is either a science or is balderdash". Based on such statements, one could easily conclude that his agenda is scientistic in the strong sense of the term.

From a contemporary perspective, the notion that philosophy is or should be a science may seem misguided or even dangerous (Colapietro 1998, p. 248). In particular, a strictly scientific conception of philosophy would appear to exclude a

1 I refer to Peirce's texts using abbreviations. *CP x.y* refers to *The Collected Papers of Charles Sanders Peirce*; *v* indicates volume number, *p* paragraph number. *EP v:p* refers to *The Essential Peirce: Selected Philosophical Writings*; *v* indicates volume number, *p* page number. *MS m:p* refers to an original manuscript; *m* indicates manuscript number, *p* page number (a $_b$ after the page number indicates that it has been obtained by the author by counting manuscript pages, a $_v$ indicates a variant page, while a $_d$ indicates that Peirce appears to have discarded the page in question). *NEM v:p* refers to *The New Elements of Mathematics by Charles S. Peirce*; *v* indicates volume number, *p* page number. *RLT p* refers to *Reasoning and the Logic of Things*; *p* indicates page number.

large part of that which is commonly called "philosophy". The fact that Peirce sees the practice of natural scientists as an exemplar for all scientific inquiry does nothing to reduce such fears. This worry is succinctly expressed by Vincent Colapietro (1998, p. 251): "to assimilate too thoroughly the work of philosophers to that of scientists, especially natural scientists (as Peirce would have us do) tends to occlude the character of philosophy as an ongoing quest for critical self-consciousness, one undertaken by participants in the intersecting practices dominant in their historical time and, in some form, perhaps even discernible throughout human history".

Yet, Peirce also maintains that the philosophical point of view differs from that of mathematics and the so-called special sciences, which include the natural disciplines. This contention may appear to clash with the scientific approach; but in fact, Peirce identifies philosophy as a distinctive type of science, delimited by its methods of observation and objects of knowledge.

In this article, I first outline what Peirce means by science in general. This enables us to see what his requirement that philosophy should be pursued in a scientific spirit really entails. Next, I pinpoint what he sees as the central distinguishing marks between the science of philosophy and the other sciences. I think that the picture of the relationship between science and philosophy that emerges from these considerations is a largely consistent and a quite interesting one – which is not to say that it is without faults, of course. Although I argue that Peirce's conception of philosophy is more inclusive than his natural scientific stance immediately suggests, it can still be criticised for excluding too much. Therefore, I conclude my discussion with a brief consideration of his apparently negative view of the possible contribution of literature to philosophy.

1. Peirce's General Conception of Science

In numerous writings, Peirce identifies three basic views of science by distinguishing different uses of the word "science":

(1) According to the classic view, "science" denotes certain knowledge (EP 2:372 [c. 1906]). In other words, the term refers to the demonstrative character of certain cognition, or "knowledge through principles" (MS 618:2 [1909]; MS 339:267 [1905]). This conception has very little to do with the actual practice of scientific inquiries.

(2) On the other hand, "science" can also mean systematised knowledge (EP 2:372 [c. 1906]; MS 618:2–3 [1909]). Such an understanding of science is common among non-scientific people, and is often repeated in dictionary definitions (MS 965). According to Peirce (EP 2:372

[c. 1906]), this sense of "science" can be traced to Samuel Taylor Coleridge.

(3) Finally, Peirce claims that for practising scientists "science" means "the concrete body of their own proper activities" (EP 2:372 [c. 1906]). This idea of science is more focused on the methods used than on the knowledge acquired; but genuine scientific investigation cannot be codified as a given methodological toolbox (cf. CP 6.428 [1893]). Rather, the emphasis is placed on the *life* of inquiry.

It is obvious that Peirce opts for the third conception; it is, in fact, connected to his own definition of *science of discovery*, which he distinguishes from *practical science*, inquiry aimed at utilities, and *science of review*, or Coleridgean science. In the following discussion, "science" will be understood in the heuristic sense.

According to Peirce, there is one thing that is *absolutely* indispensable to genuine research, namely the unfaltering *desire* to know and learn – the *scientific spirit* or *interest* (CP 6.428 [1893]; MS 860:2_b [c. 1896]; CP 1.235 [c. 1902]; MS 693:48 [1904]; MS 618:3 [1909]; MS 326:6 [late]; MS 601:25_v [late]). In a sense, it all follows from the simple principle that "in order to learn you must desire to learn and in so desiring not be satisfied with what you already incline to think" (RLT 178 [1898]). Indeed, as Susan Haack (1997, p. 241) has pointed out, this is Peirce's "First Rule of Reason" – even more fundamental than his famous maxim "do not block the way of inquiry".

Purportedly, the moral attributes that are "the most *vital* factors in the method of modern science" (CP 7.87 [1902]) are properly expressed by the consequences of the first rule. One of these is the imperative that forbids the obstruction of inquiry. Another implication is that one ought *honestly* to concede that one's knowledge has flaws (MS 860:2_b [c. 1896]; CP 1.13 [c. 1897]). The desire to learn the truth entails a willingness to revise one's beliefs in the light of evidence, and the readiness to have one's errors corrected by critical examination (MS 339:267 [1905]). This requires *sincerity*, on the one hand, and *diligence*, on the other.

> Science consists in the sincere and thorough search for truth according to the best available methods. Its only quite indispensable condition is the absolute single hearted energy with which it works to ascertain the truth, regardless of what the character of that truth may be. It is not science if is not an intelligently directed research. But it will come to be so if it is absolutely sincere and highly energetic. These dispositions will generate the intelligence required. (NEM 4:xix; cf. NEM 4:227 [1905–6]).

In a sense, scientific intelligence is a product of inquiry, rather than the other way around. This claim would be rather peculiar, if we were to understand "intelligence" to refer merely to the personal mind. Clearly, Peirce must hold that the intelligence of scientific research is largely *social*, an attribute of the community that can be said to constitute *a* science.

What does this alleged sociality of science entail? On the one hand, Peirce argues that it is a *sociological* fact that science entails "the total principal industry of a social group", which is bound together by a common object of interest, shared understanding of methods, and a devotion to the pursuit (MS 655:16 [1910]; cf. NEM 4:227 [1905-6]; NEM 3:232 [1909]). The studies of a single human being are not science; science involves a wider *communicational* community (MS 1334: 12-13 [1905]). And while cooperation and the open sharing of results are important manifestations of the social nature of science, Peirce notes that these do not suffice fully to spell out the concrete social nature of scientific inquiry; true science requires the possibility of critical examination by peers (MS 615:10 [1908]). No discovery or item of knowledge – no matter how true – can be considered as a part of science before it is open to the "kind but searching and inflexible" criticism of the relevant social group (MS 615:9-10 [1908]).

On the other hand, the community of inquirers is not restricted to the colleagues of one's generation; it is, in a significant sense, *unlimited*. Obviously, many scientific practitioners never find the truth concerning the matters they investigate. The genuine scientist is not thrown off track by this fact; the group within which he or she works is a *goal-directed* community, ideally united by a shared desire to find the truth concerning certain matters. Given that many questions will never be answered in one's lifetime, it is primarily for the benefit of *future* inquirers that such an unselfish scientist labours, fuelled by a hope that truth is something that could be discovered, given enough time and effort.

> [...] the method of modern science is social in respect to the solidarity of its efforts. The scientific world is like a colony of insects, in that the individual strives to produce that which he himself cannot hope to enjoy. One generation collects premises in order that a distant generation may discover what they mean. When a problem comes before the scientific world, a hundred men immediately set all their energies to work upon it. One contributes this, another that. Another company, standing upon the shoulders of the first, strike a little higher, until at last the parapet is attained. (CP 7.87; cf. CP 6.3 [1898])

It is because of its orientation toward the future that science possesses a living character. Furthermore, while the life of inquiry involves a sincere striving for true knowledge, it is essential for the very being of science that it is not conceived to be immune to error; as "a living thing, animating men, [science] need not be free from error, – nor can it be" (MS 965).

As we have seen, the Peircean scientific ethos requires a willingness to accept individual limitations. Now, if we look at the matter from a different perspective, we find that it is precisely this *fallibility* of human knowledge that feeds the growth of science; scientific inquiry "cannot exist as long as people think they know already or think they have an infallible teacher" (MS 965). Being fallible, human beings naturally tend to combine their efforts in their pursuit of truth. Consequently, Peirce asserts that "the most essential element of the spirit of inquiry is a swiftness to see that you have been in the wrong" (MS 860:2_b [c. 1896]). It is but a different way of saying that the essence of science is found in the passion to learn.

Although there would be much more to say about Peirce's general conception of science, we may halt here and consider what this approach would entail for philosophical inquiry. If philosophy is to be a science in the Peircean sense it ought to be driven by a genuine desire to learn, be a sincere pursuit of truth, be a social rather than an individual activity, be future-oriented, and involve a recognition of the fallibility of the whole undertaking. We may consider these requirements in turn.

To begin with, it may be noted that the demand that philosophy ought to be fuelled by a desire to learn is quite uncontroversial, perhaps even trivial. Presumably, most philosophers at least claim they are driven by such a passion. However, Peirce's viewpoint is not quite so bland. When combined with the second demand, that philosophy should involve a sincere pursuit of truth, it does in fact delimit the field in certain ways. Peirce himself criticises "seminary" philosophers, who are mostly interested in defending a certain creed (see, e.g., CP 1.128 [c. 1905]). In our day and age, this scientific conception would perhaps exclude certain so-called postmodernists, but also the kind of neopragmatism that maintains that the aim of philosophy is not to strive for truth, but rather to keep up the conversation. Peirce would certainly disapprove of Richard Rorty's claim that the only constraints on philosophical investigation are conversational (see Rorty 1980, pp. 389–394; Colapietro 1998, pp. 265–267).

The requirement that philosophy ought to be a social mode of activity, rather than an individual pursuit will, perhaps, prove to be more problematic, or at least it will seem to be in conflict with the traditional view of the philosopher as a solitary seeker for wisdom. On the other hand, the call for public criticism and peer review is hardly controversial; at least, it is the model that most of contemporary philosophy strives to follow. The more interesting aspect of Peirce's stress on the social character of science is related to the notion that scientific claims are future-oriented and fallible. Albeit anti-foundational, Peircean philosophy may well be cumulative.

Some of Peirce's pragmatistic statements could easily be interpreted as a variant of anti-metaphysical positivism – perhaps even be taken to suggest that the mission of philosophy is to eliminate "pseudo-problems", thus paving way for proper, empirical science. This is an erroneous conclusion; in fact, a closer look at Peirce's characterisations of philosophy will reveal that he conceives of philosophy as a unique mode of inquiry.

2. Philosophical Common Sense

In Peirce's classification of science, philosophy[2] is placed between mathematics and special science. Of these, mathematics is purportedly not a positive discipline. That is, it is a hypothetical science and not based on positive experience, although it involves operations on and observations of imaginary diagrams. Philosophy, in contrast, is a positive science, a quality it shares with the special sciences. However, philosophical investigation is experiential in a particular sense.

Reflecting upon experience from the point of view of inquiry, Peirce distinguishes two different *scientific* senses of "experience". In the special sciences, experience is "that which their special means of observation directly bring to light, and it is contrasted with the interpretations of those observations which are effected by connecting these experiences with what we otherwise know" (CP 7.538). However, in the philosophical sense, experience is not something that requires specifically developed resources of observation.

> [...] in philosophy there is no special observational art, and there is no knowledge antecedently acquired in the light of which experience is to be interpreted. The interpretation itself is experience. (7.527)

In other words, the "experience" of philosophy is everyday experience, which is constantly had, and which therefore requires no other means of observation than the natural cognitive capabilities of human beings. However, this does not mean that the common experience, which is the proper object of observation of philosophy, would be under direct conscious control; in fact, the opposite applies. Peirce often defines "experience" as "the cognitive resultant" of life. That is, experience is a *determination* of belief and cognition that life forces upon a human being; one "may lie about it; but one cannot escape the fact that some things *are* forced upon his cognition". (CP 2.138 [c. 1902])

Although interpretational or inferential, the cognitions are experiential because they are *compulsive*. We are not simply fed atomic experiences, from which we then build reasonable conceptions by interpretations; rather, there are interpretations we cannot avoid making, *as they occur*. We look around and see things

2 In order to simplify matters, I will just speak of "philosophy", and not pay attention to Peirce's division of the field of philosophical inquiry into phenomenology, aesthetics, ethics, logic or *semeiotic*, and metaphysics.

such as people, chairs, and computers. At first, these experiences seem to be simply given; and so they are, at the very moment of their occurrence, as these things that are undeniably there whether we want to or not. It is only in later reflection that the interpretational character of the experiences can become evident. This character of *opposition* is typical of all experiential objects. Even imaginations present a certain degree of resistance, once they have been imagined.

Now it may reasonably be asked whether Peircean philosophy is truly scientific in the Peircean sense, as it appears to place much weight on the observations of an individual. Of course, philosophy may fulfil a part of the social requirements by making its results public and openly criticisable; but one might still contend that its objects of study are too private to form the basis for a truly scientific community. Peirce would deny the validity of such counter-arguments on the basis that experience in the broad sense is felt to be collective.

> The course of life has developed certain compulsions of thought which we speak of collectively as Experience. Moreover, the inquirer more or less vaguely identifies himself in sentiment with a Community of which he is a member, and which includes, for example, besides his momentary self, his self of ten years hence; and he speaks of the resultant cognitive compulsions of the course of life of that community as Our Experience. (CP 8.101 [1900])

Moreover, Peirce argues that a person is not quite as individual as we are inclined to think. In fact, human ideas and cognitive competences are largely social. Given this continuity between person and community, "a man is not radically devoid of the power of saying what every scientific intelligence must observe, if he has the power of saying what he observes himself" (NEM 4:ix). This kind of observation is not flawless; in fact, one could argue that it is *highly* fallible, as it is directed toward the very means of observation and cognition themselves and requires rather bold generalisations.

Admittedly, there is a sense in which the experience examined by the philosophical sciences is simply given. Being "irresistibly forced upon us in the course of life" (MS 1336:1), such experience is beyond conscious control. However, while we are not in charge of experiences *as* they happen, their contents can later be reviewed critically; more specifically, we can examine the factors involved in interpretation.

Yet, Peirce argues that we cannot begin by a wholesale doubt of all of our cognitions, but must start out from the beliefs and prejudices that we actually have. Any one of our beliefs can be erroneous, but to assume that all of our common-sense beliefs are mistaken is impossible; that would simply leave us with nowhere to stand and lead to impotent despair. In other words, the common-sense knowledge we possess must be taken as a starting-point for philosophical inquiry (CP 1.129 [c. 1905]). Such beliefs are mostly *practically* indubitable. For instance, it is hardly possible to live as a human being and deny the existence of a reality external to our thoughts and representations of it. This does not mean that the belief in question would serve as a certain foundation, but merely that it cannot be doubted in view of the habits of cognition that have been forced upon us.

Philosophy is in this regard neither privileged nor disadvantaged in comparison to the special sciences. According to Peirce, positive science can only rest on experience, but he adds that "experience can never result in absolute certainty, exactitude, necessity, or universality" (CP 1.55 [c. 1896]). In his memorable metaphor, science is walking on a bog, and can only say that the ground it stands upon seems to hold for now (RLT 176-177 [1898]). This is true of philosophy as well as of chemistry and of the discipline of history.

3. Literature as a Philosophical Laboratory

Viewed on a suitably abstract level, Peirce's programme of scientific philosophy possesses considerable appeal. It places the philosophical endeavour within the broader context of science, but avoids the reductionism – even nihilism – typical of many scientistic and "descriptive" approaches, which cast across-the-board doubt on the validity of any special knowledge claims of philosophy. Unsurprisingly, things become much more complicated when we turn to more specific areas of philosophical inquiry, and enquire whether the Peircean conception could accommodate these.

It is not possible to scrutinise the detailed ramifications of Peirce's programme within the confines of this article, but we may conclude our examination with a few remarks on one area that seems to be categorically excluded from its purview. Namely, Peirce occasionally contrasts the scientific method to "studying in the literary spirit", and in fact suggests that there is a need "to rescue the good ship Philosophy for the service of Science from the hands of lawless rovers of the sea of literature" (EP 2:352 [1905]). Accordingly, it seems that Peirce would advocate a sharp distinction between philosophy and literature, an opinion that some of our contemporaries certainly share, but which no doubt would be condemned by the increasing number of intellectuals – notably neopragmatists – who argue that philosophy *is* nothing but a form of literature.

Here, then, appears to be one tangible but controversial upshot of Peirce's scientific programme: the elimination of the literary character from philosophy. However, we should take note of the context of his condemnation; Peirce's most vehement attack on the literary approach is connected to the emergence of the pragmatist movement in the early 1900s. In short, he felt that the term "pragmatism" was being hijacked by "pirates", especially by writers in certain literary journals (see EP 2:334 [1905]).

In fact, Peirce's anti-literature attitude is not as strict as it may appear at first blush. He is no stranger to employing a variety of literary forms, such as dialogues, in his philosophical writings. Moreover, Peirce is in fact more concerned with the proper means of expression than most philosophers, as his writings on the rhetoric of science show (see, e.g., EP 2:325–330 [1904]). True, most of his efforts along this line stress the importance of devising proper technical terminologies, and thus serve to underline the difference between philosophy and literature; but he also concedes that certain indefinite truths or concepts may need to be expressed metaphorically. Thus we find Peirce asserting – *against* those who take physical science to be the only route to truth – that poetry is capable of communicating truths that elude more precise expression (MS 634:18 [1909]; cf. CP 1.315 [1903]). Moreover, not "only metaphysics, but logical and phaneroscopical [i.e., phenomenological] concepts need to be clothed in such garments", for "a pure idea without metaphor or other significant clothing is an onion without a peel" (EP 2:392 [c. 1906]). Indeed, Peirce maintains that it is necessary that the truth should have figurative garb; consequently, it would appear that he cannot be taken to call for the elimination of *all* literary aspects of philosophy.

Given these considerations, it seems safe to say that Peirce's disapproval of the literary spirit should be understood as a part of his criticism of the kind of philosophy that has no interest in furthering inquiry – that is pursued for other kinds of motives. So interpreted, his scientific programme would not entail a full-scale refusal to acknowledge connections between philosophy and literature. At least, to the extent that a literary work can be said to contribute to the development of knowledge, it can be said to partake of the nature of science in the broad sense. As fiction functions as a trainer of imagination, it may pave the way for scientific inquiry, for as Peirce observes, a "pretty wild play of the imagination is [...] an inevitable and probably even a useful prelude to science proper (CP 1.235 [c. 1902]).

Nonetheless, in spite of these clarifications and qualifications, it may be argued that one of the shortcomings of Peirce's conception of philosophy is its neglect of the philosophical potential of literature. Earlier, we briefly noted that philosophy includes the critical inspection of the stream of our experiences, typically in the form of interpretations. However, if this flow of experiences is beyond criticism as it happens, we may well ask by what means philosophy can pursue critical examinations of it. Peirce addresses this question only partly in his writings, but he does suggest that philosophers ought to employ thought experiments; or, as he sometimes expresses the matter, draw imaginary diagrams in the mind, in order to see what would follow under various circumstances. In fact, this "laboratory method" should be added to the central components of Peirce's scientific conception of philosophy; albeit the experimental approach is not as vital a part of science as the desire to learn and the social impulse, it is no doubt one of the features of "the successful sciences" that he feels that philosophers ought to adopt.

Now what I would like to suggest is that fictional literature – and perhaps other forms of art as well – can play an important role as resources for thought experiments. No doubt, this would be most useful in ethics. According to Peirce's admittedly rather sketchy account of the normative field, the main function of ethical investigation is the critical assessment of deliberate action (MS 634:130 [1904]). Yet, our capacity to control our actions is limited; when we act, we often feel free to choose, but our conduct is in fact to a large extent determined by natural dispositions and ingrained habits. This is not a deplorable situation, but rather a fact of life; we are simply not able to completely step outside of ourselves. Yet, human beings possess the capacity to reflect on ideals, and can to a limited extent modify and control their own habits, using imaginary experiments to predict the results of various courses of action (see EP 2:246 [1903]). Such experimentation is in fact a prerequisite for rational development; "where there is no self-control, nothing but the normal is possible" (CP 4.540 [1906]). Obviously, this kind of self-examination is nothing extraordinary; rather, it is a part of daily human experience. Peircean ethics can be seen as a means for improving this natural ability, offering a space where various kinds of higher-level goals and ideals, including possible future selves, can be examined and criticised systematically. Importantly, fiction allows us to manipulate and play with elements of common experience, and can thus function as a kind of virtual laboratory.

I realise that this view of the philosophical possibilities of literature is hardly original; but I think that the picture sketched above would fit quite nicely within the frame of Peirce's science. If we allow Peircean philosophy to expand in this direction, it will be possible to dispel at least a part of the fear that its scientific outlook is too narrow to embrace the development of critical self-consciousness.

University of Helsinki

References

Brunning, J. and P. Forster (eds.) (1998), *The Rule of Reason: The Philosophy of Charles Sanders Peirce*, Toronto, University of Toronto Press.

Colapietro, V. M. (1998), "Transforming Philosophy into a Science: A Debilitating Chimera or a Realizable Desideratum?", *American Catholic Philosophical Quarterly* LXXII (2), 245–278.

Haack, S. (1997), "The First Rule of Reason", in J. Brunning and P. Forster (1998), pp. 241–261.

Ketner, K. L. (1987), "Charles Sanders Peirce: Introduction", in J. J. Stuhr (1987), pp. 13–25.

Peirce, C. S. (1931-58), *Collected Papers of Charles S. Peirce*, 8 vols., ed. by C. Hartshorne & P. Weiss (vols. 1-6), and A. Burks (vols. 7-8), Cambridge, Mass., Harvard University Press.

Peirce, C. S. (1976), *The New Elements of Mathematics*, 4 vols., ed. by C. Eisele, The Hague, Mouton.

Peirce, C. S. (1992), *Reasoning and the Logic of Things: The Cambridge Conferences Lectures of 1898*, ed. by K. L. Ketner. Cambridge, Mass., Harvard University Press.

Peirce, C. S. (1992-8), *The Essential Peirce: Selected Philosophical Writings*, 2 vols., ed. by N. Houser & C. Kloesel (vol. 1), and The Peirce Edition Project (vol. 2), Bloomington, Indiana University Press.

Rorty, R. (1980), *Philosophy and the Mirror of Nature*, Oxford, Blackwell.

Stuhr, J. J. (ed.) (1987), *Classical American Philosophy: Essential Readings and Interpretive Essays*, New York, Oxford University Press.

Phenomenology – a Challenge to Science

Juha Himanka

This presentation has three parts. *First* I consider Edmund Husserl's view on the relation between philosophy and science. In the *second* part, I present what follows when Husserl puts these views in action. I use Husserl's investigation of the earth as an example. In the *last* part, I reflect on what we should do when there is a real conflict between the results of a philosophical and a scientific investigation.

1. Phenomenology and Science According to Husserl

My background in this presentation is Edmund Husserl's philosophy. There are many interpretations of the philosophical approach that Husserl called phenomenology. This is no wonder, as the *Husserliana*, Husserl's collected works, actually contains a very wide range of different approaches. I think that, in the end, one has to acknowledge that there are different motivations and different attitudes in play within Husserlian thought. The methodological step that starts the phenomenological work, however, is actually a change of attitude. In my eyes, the final approach to *Husserliana* means that we locate these reductions and learn to see how the change of attitude takes place within Husserl's texts.[1]

Husserl calls these attitudes between which the change takes place natural and phenomenological. The natural attitude includes the theoretical and experimental approach of the sciences. In reduction, we overcome this scientific point of view – but this is done in order to reach an even more scientific viewpoint, the point of view of rigorous science (*strenge Wissenschaft*).

What is strange in this approach is that, according to Husserl, we should distance ourselves from the scientific approach in order to reach this even more scientific viewpoint. Husserl states this in a radical manner in his lectures on the *Idea of Phenomenology* (1999).

In the first lecture, Husserl makes two radical claims about scientific knowledge:

(1) There is no difference between levels of knowing:

1 Himanka (1999).

"Exact knowledge is no less enigmatic than in-exact knowledge, scientific no less than prescientific" (ibid., p. 20).
"The most rigorous forms of mathematics and mathematical natural science here have not the slightest advantage over any actual or alleged knowledge belonging to common experience" (ibid., p. 21).

(2) Philosophy or phenomenology differs radically from knowing and science or the natural attitude.
"In comparison to all positive knowledge, philosophy, –, lies in a new dimension; and to this new dimension corresponds a fundamentally new method which is to be contrasted with the 'natural' method. [...] Anyone who denies this has failed to understand the peculiar level at which the problems of the critique of knowledge must be posed, and thus has failed to understand what philosophy actually wants to accomplish – and should accomplish – and what gives philosophy, as opposed to all positive knowledge and science, its proper character and authority" (ibid.).

Husserl does not want to start philosophy from the highest forms of knowing: phenomenology does not "engage in theory or mathematical construction" (ibid., p. 43). Instead, "*[p]henomenology carries out its clarification in acts of seeing, determining, and distinguishing sense*" (ibid.). Phenomenology tries to reach the prescientific and pretheoretical level that is supposed by sciences but cannot be reached with their method: phenomenology "ends where objectifying science begins" (ibid.).

2. The Example

Within Husserl's *corpus*, the tension between the phenomenological and the prevailing view of our culture is perhaps most clearly apparent in his late text on the earth, "Grundlegende Untersuchungen zum phänomenologischen Ursprung der räumlichkeit der Natur" (1940) – in what follows, I will refer to this text as *Investigations*. This tension is situated between *our* experienced earth and the experimental and theoretical earth of the scientists, between the phenomenological and the natural attitude.

Both of these views on the earth – phenomenological and scientific – are right in their own right, but their justifications are different. When justifying either of these two options, one also has to put forward one's own view on how reality becomes real or what makes real real. The main possibilities to choose from are concrete phenomenology, where the tests for reality are experience and presence, and theoretical science, where the tests are experiments and calculations.

Our task is now to follow through Husserl's reduction in the *Investigations*. The first step is to distance ourselves from the natural views we are used to. Husserl calls this phase epoché.

2.1. Step one: epoché

Our point of view within the natural attitude towards the earth is called Copernican. The main thesis of this theory is that the earth is a physical body, and as a body it can move or rest. In his *Investigation*, Husserl attacks this view directly: as we never see or experience the earth as a body, how can we state that it is a physical body? Here we need to elaborate a bit further than Husserl. I state three theses of natural attitude in relation to the earth and question them within the scientific approach.

Thesis I
It is a fact that the earth revolves around the sun; or that Copernicus's view is correct and Ptolemaios's is wrong.

Question 1
Fred Hoyle states in his book *Nicolaus Copernicus* (1973, p. 79) that "Today [1973] we cannot say that the Copernican theory is 'right' and Ptolemaic theory 'wrong' in any meaningful physical sense. The two theories are [...] physically equivalent to one another."

Thesis II
Since Foucault's experience in 1851 it has been a fact that the earth rotates about its axis.

Question II
Ernst Mach writes in his *Science of Mechanics* (1960, p. 283): "If the earth is affected with an *absolute* rotation about its axis [...] the plane of Foucault's pendulum rotates. [...] This, indeed, is the case if we start *ab initio* [*von vornherein*] from the idea [*Vorstellung*] of absolute space. But if we take our start on the basis of facts, we shall find we have knowledge only of *relative* spaces and motions."

Thesis III
Someone has seen that the earth moves absolutely.

Question III

Pierre Kerszberg (1987, p. 196) writes: "The motion of the earth, as it revolves about its axis or around the sun, is never perceived as such." Now that we have questioned our natural attitude, that is, accomplished epoché, we might manage to complete a reduction and switch to the phenomenological attitude.

2.2. Step two: reduction

With the epoché, we question and set aside the views of the natural attitude, in this case the Copernican viewpoint. What does phenomenology have to offer in their place? What phenomenology offers is (1) original and (2) transcendental. Instead of concentrating on the advanced, exact and objective, Husserl focuses on the original and transcendental.

Original

According to Husserl, the source of the original views is experience. It is Husserl's conviction that the first step to reality is always taken in experience. For Husserl a reality without experience, for example the sun before life, is not thinkable. Also the earth is given originally in experience.

How do we experience the earth? In his investigations, Husserl tries very hard to find a way to experience the earth as a physical body. He imagines the possibilities of living on the moon and being born in a moving vehicle, etc. Husserl knew that he himself had never experienced the earth as a physical body. The result of the investigations is, however, much more radical: he cannot imagine the possibility of experiencing the earth as a physical body.

From this follows the situation:

(1) Original or primitive level: the earth is not a physical body as it is never experienced as such.
(2) Copernican or advanced level: the earth is a physical body.

The task is now to find a way to link these two levels. The first level is the level of the life-world; in our lives, we never actually experience the earth as a body. The second level is the level of natural sciences, experiments and theories. Here we can point out what Husserl means by saying that there is a crisis within European Sciences. The crisis is the absence of a link between life and science. More and more, modern sciences tend to deal with a reality that has very little to do with the reality of our lives.

Husserl tries to find the link between the original or experienced earth and the scientific earth. Unfortunately, Husserl fails. In order to see this, we need to take the transcendental perspective.

According to Husserl, a physical thing is something that has a place and could move. This possibility of movement only has sense in relation to something. Movement is relative. The tension between phenomenology and science asks whether there is only one point of reference for us all – the earth – or could we really choose freely whatever reference points of our own.

In the *Investigations*, Husserl shows that ultimately this in relation to which the movement could take place can be neither another physical body nor my own body. This something else has in all human cultures been the earth. Movement and rest have sense only in relation to that something called the earth. It follows that in order for the earth to be a thing and for the possibility of movement, there has to be another earth before our original earth. The earth could move only in relation to an even more original earth. According to Husserl's result, such an even more radical sense of originality is not possible to us humans. It follows that the earth will never really enter into the field of physical bodies, into the field of movement and rest.[2]

What is the role of this kind of transcendental argument? Let me take a short example. Here is a thing: a pen. When we follow Husserl's third logical investigation, we notice that we have here the following elements as parts of this thing: brightness, colour and surface. Husserl calls these parts moments. Moments are a particular kind of parts, as they have essential relations. I quote Robert Sokolowski (1967, p. 540): "I cannot disengage brightness from color, I cannot consider color without locating it within a certain surface, and I cannot consider surface without seeing it as a moment of an extended thing". The relationships among moments are rigorously determined: "Brightness cannot be immediately blended with surface, it must be mediated by color."

It is pretty difficult to argue against this kind of transcendental argumentation. Whatever results theories of colours will achieve, the basic conditions of seeing colours will not change. From the point of view of Husserl, his argumentation of the earth situates in this transcendental level. The earth is situated before the possibility of movement as clearly as brightness and surface must be mediated by a colour.

In his *Vienna Lecture*, Husserl (1970, p. 295) points out that Einstein's innovations do not really concern the space and time of our experience. The same already holds for the Copernican view: although we know that the earth moves, we do not experience it moving under our feet at the speed of a jet plane.

---

2 Himanka (2005).

3. Reflections

I have outlined the following situation: Husserl considers the original earth of our experience and discovers that it is not and cannot be a physical body. It follows that the earth cannot move or rest.

On the side of the sciences, it is crystal clear that the earth must be a physical body. We can measure its mass and its size, and we can locate it in relation to the fixed stars. Husserl would not argue against this. In the *Investigations* (p. 230) he actually writes that "we do not even touch upon physics". I think he meant that as physics deals with movement and rest, it cannot deal with the earth. The earth in the original sense is situated outside the scope of physics. Phenomenology considers the concrete earth, while the sciences consider a theoretical entity they – for historical reasons – also call the earth.

It is clear form the text of *Investigations* that Husserl knew he would be laughed at. He knew that his result would be seen as "unbelievable philosophical hybris". Yet, he saw no possibility to "back down on the consequences" of his clarification. When I started to read this manuscript, I thought there must a mistake in the argument. But the more I studied the text, the more convinced I became that the actual argument is phenomenologically valid. If we accept the phenomenological view of the original reality of the real, it follows that the earth does not move. It now seems to me that if we want to deny Husserl's result, we have to start from the starting point that already horrified Plato: we do not live in a real reality (*Sophist* 246a–249c).

It is in this sense that phenomenology challenges science. It asks for the meaning of science that denies or overlooks our reality. Husserl's aim, however, was not to separate phenomenology from the sciences. His aim was just the opposite: to show the connection between our lived experiences and the theoretical reality of the sciences. That I see to be the real challenge for both philosophy and science.

University of Helsinki

References

Farber, M. (1940), *Philosophical Essays in Memory of Edmund Husserl*, Cambridge, Mass., Harvard University Press.

Himanka, J. (1999), "Reduction *in concreto*, Two Readings of the *Idea of Phenomenology*", *Recherches husserliennes* 11, 51–78.

Himanka, J. (2005), "Husserl's Argumentation for the Pre-Copernican View on Earth", *Review of Metaphysics*, LVIII, 621–644.

Hoyle, F. (1973), *Nicolaus Copernicus*, Suffolk, Heinemann.

Husserl, E. (1940), "Grundlegende Untersuchungen zum phänomenologischen Ursprung der räumlichkeit der Natur", in M. Farber (1940), pp. 307-325.
Husserl, E. (1970), "The Vienna Lecture", in Husserl (1970a), pp. 269-296.
Husserl, E. (1970a), *The Crisis of European Sciences and Transcendental Phenomenology*, Evanston, Northwestern University Press.
Husserl, E. (1999), *The Idea of Phenomenology*, Dordrecht, Kluwer.
Kerszberg, P. (1987), "The Phenomenological Analysis of the Earth's Motion", *Philosophy and Phenomenological Research* XLVIII, 190-217.
Mach, E. (1960), *The Science of Mechanics: A Critical and Historical Account of Its Development*, La Salle, Ill., The Open Court.
Plato (1892), "Sophist", in *The Dialogues of Plato, Vol. 4, translated into English with Analyses and Introductions by B. Jowett*, 3rd rev. ed., London, Oxford University Press, pp. 339-407.
Sokolowski, R. (1967), "The Logic of Parts and Wholes in Husserl's *Investigations*", *Philosophy and Phenomenological Research* XXVIII, 537-553.

The Constituent Function Analysis of Functions

Ingvar Johansson

The emergence of modern physics and evolutionary biology brought with it a general ban on talk of final causality in the natural sciences. No natural science ought to put forward hypotheses about entities that strive toward a certain goal, since such strivings can only belong to beings with consciousness. Aristotle's view that the function of acorns is to become fully grown oaks, and that acorns actually strive to become oaks, came to be regarded as an anthropomorphization. Nonetheless, even after Darwin biologists and medical scientists have continued to talk about functions in a seemingly pre-Darwinian way. What to say about this? Do not functions presuppose goals, and are not goals subjective, be they person-subjective or culture-subjective?

1. Today's Philosophy of Functions

In the philosophy of science, the mainstream view of the 20th century and of today is that all function talk within biology and medicine can, without loss of scientific content, be reinterpreted as being only about efficient causality in general or, more specifically, about natural selection as a purely causal phenomenon. This reinterpretation is thought to take away only a shell of unscientific purpose talk, and its result has been called neo-teleology (Cummins 2002) even though it might just as well be called non-teleology. There are, however, critical voices such as Robert Cummins (1994; 2002), Christopher Boorse (1976; 2002), and Richard N. Manning (1997) who argue that function talk cannot at all be reduced to evolutionary talk. Cummins has put forward an alternative analysis of functions that I will call, like Peter Melander, the *intrasystemic role analysis* (1997, p. 51). Boorse (2002, p. 64) calls it the causal role analysis of functions, and he calls his own alternative the *general goal-contribution analysis* of functions; Manning's contribution to the discussion is mainly critical. For an attempt at an overall classification of philosophical analyses of functions, see Mark Perlman (2004).

I agree with many of the criticisms of the mainstream, neo-teleological view of functions, and the rejection of this view will in what follows be taken for granted. Nonetheless, I find both Cummins's and Boorse's own alternatives wanting.

Therefore, I will here put forward what I hope is a better alternative; it will be called the *constituent function analysis* of functions.

According to Cummins (1994, p. 67), "[t]o ascribe a function to something is to ascribe a capacity to it which is singled out by its *role* in an analysis of some capacity of a *containing system*" (italics added). That is, the function of an entity is its intrasystemic role in a larger system that carries out a certain activity. The function of the heart is to pump blood because this is its intrasystemic role in the activity of the circulatory system.

Cummins (2002, p. 158) criticizes both teleological and neo-teleological analyses of functions and wants them to be replaced by his proposed brand of functional analysis: "While teleology seeks to answer a why-is-it-there question by answering a prior what-is-it-for question, [intrasystemic role] functional analysis does not address a why-is-it-there question at all, but a how-does-it-work question".

According to Cummins, a *functional analysis* of the heart must be kept distinct from an *evolutionary analysis* of the heart. Only the latter is concerned with the problem why there are hearts. I agree, and the view that I will put forward comes close to Cummins's, but my complaint is that he never answers the following questions:

- In what significant sense does the concept of *role* differ from that of *function* itself? What is the difference between saying that "the role of the heart is to pump blood" and "the function of the heart is to pump blood"? That is, is not Cummins's analysis circular? It takes away the explicit concept of function only to introduce an implicit function concept, that of role.
- Must not the "containing system" spoken of be a *functional* whole? In a purely causal system such as the solar system, there are no roles for the planets. That is, again, is not Cummins's analysis circular? It takes away the explicit concept of function, but introduces instead an implicit function concept in its concept of containing system.
- Are there mind-independent roles in nature? That is, on Cummins's analysis can there be objectively existing functions?

Boorse is in favor of classical teleology, and he treats his concept of goal-directedness as a primitive concept. He analyzes "functions as causal contributions to goals of a goal-directed system," and he "takes goal-directedness to be an objective, non-mental property of all living organisms" (2002, pp. 63-64). Function statements in this sense are, he claims: "literally true throughout the whole biological domain, not merely metaphorically true, heuristically useful or the like" (ibid.). However, I cannot see that he ever succeeds in explaining the presumed objectivity of the goal-directedness spoken of.

I regard it as a fact that we often spontaneously *perceive* living organisms as being engaged in goal-directed behavior. For instance, when I hunt a fly I perceive it as fleeing from me. Such perceptual facts, however, cannot prove the objective existence of goal-directed behavior in animals. In sunsets we perceive the sun as moving, but this does not prove that the sun really is moving, as Copernicus made clear. When Boorse (2002, sect. 2) discusses the objectivity of goal-directedness, he simply bypasses this problem of the possibility of illusory perceptions of goal-directedness.

In my view, neither Cummins nor Boorse has managed to establish that his own philosophical analysis of functions gives rise to a concept of function that supports the view that functions can be objectively investigated by science.

Before I embark on my central, ontological task – to show that there is indeed a kind of objectively existing function, which I call "constituent function" – I will say some words about the semantics of the concepts of causes, goals, and functions. If it is not regarded as trivially true that these concepts are different and not semantically reducible to each other, then at least this fact becomes clear when their possible referents are related to time.

2. Causes, Goals, Functions, and Time

Whatever time is, it has to be represented as a line, be it straight, circular, or curved in some other way. For the purposes of my remarks, we need to distinguish between (i) a certain chosen instant in time, (ii) the next moment, (iii) a more distant future moment, and (iv) the time interval between the instant and the future distant moment. The problem of how to apply the concept of the next moment to continuous time – in continuous time, between any two time points there is necessarily a third time point – will be disregarded; it is of no importance for my argument.

In everyday talk as well as in philosophical discourse, the concept of cause seems to be used in such a way that the effect of a cause is either (i) simultaneous with the cause ("His presence causes irritation"), or (ii) it comes into being in the next moment ("His cutting the rope was the cause of the flag's falling").

Goal talk has another relation to time. One may have a goal that (i) is instantly realized; for instance, showing someone that you are hearing and understanding what she is saying. And one may have a goal that (ii) is possible to realize in the next moment, e.g., opening the door. In this respect, there is a similarity between goals and causes. However, there are also goals that (iii) can be realized only in a distant future moment. For example, my goal now is to have a finished paper by next week. When such a goal exists, there is a directedness (intentionality) towards a future distant moment. A time jump from "now" to "future" is in some sense involved since "the next moment" is not taken into account. Finally, there are goals that (iv) directly encompass a whole time interval. All activities that are regarded as ends in themselves have this character. If my goal is to listen to music, then my goal has an implicit relation to a time interval.

Goals, I argue, inhere in goal-bearers (persons) in about the same way as monadic qualities such as shape and mass inhere in quality bearers (things). Many human goals, in the sense of "prior intentions",[1] are of kind (iii), i.e., they are directed towards something in a distant future moment of time. This might seem to be the case even with things such as goal-seeking cruise missiles, but this is a wrong impression. When such a missile is heading towards its target, there is no goal (towards a future hitting of the target) that inheres only in the missile. Without a more or less continuous fit between "the map within" and "the world around" the missile in itself has no goal whatsoever, even though the people who programmed it had. A cruise missile that is fired from the wrong place has no goal at all, but a human being can retain a goal to travel to a certain place quite independently of where on earth he happens to be situated. Even more, a human being can have a goal without knowing how to realize it, but no present-day artificial goal-seeking device can be ascribed such a goal.

What then about function talk? I think one example is enough to sustain my point that it differs from both causal talk and goal talk. The function of the heart is to pump blood. At every moment that a heart is functioning, it has a relation to the same or to the next moment (of the sort which obtains in the case of causes), but it is not able (in contrast to what is the case with humans with goals) to be directly aiming at a distant moment. Even though a function can exist at a single point of time, an actual functioning is necessarily extended in time. Essentially, if only implicitly, function talk makes reference to whole time intervals.[2]

This brief excursion into conceptual analysis will have to do. If I am right, then every presumed semantic reduction of the intension of "function" to that of "cause" or that of "goal" does violence to at least one of these concepts' relation to time. Let me now turn to ontology. Do the three concepts in question have an extension? Certainly, there are goals. First-person introspection reveals that I have goals. But are there causes and functions? In what follows – for the sake of both simplicity and (I think) truth – I will assume that the mind-independent world contains causes and causal relations. Nonetheless, nothing in my argument hinges on this assumption. What is at stake is the ontology of functions. This brings us to the next section.

---

1 Searle (1983, pp. 84f.).

2 See also Johansson (2004, pp. 98f.).

3. Constituent Functions and Intrinsic Functions

Let me start by introducing two distinctions: first between *subjective* and *objective* functions of material entities, and then between *intrinsic* and *constituent* functions.

Objective functions are functions that have in some sense a mind-independent existence, whereas subjective functions are functions that are wholly projected onto material entities by human beings (or by some other beings with minds). If the entities in question are designed, the projection comes with the design. Surely, there are subjective functions, both culture-subjective and person-subjective. Most functional *artifacts* can be regarded as having culture-subjective functions. But I can also myself assign a new function to an already existing culture-subjective functional artifact and thereby create a person-subjective function. For instance, I can start to look upon one of the books on my desk as having the function of killing flies and mosquitoes. John Searle (1995, p. 14), who also opposes the reduction of functions to causes, claims that *all* functions are subjective, primarily culture-subjective.

I will by stipulation define an intrinsic function as being objective and monadic. An intrinsic function inheres in an entity in the way a quality like mass is assumed to inhere in Newtonian corpuscles, or in the way personal human goals are assumed to inhere in individual persons. A constituent function, on the other hand, is a relational entity. It is a feature of something that is a spatial part of a larger entity. Instead of the term "constituent function", one might also profitably use terms such as "component function" and "part-to-whole function". All these terms indicate that constituent functions are relational in character. The question whether constituent functions exist subjectively or objectively will be answered below. Its answer will expose a complication in the subjective-objective distinction.

It is clear that there are subjective functions. I will now argue that it is no less clear that there are constituent functions also. However, no attempt will be made to answer the question whether or not there are intrinsic functions. My argument in favor of the existence of constituent functions needs the *concept* of intrinsic functions, but not their *existence*.

A sufficient and necessary condition for something's being a constituent function is the following:

- *F* is a constituent function borne by *B* if and only if:
  (a) there is a functional whole *A*;
  (b) *B* is both a spatial part and a subunit of *A*;
  (c) *B F*'s in relation to some other entities (*X*, *Y*, *Z*) that are relevant for *A*.

This requirement can be directly applied to our prototypical example of the function of the heart, and by implication, also to many other traditional examples. Here comes a constituent function of the heart:

- To pump blood (*F*) is a constituent function borne by the heart (*B*) since:
  (a) there is a functional whole, the circulatory system (*A*);
  (b) the heart (*B*) is both a spatial part and a subunit of the circulatory system (*A*);
  (c) the heart (*B*) pumps blood (*F*) in the blood vessel system (*X*).

This requirement must not be interpreted as a definition of the concept of function, even though it may be regarded as being not only a requirement but an unfolding of the intension of the concept of constituent function. As a definition of "function" it would be circular, since this concept also appears in the definiens, both explicitly (in "functional") and implicitly (in "subunit of *A*" and "relevance for *A*").

In my opinion, "function" is a primitive undefinable concept. Physics works with seven undefined quantities, the so-called base SI-units: length (metre), mass (kilogram), time (second), electric current (ampere), thermodynamic temperature (Kelvin), luminous intensity (candela), and amount of substance of kind K (mole). As "length" comes with the idea that there can be different lengths, and "substance" comes with the idea that there can be different amounts of a certain kind of substance, "function" comes with the idea that a certain kind of function *F* can be realized more or less well. Even though it seems impossible to quantify well-functioning, there is a grounded similarity between quantity-talk in physics and functioning-well-talk in medicine and biology (Johansson 2004). The life sciences can regard "degree of well-functioning of function *F* (goodness)" as an undefined concept on a par with the SI-units of physics.

The essence of the definition is that *B* has its constituent function only as part of, and in relation to, a whole *A* that already has a function. Constituent functions are inherited functions. This fact, of course, gives rise to this question: Should the functional whole *A* be allowed to have (1) a subjective, (2) an objective-intrinsic, or (3) a constituent function? In my opinion, all three alternatives can be allowed, but, unfortunately, all three may also give rise to the false impression that they are problematic:

(1) If *A*'s function is subjective, then are not the corresponding constituent functions also subjective? That is, are we not then allowing in the life sciences the kind of subjectivity that the natural sciences should avoid?
(2) If *A*'s function is objective-intrinsic, then do we not have to try to establish that nature truly can contain intrinsic functions? That is, do we

not then have to offer an account of what an intrinsic function is? Would this not threaten to require a rehabilitation of classical teleology?

(3) If A's function, too, is a constituent function, then are we not facing a vicious infinite regress, in which B gets its functionality from A_1, which gets its functionality from A_2, which gets its functionality from A_3, and so on?

The answer to the first question is that a constituent function can have, appearances notwithstanding, a kind of objectivity even when its functionality is inherited from a functional whole with a subjective function.

Artifacts (screwdrivers, nails, cars, radios, etc.) are entities with subjective functions. When, in some time interval, such a material entity realizes its function, a material process occurs independently of man, but without man there is no function and functioning. Think of a house and its doors. In itself, the house and its parts make up a pure material structure that lacks functionality. However, as soon as the house is ascribed its usual house-function, *it becomes an empirical question* whether or not the door-parts have a function. If one finds that the doors have a function in relation to the function of the whole house, e.g., to make it possible for people easily to move between its rooms, then, second, it also becomes an empirical question whether each door performs this function in a bad or in a good way. Therefore, the constituent function of the doors is not a mere projection – in spite of the fact that the function of the whole house is a subjective function. Of course, one might say that objective-constituent functions (as I will now call them) are not as objective as objective-intrinsic functions (if such there are); but they deserve nonetheless to be called objective. From a scientific point of view, they are open to normal empirical and theoretical investigations.

Constituent functions are not the only kind of entities that have this kind of objectiveness. The same is true of actions with means-end rationality. Let me explain. Is it rational never to lie? Kant claimed, famously: Yes, it is absolutely and categorically rational never to lie. According to Kant, each act of attempted truth-telling is an *intrinsically rational* action. Today, however, most philosophers would say that the concept of intrinsic rationality has no extension, but that it depends on one's goals in life whether or not it is rational to tell the truth or a lie in a given situation. The question whether it is rational never to lie has been transformed into the question whether lying can be a rational means to achieve some pre-given subjective goal. Science cannot investigate what the subjective goals ought to be; but as soon as such a goal is given, it can be scientifically investigated what actions are rational from the point of view of means and to ends. Is it rational to be a non-smoker? If it is a question of intrinsic rationality it seems to be unintelligible or to have no answer; but if it is a question of means-ends rationality it can be objectively investigated as soon as some end is specified, be this end rational or irrational. Game theory is a science wholly devoted to the study of complicated means-end rationalities.

Actions that are means-end rational are relational in character, just like constituent functions. What has become known as means-end rationality could equally well have been called means-to-end rationality, and what I call constituent functionality might also be called part-to-whole functionality. Doors have a part-to-whole function in relation to the houses of which they are parts, and hearts have such a function in relation to the circulatory systems of which they are parts.

Thus my answer to question (1) brings with it an easy solution to (2). If there can be objective-constituent functions in relation to wholes with subjective functions, there can of course be objective-constituent functions in relation to wholes with objective-intrinsic functions as well. Therefore, one can investigate constituent functions scientifically without bothering about whether the functional whole in question has a subjective or objective function. Both believers and non-believers in intrinsic functions can equally well study constituent functions.

What, finally, of question (3)? Can any function whatsoever be an objective-constituent function? The answer is simple: No, it cannot. Without a starting point, a pre-given functional whole (recall Aristotle's "unmoved mover"), the concept of constituent function loses all its sense. The introduction of constituent functions does not make it possible to delete completely the traditional alternatives in the philosophy of functions. Every meaningful regress of constituent functions has to be finite and end in either a subjective or in an intrinsic-objective functional whole. Nonetheless, a constituent function can be an objective-constituent function in relation to a whole whose function is merely in its turn an objective-constituent function.

Let me summarize. An objective-constituent function can derive its functionality either from a subjective function, an objective-intrinsic function, or another objective-constituent function. Therefore, the life sciences can study objective-constituent functions while disregarding the question of classical teleology. However, philosophy cannot do the same.

In the final section, I concretize the regress problem.

4. The Objective-Constituent Function of the Heart

Nothing in the arguments put forward excludes the possibility that an entity is at one and the same time the bearer of a subjective, an objective-intrinsic, and an objective-constituent function; it may also be the bearer of several objective-constituent functions simultaneously. However, let us assume that the human heart is the bearer of only one function, namely the one already presented:

- To pump blood (*F*) is a constituent function borne by the heart (*B*) since:
  (a) there is a functional whole, the circulatory system (*A*);
  (b) the heart (*B*) is both a spatial part and a subunit of the circulatory system (*A*);
  (c) the heart (*B*) pumps blood (*F*) in the blood vessel system (*X*).

Let us next assume that even the circulatory system has only a constituent function: to transport substances. We then get:

- To transport substances (*F*) is a constituent function borne by the circulatory system (*B*) since:
  (a) there is a functional whole, the human body (*A*);
  (b) the circulatory system (*B*) is both a spatial part and a subunit of the human body (*A*);
  (c) the circulatory system (*B*) transports substances (*F*) between several bodily systems (*X*, *Y*, *Z*).

What now about the functional whole of the human body? What is its function? Three kinds of contemporary proposals are:

(1) The human body as a whole has the *culture-subjective* function of preserving its own life.
(2) The human body as a whole has the *objective-intrinsic* function of preserving its own life.
(3) The human body as a whole has the *objective-constituent* function of (a) converting some substances into some other substances in the ecological system of the earth, or (b) promoting the existence of the human species.

The main criticism of the first proposal is that it looks too conventional, and the main criticism of the second one is that it makes death unexplainable. Often, much too often, death comes through accidents, diseases, killings, and murder, but there is also a kind of death that appears just as natural as life. Of course, all persons who make proposals of the third kind have to face the regress I have mentioned. Ecologically minded persons who put forward (3a) have to answer the question: What is the function of the ecological system of the earth? And in case of (3b) the question becomes: What is the function of the human species? Does the human species have an objective-intrinsic function? Be this as it may.

The point of this paper, applied to the heart and the circulatory system, is that we can in an objectively scientific way study the functions of both of them without bothering about the trilemma stated for the human body as a whole. Independently of whether or not self-preservation is itself an intrinsic function of the human body, a merely projected function, or an objective-constituent function in a larger whole, the human body contains many *objective-constituent* functions. This fact suffices to make also non-evolutionary studies of functions a tractable scientific enterprise. Furthermore, functions might very well provide the ground for taxonomic work. One might order the parts of the human body according to their function and create a *functional anatomy*.[3]

Philosophy can, I think, meet "the function challenge" that the life sciences have posed.

Saarland University, Umeå University

Acknowledgments

The paper was written under the auspices of the Wolfgang Paul Program of the Alexander von Humboldt Foundation, the European Union Network of Excellence on Medical Informatics and Semantic Data Mining, and the Volkswagen Foundation under the auspices of the project "Forms of Life". I would like to thank the participants in the discussion in Helsinki, but first and foremost, all the co-authors of (Johansson, Smith, et al., 2005), in particular Barry Smith and Katherine Munn.

References

Ariew, A., Cummins, R. and M. Perlman (eds.) (2002), *Functions. New Essays in the Philosophy of Psychology and Biology*, Oxford, Oxford University Press.

Boorse, C. (1976), "Wright on Functions", *Philosophical Review* 85, 70-86.

Boorse, C. (2002), "A Rebuttal on Functions", in A. Ariew et al. (2002), pp. 63-112.

Cummins, R. (1994) [1975], "Functional Analysis", in E. Sober (1994), pp. 49-69.

Cummins, R. (2002), "Neo-Teleology", in A. Ariew et al. (2002), pp. 157-172.

Johansson, I. (2004), "Functions, Function Concepts, and Scales", *The Monist* 86, 96-115.

Johansson, I., Smith, B., Tsikolia, N., Elsner, K., Ernst, D., Munn, K., and D. Siebert (2005), "Functional Anatomy: A Taxonomic Proposal", *Acta Biotheoretica* (forthcoming).

Manning, R. N. (1997), "Biological Function, Selection, and Reduction", *British Journal for the Philosophy of Science* 48, 69-82.

Melander, P. (1997), *Analyzing Functions. An Essay on a Fundamental Notion in Biology*, Stockholm, Almkvist & Wiksell.

Perlman, M. (2004), "The Modern Philosophical Resurrection of Teleology", *The Monist* 86, 3-51.

---

3 See Johansson, Smith, et al. (2005).

Searle, J. (1983), *Intentionality*, Cambridge, Cambridge University Press.
Searle, J. (1995), *The Construction of Social Reality*, New York, The Free Press.
Sober, E. (ed.) (1994), *Conceptual Issues in Evolutionary Biology*, Cambridge, Mass., MIT Press.

Thomas Kuhn – Last Logical Empiricist or First Post-Empiricist?

Jouni-Matti Kuukkanen

The received wisdom for decades was that Kuhn and other historical philosophers of science are arch enemies of logical empiricism. Recently, however, various philosophers have challenged this conception. In their opinion, Kuhn held views that were either remarkably similar to late logical empiricism or even extensions of its doctrines. In this essay I examine the question of whether Kuhn was one of the first who looked beyond logical empiricism or whether he was one of the last flag bearers of that school. I will eventually settle on the side of the former view, but not for the reasons inherent in the old orthodoxy. My grounds for seeing Kuhn as a non-logical empiricist are firstly that, no matter whatever parallels we can point out between these two traditions, the genesis of Kuhn's view is entirely different. There is no chance that Kuhn had adopted and redeveloped tenets of logical empiricism. Secondly, I admit that we can spot similarities, but even if this is so, this should not blind us from noticing radically different and interesting theorising in seed in Kuhn's philosophy. For example, Kuhn did not reach the level of sophistication with his view on meaning, but quite a few of his insights have been captured successfully and developed further during the last ten years in the cognitive-historical studies that take advantage of theoretical innovations developed in cognitive science for the benefit of understanding the history of science. I also believe that a cognitive-historical interpretation of some of Kuhn's insights can improve our understanding of scientific change, and is, therefore, an important object of study.

I have narrowed my focus here on Kuhn's views on meaning and meaning change. That is mainly because the strongest case for Kuhn being the last logical empiricist can be made on the basis of meaning theorising. Furthermore, Kuhn's view on meaning and meaning change is responsible for the infamous problem of incommensurability. What is more, these views contain elements of an interesting theory of meaning and scientific change. I start by taking a brief look at the revisionist interpretation of the relationship between Kuhn and logical empiricism. After that, I argue that the genesis of Kuhn's ideas of meaning is in the study of history, not in philosophy. Thirdly, I show how the cognitive-historical viewpoint can explicate Kuhn's insights and then comment on the usefulness of the emerging view in respect of the understanding of scientific change.

1. Kuhn as the Last Logical Empiricist

Since the 1960s it has been customarily noted in numerous articles and books how markedly Kuhn's theses stand in contrast to the doctrines of logical positivism and empiricism. It is enough here to remind ourselves that Kuhn argued against neutral observation, the invariability of meaning, the commensurability of scientific theories and the cumulativity of scientific knowledge. Despite this list, which undoubtedly can be said to be anti logical empiricist, a number of philosophers have insisted that Kuhn is, after all, not so different from the late Carnap, who was arguably the most eloquent defender of logical empiricism.

Jane English (1978) was probably the first who gave a detailed analysis of the parallels between Kuhn and Carnap. English labels Kuhn's and Feyerabend's view of meaning change as contextualist in the sense that every substantial change in theory at least changes the meanings of terms that occur in revised sentences. According to English, this makes intertranslation, contradiction and progress as increasing verisimilitude, impossible. English's main point is that, in contrast to what has been thought, Carnap's view faces these same objections as the meaning change view. George A. Reisch (1991, p. 264) explains Carnap's appreciation of Kuhn's *Structure of Scientific Revolutions* in a letter on the possible publication of *Structure*[1] in 1960 by the similarity of their views and claims "that logical empiricism, insofar as that program rested on Carnap's shoulders, was not substantially upstaged by Kuhn's book". That is, both Carnap and Kuhn adopted the distinction between radically different theories, languages or conceptual frameworks, which change during the revolutionary period. They are incommensurable and can therefore be evaluated only pragmatically. John Earman (1993) writes that many of the post-positivist themes are extensions from Carnap and other logical positivists and empiricists. Such themes are at least the non-existence of neutral facts, incommensurability in the form of failure of intertranslatability and paradigm as analogous to Carnap's linguistic framework. Also, according to Earman, a holistic view of meaning emerges from logical positivism. Gürol Irzik and Teo Grünberg (1995) conduct a comparative analysis between Carnap and Kuhn, and they likewise want to challenge the conventional wisdom that the post-positivist philosophy of science constituted a rupture without too many affinities with logical positivism. They attempt to show that Carnap's linguistic frameworks function in the same way as Kuhn's lexical structures; that Carnap's scientific theories resemble, at least in some respect, Kuhn's paradigms; and Carnap's and Kuhn's characterisations of scientific revolutions are nearly identical. Furthermore, they

1 The following abbreviations are used in reference to Kuhn's books: *Structure of Scientific Revolutions* (*Structure*), *Essential Tension* (ET) and *Road Since Structure* (RSS).

agree with English and disagree with Earman that the Kuhnian type of semantic holism is inherent in later Carnap and that semantic incommensurability is a direct consequence of it. Michael Friedman (2003) in turn agrees with Reisch that both Kuhn and Carnap accepted the philosophy of linguistic frameworks, and deepens this analysis by studying the influence of neo-Kantianism. Finally, Alexander Bird (2000) presents Kuhn as the last empiricist rather than the first post-empiricist. Bird argues that Kuhn failed to break entirely with the preceding tradition and consequently kept many theses from logical empiricism. Some of these are, for instance, intensionalism in reference determination, meaning holism, and scepticism towards theoretical entities. In terms of meaning and meaning change, Bird (2002, p. 14) introduces a catchy phrase that "Kuhn's incommensurability thesis can be seen as positivism plus the theory-dependence of observation".

The most interesting revisionist claim is that semantic holism is more or less just an extension of views on meaning held by logical empiricism. This relies on the idea that Carnap's and Kuhn's meaning theories were fundamentally very similar. Both thought that meaning is dependent on theory in some way. Kuhn, however, went one step further by extending semantic holism, not only to theoretical, but also to observational expressions. It is not always clear though whether these various writers are ready to claim that there is actually a historical continuum from logical empiricism to Kuhn.[2] In any case, they all want to read Kuhn's philosophical impact as an addition to the philosophy of logical empiricism. And it is true that this orientation is not totally unreasonable. Let us remember how Carnap had to liberalise the logical positivists' strict requirement for scientific language. His departure from the meaning theory of logical positivism was mediated by problems with dispositional terms, which led to an admission that not all expressions can be given a full definition by observational terms. Carnap's (1953) suggestion in "Testability and Meaning", first published in 1936, was that the job could be done by reduction sentences. We may regard this as the first step in the liberalisation. Later in "The Methodological Character of Theoretical Concepts" (1956), Carnap came to the conclusion that definition by reduction sentences was not enough, and accepted theoretical language as legitimate scientific language. This counts as a second step in the liberalisation. These moves functioned together as a Trojan horse bringing in the principle of the theory-dependence of meaning. If we now take away the idea that observation fixes the meaning of at least observational terms, we break the foundation of the meaning theory of logical positivism. And this is just what Kuhn and other historical philosophers seemed to

---

2 Some, such as Reisch, Irzik and Grünberg attempt to show merely analogies, while Earman and Bird argue that they shared at least partially the same tradition and, therefore, certain of Kuhn's views can be traced to logical empiricism. See also Sankey (1997; 2000) on similar comparisons.

have done by showing that observation is not theory neutral. This is then the third and last step in the liberalisation of the criteria for scientific language. By this move, the meaning theory of logical positivism appears to be turned upside down. Instead of determining meaning unambiguously and permanently by neutral observation, all meaning is at the mercy of theory. Once we arrive at semantic holism meaning change, incommensurability and problems with referential determination follow.[3]

The fashionable response to practically all the problems that derive from semantic holism has been a causal theory of reference, whether pure or more recently in some modified form, allowing also a limited form of descriptivism in some cases. Kuhn may even look conservative and old-fashioned, if we think externalism in semantics and epistemology have been The contributions to philosophy in the latter part of 20th century. However, at this point I want to put on the breaks! Whatever the improvements that theories of reference are able to bring, we should not rush to close the chapter on Kuhn. There is another Kuhn still to be discovered.

2. The Genesis of Kuhn's Notion of Meaning

I contend that regarding Kuhn's view on meaning and meaning change there is no historical continuum between Kuhn and logical empiricism. Even though the comparison of different views is naturally entirely legitimate, it should not hide the fact that Kuhn's thinking stems from a different origin than those with apparently similar views in logical empiricism, and that his notion of meaning, in particular, is not just a further development of that notion by logical empiricism.

It is clear that Kuhn was not familiar with logical empiricism. First of all, we should remember that he was, as Kuhn himself expressed, "a physicist turned historian for philosophical purposes" (RSS, pp. 320–321). As a consequence of this, although his interest was philosophical, he did not know philosophy very well at the beginning of his career. After Earman had attempted to show parallels between Kuhn and logical positivism, Kuhn felt compelled to answer: "Whatever role the problems encountered by positivism may have played in the background for *The Structure of Scientific Revolutions*, my knowledge of the literature that attempted to deal with those problems was decidedly sketchy when the book was written. In particular, I was almost totally innocent of the post-*Aufbau* Carnap, and discovering him has distressed me acutely" (ibid., p. 227). How limited Kuhn's

3 On this, see Sankey's (1994; 1997; 2000) studies on the problematic and his formulation of it as the incommensurability thesis.

knowledge of logical empiricism was becomes clear from his last interview.[4] Kuhn's ignorance can be seen also in references in *Structure* or in other early works. In *Structure* there are two unimportant references to philosophers that could properly be said to be logical empiricists; one is to Ernest Nagel and the other to Philipp Frank.

Kuhn had nevertheless some idea of logical positivism, but it was such a general idea that Kuhn himself later called it an "everyday image of logical positivism" (ibid., p. 306). He opposed that view because it contradicted the one that seemed to arise from studies of the history of science. It can, in fact, be argued that Kuhn's main objectives in his career were, firstly, to argue against Whig-history that interprets the past in terms of the present in favour of the view that understands the past on its own conditions; and, secondly, to change the public image of science. He saw the public image as being based on the flawed view of the history of science as cumulative progress.[5]

In the midst of all this we can discern the single most important historical experience in Kuhn's career that colours virtually the whole of his later thinking. Kuhn returns to his experience of reading Aristotle in 1947 as a student in various places officially in writing[6] as well as unofficially in speaking.[7] He tells us how, to him, Aristotle's physics seemed not only totally wrong but also absurd. Kuhn thought this was curious as Aristotle has managed to leave a much more lasting legacy in many other fields, such as in biology or in political behaviour. A sudden realisation changed all this. Kuhn realised how he had been reading Aristotle from a Newtonian perspective. In order to understand Aristotle's world one needs to read his works on their own conditions, not in relation to modern science. After that, he felt that he could finally understand Aristotle, "strained metaphors became naturalistic reports, and much apparent absurdity vanished", and that he had learned to think like an Aristotelian physicist. Kuhn calls this moment his "first scientific revolution". This new, "out-of-date", way of reading made history "consequential" and had a "decisive effect" on Kuhn's view of science. (ET, pp. xi–xii.)

We can also find other extremely revealing passages from Kuhn. In an interview he says: "I had wanted to write *The Structure of Scientific Revolutions* ever

---

4 Kuhn says, "I read some von Mises; I certainly read Bridman's *Logic of Modern Physics*; I read some Philipp Frank; I read a litte bit of Carnap, but not the Carnap that people later point to as the stuff that has real parallels to me" (RSS, pp. 305–306).

5 On this, see Sharrock and Read (2002), especially "Introduction".

6 See "Preface" (ET), "A Function of Thought Experiment" (ibid.), "Possible Worlds in History of Science" (ibid.), "What are Scientific Revolutions" (RSS), and also "A Discussion with Thomas S. Kuhn" (ibid.).

7 Kuhn's daughter Sarah Kuhn says the importance of this experience cannot be overstated as his children heard about it from time to time. Andresen (1999, p. 56).

since the Aristotle experience. That's why I had gotten into history of science" (RSS, p. 292). Furthermore, he says that "I didn't know enough about meaning, so I was leaning hard on gestalt switches; I think I talked about meaning change in *Structure*, but I looked to find the passages recently, and I was surprised at how few of them are" (ibid., p. 298). If, now, we put all what was said above together, we get a picture of the young Kuhn not really knowing what was going on in the philosophy of his day but who then awakened to an understanding of past science in its own right. He became further convinced that this realisation would change the view of the history of science, as well as the public image of science. Understanding past science required deciphering old texts hermeneutically and respecting the fact that the terms of past science have not the same meaning as in modern science.

This is, in short, the root of Kuhn's meaning talk. It stems from an attempt to make sense of old texts that, at first sight, look incoherent and absurd. Kuhn's notion of meaning implies two basic components that are carried forward throughout his career. The first is that Kuhn is talking about meaning, either in the psychological or sociopsyhological sense. He is interested in what people thought or believed. The presumption is that people expressed their views in their language, and that's why studying "meanings" is so important. The second component is that Kuhn strongly believed that meaning has to be holistic in some sense. If a person understands one meaning, s/he has to understand many, because meanings are somehow dependent on each other.

There is, thus, a crucial difference in terms of the genesis between Kuhn and Carnap. Carnap succumbed to his position after all the difficulties with the Received View of logical positivism. Kuhn's "meaning theory" and its holistic flavour are, in turn, positive constructions that arose from historical studies and were assumed to be necessary preconditions for understanding historical texts. It would be inappropriate to suppose or imply that Kuhn continued from where logical empiricism had arrived at. To suggest a clear continuity in meaning theory would suggest that Kuhn and Carnap, for example, shared the same theoretical body and were led the same way towards a certain conclusion, only for Kuhn to end up with a more radical view. The proper context is that Carnap was led to his view by careful examination of the Received View, while Kuhn arrived at a somewhat similar view from his hermeneutical objective by reading and interpreting historical scientific texts. This is also something that Kuhn himself remarked on: "the view that emerges in *Structure* is not the same as the Carnap view, but it's interesting that coming from what were partially different [...] Carnap staying within tradition had been driven to this – I had rebelled already and come to it from another direction, and in any case we were still different" (RSS, 306). Next I will show why the views "were still different".

3. Kuhn's View of Meaning

What is wrong with the new predominant conception that interprets Kuhn as the last logical empiricist with a holistic account of meaning (and often also with a descriptive theory of reference) is that the view is usually a result of reading some of the early works by Kuhn but then overlooks most of Kuhn's later attempts to explicate his notion of meaning. I believe that, by following Kuhn's career, we can actually find progress in his thinking, even if we do not find a sophisticated meaning theory. What is important is that his insights offer ingredients for further development. There are two aspects to which I pay attention in this limited space. One regards the speculation on how conceptual and empirical knowledge[8] are organised and stored in our mind. The other aspect is the claim that the classical account of concept definition is untenable and the suggestion of the family resemblance account as a replacement.

Kuhn had already written about conceptual schemes in *The Copernican Revolution*. They are said to be frameworks for the organisation of knowledge in order to guide future research and to aid the recall of data (Kuhn 1985, pp. 37–41). Throughout *Structure*, Kuhn talks about conceptual webs or interrelated concepts that change in scientific revolutions. This preoccupation of how scientific knowledge is organised also continues later in his career. Firstly, Kuhn becomes interested in how kind terms form hierarchical taxonomies. Soon, "lexicon" becomes the central term in his philosophy. He asks us to imagine that each individual referring term, i.e. taxonomic term, is "a node in a lexical network from which radiate labels for the criteria that he or she uses in identifying the referents of the nodal term. Those criteria will tie some terms together and distance them from others, thus building a multidimensional structure within the lexicon" (RSS, p. 52). Kuhn connects the notion of conceptual scheme to his new theoretical notions: "I might more appropriately speak of concepts than of words. What I have been calling a lexical taxonomy might, that is, better to be called a conceptual scheme". A conceptual scheme, Kuhn understands not as a set of beliefs, but as "a particular operating mode of a mental module". Mental module is thus some kind of cognitive structure in people's minds that organize and store the community's kind terms taxonomically. It is also a prerequisite to having beliefs, and supplies and bounds the set of beliefs that are possible. (Ibid., p. 94.) It has a communal

8 To avoid terminological misunderstanding, it is important to notice that I use here the word "knowledge" more freely than traditionally in epistemology. "Knowledge" does not mean here something like a true justified belief. Because both Kuhn and cognitive scientists talk of "knowledge" without implication of truth, the closest translation philosophically is "set of beliefs". It is another matter that they also assume that such a set has a specific structure.

equivalent, namely "lexical structure". Members of a language community share a homology of lexical structure. (Ibid., p. 104.)

The idea of conceptual scheme aroused famously vigorous opposition by some philosophers, most notably by Donald Davidson. It now seems, however, that there is a reason to take a second look at conceptual schemes in philosophy. For years cognitive sciences have taken an interest in how information is stored and organised in our minds. Since the beginning of the 1960s it has been noticed that information cannot be taken merely as a list of unorganised propositions. That supposition is psychologically unrealistic, and does not reflect the ease by which information can be retrieved from memory and how the mental lexicon can be accessed. After Collins and Quillan's introduction of semantic networks as theoretical notion and an alternative to predicate logic in "knowledge representation", many rivals or modifications have appeared that all assume an interrelational information structure.[9]

Paul Thagard argues that philosophers have previously understood conceptual schemes as being entirely abstract, but that they can now be explicated specifically. One interesting application to the history of science is made by Thagard (1992) himself. Also the historian Peter Barker notes that for a long time philosophers have talked about conceptual schemes without being able to say anything specific about their structure. He argues that the frame model, which he has used to describe the history of science, together with Hanne Andersen and Xian Chen (1996; 1999; 2003), gives us a tool for representing conceptual schemes in a concrete way. Barker thinks that, unlike earlier philosophical attempts to understand scientific change, the frame model applies to history without either falsifying the history or distorting the model. Furthermore, he is able to cite a few successful applications, such as the discovery of nuclear fission, some classification revisions made by ornithologists after discovery of South American waterfowl called "Screamer" and the Copernican revolution. Especially interesting is the second case above because the authors predicted the nature of the change beforehand and managed to verify it later by actual historical data.[10]

There is another insight in Kuhn that appears to gain support from recent research in cognitive science. Kuhn introduced Wittgenstein's idea of family resemblance to the philosophy of science in *Structure*. That is, Wittgenstein's famous idea that the application of the term "chair", for example, does not require grasping some set of attributes that all chairs and only chairs have in common. There simply is not such a set to be found; therefore, the recognition and the naming of objects is possible by noticing the family resemblance between objects

9 See, e.g., Stevenson, especially 2nd chapter, for introduction to knowledge representation.

10 Barker (2001), pp. 244–245.

that fall under the concept. The main contribution of the writings on meaning by Kuhn in the 1970s was the specification of the family-resemblance account and the development of a theory on the learning of concepts.[11] According to Kuhn, objects are grouped in similarity sets that form natural families, such as "swans", "geese" and "ducks". They are organised into natural families not only by their internal similarity to each other, but also by their difference to other families. Kuhn assumes that learning is ostensive and pre-linguistic, and does not require the conscious application of criteria for categorising.

What emerges from this is a definition of concept by family resemblance, in contrast to the classical account of concept as a set of necessary and sufficient conditions. Interestingly, the family resemblance notion can be seen receiving empirical support from cognitive psychology. The pioneering work was done by psychologist Eleanor Rosch whose studies showed that human beings regularly grade instances of a concept as bad, better and best examples. Some instances are thus better examples of a concept than others. The first studies showed that even in colour concepts some samples are thought to be better than others. Later studies, for example, have discovered that the same phenomena is found with the categorisation of tools, clothing, furniture, animals, trees, fish and birds.[12]

The most interesting application of the family resemblance notion comes from the previously mentioned studies by Andersen, Barker and Chen. They have utilised Lawrence Barsalou's frame model to represent the Kuhnian family resemblance account. In the model, concept is represented as having attribute-value structure. The idea of family resemblance is illustrated by the possibility that instances of concepts can have different values for the same attributes. For example, the frame for "bird" could have an attribute "main mean of movement". While most birds have the value "flying" for this attribute, an ostrich would have the value "walking" or "running", and yet it would be counted as a bird. And remarkably, Kuhn's taxonomic trees of kind terms can be built in the frame model. There is no space here to go on to discuss their model in any detail. However, the authors are convinced that as long as we are concerned with human concepts, family resemblance seems to be the only realistic option.

Where does this leave us with Kuhn's notion of meaning? After all, it is not clear whether Kuhn wanted to develop a meaning theory or not. His views on this oscillated throughout his career.[13] At the end of it Kuhn characterised his position

---

11 The most important essay is arguably "Second Thoughts on Paradigms" (ET).

12 See Andersen (1996, p. 352).

13 Compare Kuhn's (1977, p. 506) declarations in discussion in 1974, "Commensurability, Comparability, Communicability" (RSS, p. 56), "The Road since Structure" (ibid., p. 92) and "Afterwords" (ibid., p. 229). Kuhn travels from confidence about meaning to

as follows: "to the extent that I'm concerned with language and meanings at all [...] it is with the meaning of a restricted class of terms. Roughly speaking, they are taxonomic terms or kind terms"(RSS, 92). Responding to Hacking's paper, which tries to eliminate all talk of meaning from Kuhn's philosophy, Kuhn says similarly that "with respect to kind terms, aspects of a theory of meaning remain at the heart of my position" (ibid., p. 229) This more limited conception of meaning that Kuhn saw so important has to do, in other words, with organisation and definition of kind terms, as discussed above.

4. Conclusion

I believe that the proper origin for Kuhn's notion of meaning is the study of history. Latterly Kuhn clearly became more and more interested in a philosophical explication of his views. This transformed the early rough holistic notion of meaning to characterisation of kind terms through similarity and dissimilarity relations in lexicons and lexical structures. I have attempted to show here that Kuhn's ideas of information organization and concept definition rest on the firmer ground than has been recognised. Now it can be argued that there is empirical evidence for Kuhn's insights, which should be of interest to anyone who feels inclined to naturalistic philosophy. Furthermore, applications to the history of science have been promising. These aspects have not, and would not, be noticed if we pay too much attention to Kuhn's assumed connection to logical empiricism. And so, in answer to the question at the beginning, Kuhn was, if not the first, then one of the first post logical empiricists in anticipation of something what is now much more specifically understood.

University of Edinburgh

References

Andersen, H. (1996), "Kuhn's mature philosophy of science and cognitive psychology", *Philosophical Psychology* 9, 347-363.

Andresen, J. (1999), "Crisis and Kuhn", *Isis* 90, Supplement, 43-67.

Barker, P. (2001), "Incommensurability and Conceptual Change During the Copernican Revolution", in P. Hoyningen-Huene and H. Sankey (2001), pp. 241-273.

---

uncertainty, becoming more assured yet again, until finally settling on with a more limited concern of meaning.

Barker, P., Chen, X., and H. Andersen (2003), "Kuhn on Concepts and Categorisation", in T. Nickles (2003), pp. 178–212.

Bird, A. (2000), *Thomas Kuhn*, Chesham, Acumen.

Bird, A. (2002), "The Road Since Structure – Kuhn's Wrong Turning", *Studies in History and Philosophy of Science* 33, 443–463.

Carnap, R.(1953). "Testability and Meaning" (reprinted with omissions), in H. Feigl and M. Brodbeck (1953), pp. 47–93.

Carnap, R. (1956), "The Methodological Character of Theoretical Concepts", in H. Feigl and M. Scriven (1956), pp. 33–76.

Chen, X., Andersen, H. & P. Barker (1998), "Kuhn's theory of Scientific Revolutions and Cognitive Psychology", *Philosophical Psychology* 11, 5–28.

Earman, J. (1993), "Carnap, Kuhn, and the Philosophy of Scientific Methodology", in P. Horwich (1993), pp. 9–37.

English, J. (1978), "Partial Interpretation", *Journal of Philosophy* 75, 57–76.

Feigl, H. and M. Brodbeck (eds.) (1953), *Readings in the Philosophy of Science*, New York, Appleton-Century-Crofts.

Feigl, H. and M. Scriven (eds.) (1956), *Minnesota Studies in the Philosophy of Science, Vol. II*, Minneapolis, University of Minnesota Press.

Friedman, M. (2003), "Kuhn and Logical Empiricism", in T. Nickles (2003), pp. 19–45.

Horwich, P. (ed.) (1993), *World Changes: Thomas Kuhn and the Nature of Science*, Cambridge, Mass., MIT Press.

Hoyningen-Huene, P. and H. Sankey (eds.) (2001), *Incommensurability and Related Matters*, Boston, Kluwer.

Irzik, G. and T. Grünberg (1996), "Carnap and Kuhn: Arch Enemies or Close Allies?", *British Journal of Philosophy of Science* 46, 285–307.

Kuhn, T. (1970), *The Structure of Scientific Revolutions*, 2nd enlarged edn., Chicago, University of Chicago Press.

Kuhn, T. (1977), "Discussion", in F. Suppe (1977), pp. 500–518.

Kuhn, T. (1977a), *The Essential Tension. Selected Studies in Scientific Tradition and Change*, Chicago, University of Chicago Press.

Kuhn, T. (1985), *Copernican Revolution*, New York, MJF Books.

Kuhn, T. (2000), *The Road since Structure*, edited by J. Conant and J. Haugeland, Chicago and London, Chicago UP.

Nickles, T. (ed.) (2003), *Thomas Kuhn*, Cambridge, Cambridge University Press.

Reisch, G. A. (1991), "Did Kuhn Kill Logical Empiricism?", *Philosophy of Science* 58, 264–277.

Sankey, H. (1994), *The Incommensurability Thesis*, Aldershot, Avebury.

Sankey, H. (1997), "Incommensurability: The Current state of Play", *Theoria* 12, 424–45.

Sankey, H. (2000), "The Language of Science: Meaning Variance and Theory Comparison", *Language Sciences* 22, 117–136.

Sharrock, W. and R. Read (2002), *Kuhn. Philosopher of Scientific Revolution*, Cambridge, Polity Press.

Stevenson, R. J. (1993), *Language, Thought and Representation*, Chichester, Wiley & Sons.

Suppe, F. (ed.) (1977), *The Structure of Scientific Theories*, 2nd edn., Urbana, Ill., University of Illinois Press.

Thagard, P. (1992), *Conceptual Revolutions*, Princeton, Princeton University Press.

Ontology of Science and Common Sense Ontology

Ólafur Páll Jónsson

When we need explanations of why our physical surroundings are the way they are and behave as they do, we look to the physical sciences for an explanation. But could it be that the very world view, which the physical sciences rely on, undermines our common sense view of the world? In other words, does the ontology of the physical sciences undermine our common sense ontology? Perhaps "undermine" is not the right word since it suggests a ranking, "conflict" is more neutral. So, I shall focus on the following question:

Question (1) Is there a conflict between the ontology of the physical sciences and common sense ontology?

Answering this question turns out to be more cumbersome than one might expect. My attempt at answering it takes us through philosophy of psychology and theories of vagueness and we will have to reconsider the notion of ontological commitment. What we get in the end is conditional answer: If logic is classical, then there is conflict, otherwise not. In working through these issues I hope to shed some light on why it can be so difficult to discuss ontological issues.

Question (1) presupposes that there are such things as *the ontology of the physical sciences* and *common sense ontology* – a presupposition that is not without difficulties. Certain disagreements among physicists are perhaps best described as disagreements about ontology. And, likewise, ordinary people may disagree about ordinary objects, properly so called, and such disagreements may be best described as ontological disagreements. But even so, I think that there is a certain common ground which physicists accept – a common ground about atoms and their immediate parts, photons and some other particles, and certain physical properties such as mass, velocity, charge etc. Likewise, the ontology accepted by ordinary people is bound to include objects such as cars, mountains, roads, telephone networks and most importantly, persons, as well as various properties such as height, colors, price, cheerfulness, reliability etc. Question (1) should, therefore, not be understood as a question about the possible conflict between two well defined ontologies but as a question about conflict between two general ontological outlooks where each outlook may comprise different and even incompatible ontologies.

1. Ontological Commitment

How can we answer Question (1)? One might suggest that there is a straight forward way of answering it: One simply compares, on the one hand, ontological theories asserting ontology of the physical sciences with theories asserting the existence of ordinary objects to see whether they postulate different entities. Following this plan, we might begin with the materialism of J. J. C. Smart (1970, p. 159), who says that there is nothing in the world except the basic particles that physics describes,

> [...] there is nothing in the world over and above those entities which are postulated by physics (or, of course, those entities which will be postulated by future and more adequate physical theories).

Here we have a clear adherence to the ontology of the physical sciences. Smart does not specify what entities there are in the world. That is a task for present and future physics. For a theory asserting the existence of ordinary objects, we might consider Lynne Rudder Baker's (1999, p. 13) practical-realist picture:

> The Practical-Realist Picture incorporates a comprehensive commonsense conception of reality, whose ontology includes medium-seized objects and animals and persons with intentional states. This commonsense conception, which is shared by all people everywhere, has its own integrity, regardless of the deliverances of the sciences. On the Practical-Realist Picture, what goes on in the world – what we want to describe, predict and explain – is governed not only by laws of nature, but also by laws of society and rules and conventions.[1]

Now, is there a conflict between the ontologies of Smart and Baker? The answer is not clear. For one thing, Smart's materialism should not *eliminate* such objects as Geiger counters or accelerators, nor should it eliminate Smart himself. Smart is not an entity postulated by physics, but he should not, for that reason alone, consider himself a fiction. Rather, Smart might think that he is identical to a mereological sum (or some such construction) of entities postulated by physics. Moreover, Smart does not think that his materialism excludes properties such as being a professor,

---

1 Baker would regard Smart as a proponent of what she calls "the physicalist picture" which she considers incompatible with her own view.

having a commitment to give a talk and being a convinced materialist, none of which is a typical physical property. These properties, Smart might say, are *in the end* nothing over and above certain physical properties.

The difficulty here is that theories, whether in physics or common sense theories about ordinary objects, do not wear their ontological commitments on their sleeves. So, before we can answer Question (1) we need to answer the following question,

Question (2) What are the ontological commitments of our theories?

where "theory" can be physics or any other scientific theory as well as a common sense theory about one's everyday surroundings.

1.1. Quine on ontological commitment

Questions about ontological commitment of theories take us back to Quine's paper "On what there is" (1971, pp. 13–14) and his criterion for ontological commitment:

> [A] theory is committed to those and only those entities to which the bound variables of the theory must be capable of referring in order that the affirmations made in the theory be true.

In addition to the above criterion for ontological commitment, Quine (ibid., pp. 16–17) argues for the view that the acceptance of an ontology is

> [...] similar in principle to our acceptance of a scientific theory, say a system of physics: we adopt, at least insofar as we are reasonable, the simplest conceptual scheme into which the disordered fragments of raw experience can be fitted and arranged. Our ontology is determined once we have fixed upon the over-all conceptual scheme which is to accommodate science in the broadest sense.

Quine's criterion has played a constructive role in ontological discourse for the past 50 years and has become an household item in analytic philosophy. And although his view about the tight relation between ontology and science is more controversial, I believe that it can be called "the received view".[2]

Quine's approach has two features which I want to draw special attention to. First, it makes explicit that one's acceptance of an ontology is not done independently of one's acceptance of theory in general; the two are inseparable. This

---

2 See, e.g., Maddy (1996).

means that matters ontological are not independent from matters of science nor in any way prior.[3] Second, and this is a feature which Quine emphasizes, withdrawing to the semantic level, i.e., discussing ontological commitments of theories rather than directly what exists, provides a common ground for the disagreeing parties. Quine (1971, p. 16) says:

> Another reason for withdrawing to a semantic plane is to find common ground on which to argue. Disagreement in ontology involves basic disagreement in conceptual schemes; [...] despite these basic disagreements, [we] find that our conceptual schemes converge sufficiently in their intermediate and upper ramifications to enable us to communicate successfully on such topics as politics, weather, and, in particular, language.

What Quine calls agreement in intermediate and upper ramifications, i.e., agreement in what we might call ordinary matters as opposed to more philosophical or scientific matters, provides, according to Quine, the common ground which is necessary for meaningful ontological debate.

It may seem that we have everything we need in order to answer Questions (1) and (2); we have (i) a criterion for ontological commitment, (ii) an account of how one accepts ontology, and (iii) an account of how ontological disagreements can be discussed. However, these elements are not enough as difficulties with vagueness and in philosophy of psychology show.

1.2. A difficulty with vagueness

If we approach the debate whether there are vague objects in the spirit of Quine, we should move to the semantic level and look at our theories and ask what their bound variables must refer to in order that their affirmations be true. This is perhaps the standard way of proceeding, but by moving to the semantic level we risk losing sight of the deep issue.

An example of how the deep issue gets lost when vagueness is discussed on the semantic level is Mark Sainsbury's "Why the World Cannot Be Vague". In this paper Sainsbury (1994, p. 63) argues that one can find no clear thesis of ontological vagueness as opposed to linguistic vagueness:

> Whether or not the world is vague appears to be a deep issue. This paper reports the results of a search for a clear and substantive thesis to the effect that the world itself,

3 Quine maintains that science is continuous with both metaphysics and common sense, a view which has been called "gradualism". See, e.g., Maddy (1996).

as opposed to our language or concepts, is vague. Such a thesis of "ontic vagueness" ought to reflect the apparent depth of the issue, and so its truth or falsehood should be a matter of controversy. I did not find such a thesis.

Sainsbury (ibid., p. 68) considers vagueness with respect to parthood and says about it:

> If "part of" is vague, [a friend of the linguistic view] can [...] believe that there are compositionally vague objects. They are objects which, relative to some object, are borderline for "part of". Being thus borderline is an ontic property, in that it applies to nonlinguistic things, but it is not itself the property of being vague, and its holding of an object is explained by vagueness in representations.

It is clear what the intended reading of the basic claim of the ontological view is, namely that there are vague objects. But Sainsbury's point, as I understand him, is that there is an unintended interpretation of the basic claim of the ontological view, a linguistic interpretation, which enables one to reap all the benefits that come with the ontological view without committing one to ontological vagueness. If Sainsbury is right, then what remains to distinguish the linguistic view, which we get from the unintended reading, from the intended ontological view, is the bare claim that there are vague objects. A claim that nothing hinges on.[4]

The ontological view of vagueness entails certain linguistic vagueness which, in turn, may seem to render any claim about ontological vagueness redundant. This is, in effect, Sainsbury's point. An example makes this clear. Suppose that Sparky is a pebble on the border of Kilimanjaro. Then the sentence

Sentence (1) Sparky is part of Kilimanjaro

is indeterminate. Let us further assume that the source of vagueness is ontological, that Sparky is a borderline case of being a part of Kilimanjaro because Kilimanjaro, the very mountain itself, is vague. This means that the ordered pair ⟨Sparky, Kilimanjaro⟩ is a borderline instance of the predicate "is part of" which, in turn, means that the predicate "is part of" is vague. We assumed ontological vagueness and now we end up with linguistic vagueness as well and, moreover, it seems that

---

4 I do not agree with Sainsbury's conclusion, although I think that his criticism of some of the formulations of the ontological view is appropriate. But that is not what is relevant right now. What is relevant is to see how Sainsbury misses sight of the deep issue by moving to the semantic level.

the philosophical work which ontological vagueness was supposed to do is, anyway, done by linguistic vagueness.[5]

The difference between the two views will not be apparent if we rest with questions about the truth value of sentences, or affirmations of theory, i.e. if we consider only whether a sentence is true, false or indeterminate. It is not even enough to ask what explains the truth status of a sentence such as (1) since the indeterminacy of (1) can be explained by the vagueness of the parthood predicate. We have to take the inquiry one step further and ask what explains the vagueness of the parthood predicate itself. And to answer this question one must settle on a particular ontology, one must say what material objects are like.

The way in which ontology permeates the discussion of vagueness is particularly clear in the supervaluationist framework. Suppose that the Sentence (1) is indeterminate. A supervaluationist account of this is given in terms of acceptable precisifications. A precisification of a singular term such as "Kilimanjaro" is a single object, a precisification of general term such as "mountain" is a set of objects. If one is going to spell out a supervaluationist account of vagueness one has to say more about what kind of entities the referents of ordinary names, such as "Kilimanjaro" and "Sparky", are in a precisification. One such account is given by Vann McGee (1998) in his paper "'Kilimanjaro'":

> For the present, let's suppose precise physical objects to be mereological sums of elementary particles. Then Kilimanjaro, being a physical object, is a mereological sum of elementary particles.[6]

Here McGee defines the acceptable precisifications in terms of physical objects which are mereological sums of elementary particles.

The indeterminacy of Sentence (1) is now explained by pointing out that in some acceptable precisifications the object that "Sparky" refers to is part of the mereological sum of elementary particles that "Kilimanjaro" refers to, in others it is not. And this is a perfectly good explanation of the indeterminacy of Sentence (1). The only difficulty with it, in the present context, is that it presupposes a particular ontology. Moreover, someone who does not accept the ontology of the supervaluationist is bound to disagree with him about what he says about ordinary language.

---

5 Mark Sainsbury makes much of this in his paper "Why the World Cannot Be Vague" (1994), especially sections II and III.

6 McGee elaborates the ontology later in his paper but that is not relevant for the present purposes. What is relevant now is to see how fundamental questions about ontology are unavoidable when discussing ordinary language.

Philosophers who disagree about vague objects also disagree about how ordinary names such as "Kilimanjaro" or "Aristotle" should be interpreted. Friends of vague objects say that names such as these should be taken at face value, that they are precise names referring in the classical sense to particular objects. But they refer to vague objects. Friends of the linguistic view say that these names are vague and that they refer only indeterminately to precise objects, say mereological sums of elementary particles. The difficulty in seeing the deep issue is not, as some say, that there is no deep issue, but, instead, that it is difficult to state one's position without begging the question against those with whom one is disagreeing. The upshot is that we might not be able to move to the semantic level independently of accepting one ontology or another. This means that the alleged virtue of Quine's approach to ontological disagreements, the common ground, is not available.

1.3. A difficulty with psychology

Another source of difficulties for Quine's approach has to do with the role of common sense theory, or theory about the objects which "the intermediate and upper ramifications" of our conceptual scheme describe. This is particularly clear in psychology where theoretical issues such as explanation of behavior are intimately connected with common sense beliefs. Lynne Rudder Baker says that her car is worth no more than the blue-book value and that this property of her car is (i) explanatory, i.e., it explains why the insurance company will not pay for it's repair, (ii) not taxonomic in any physical science, and (iii) that it does not matter that the instantiation of the property by her car is not explainable by any physical science. But whether reference to such properties as the one Baker mentions is acceptable in scientific psychology is directly related to questions about the ontological foundations of psychology. This means that when psychologists disagree about ontology, then ordinary language my not provide a common ground at all. In fact, what Quine describes as common ground turns out the be the main battleground. Again we find that the main virtue of Quine's approach is not available.

These considerations are not meant to undermine Quine's criterion. His criterion is good as far as it goes. But finding a common ground for ontological debates is difficult, and it is more difficult than Quine realized. The upshot is that in order to identify the "deep" issues in ontological disagreements and specify the terms in which such disagreements should be discussed, it is not enough to move to the semantic level. Something different is needed.

2. Ontological Stance

Quine defines ontological commitment in terms of the affirmation made by a theory. Ontological commitment of a person is then defined derivatively as the commitments of the theories which the person accepts. And these may be questionable in many ways. They may be inconsistent, i.e., a person may be committed to there being objects of a certain kind and there not being any such objects, just as one's believes in general may be inconsistent. They may lack comprehensiveness, a person may not have bothered to develop a comprehensive world view. And ontological commitments may be arbitrary, a person may believe in physics and in ghosts and in God without any attempt to bring these commitments into a unified whole. But when a person's ontological commitments are consistent, comprehensive and non-arbitrary, and the person has made some effort to make her commitments such, I shall say that she has adopted an *ontological stance.*

Let us look closer at these three ways in which one's ontological commitments may be questionable. Consistency alone is a weak condition. First, consistency is not a property of objects but theory or sets of theories. Second, someone's ontology may be wildly arbitrary and promiscuous without thereby being inconsistent. But despite being week in the above senses, consistency is a strong condition in another sense: When a theory is found to be inconsistent very little can rescue it. In that sense, inconsistency is bound to be much stronger a requirement than either comprehensiveness or non-arbitrariness.

The comprehensiveness requirement demands that the ontological commitments of different branches of science be taken into account as well as commitments of common sense theories about the everyday environment (which may be implicit and only show in our practices). What may have to be brought together are the ontological commitments of particle physics, biology, psychology, anthropology, jurisprudence, ethics and a common sense theory about one's everyday surroundings. The class of objects that one would end up with might include such entities as quarks, waves, strings, genes, species, thoughts, beliefs, fears, sensations, customs, families, laws, legal duties, virtues, moral duties, moral commitments, roads, mountains, and oceans.

The comprehensiveness requirement is demanding in the above respect, but its demands are weak in another respect. It demands that we bring together a vast and varied set of commitments, but it does not demand that we do anything in particular with these commitments. Even adding the consistency requirement does not complicate things considerably since although one's commitments may be vast and varied, these commitments need not be inconsistent. No one claims, for instance, that mental states and physical states are inconsistent; attempts to reduce the mental to the physical, or else eliminate the mental, are not driven by the need to avoid inconsistency.

The requirement that one's commitments be non-arbitrary is perhaps more demanding but, at the same time, less clear than the other demands. The non-arbitrariness requirement calls upon us to bring the totality of our ontological commitments into a unified whole. The first difficulty one meets when trying to meet this requirement is to settle on a starting point. Smart takes the physical sciences to provide such a starting point. In this respect his philosophy is entirely within the tradition of the logical positivists. Others, for instance Lynne Rudder Baker, refuses to give science such precedence over other realms. In this respect her views are in line with Wittgenstein's views: We have nothing to work with other than the various language games and there is no single language game which lays the foundation for the rest.

3. Adopting an Ontological Stance

When Smart says that "there is nothing in the world over and above those entities which are postulated by physics" he has settled on a definite starting point. He is maintaining that commitments which seem foreign to physics, say the commitments that there are persons and mountains, are *in the end* nothing but the commitments that there are those entities postulated by physics. Here Smart has solved the difficulty of settling on a starting point by a decision.[7] Having done this, Smart has to address two additional difficulties before he can maintain that his ontological commitments form a unified whole. The first difficulty has to do with translation: Smart must translate or paraphrase sentences which he affirms and which seem to commit him to entities outside of physics in such a way that they only commit him to entities of physics. The second difficulty is to explain what should be done when such translation or paraphrasing fails. Whether the resulting ontology is arbitrary or not depends then on how these two last difficulties are met.

The materialism of Smart is an example of strongly unified ontology and from it's point of view Baker's practical realist picture appears to be irreducibly arbitrary. The unity in Smarts ontology is the product of two components. First is what we might call the ontological base. This comprises the basic entities of physics and certain kinds of constructs of such basic entities, say mereological sums. Second is what we might call the reductive element. Accepted objects are

7 Perhaps there is no principled way to settle on such a starting point. Strawson (1990, p. 313) says about the two different starting points which Smart and Baker represent that "the choice between them is, ultimately, perhaps, a matter of individual temperament".

reduced to entities at the base level or mereological sums of such entities.[8] Baker's ontology, on the other hand, has nothing corresponding to the reductive element in Smart's ontology.

The gain in terms of unity in the ontology of Smart comes with a loss in intuitiveness. What needs to be reduced (or else eliminated) is not merely the mental but also the common sense world. Categories such as persons, cars, roads, governments, songs etc. are not categories of the physical sciences. No one knows how to reduce these categories to categories of physics, and very few are willing to give them up. We might say that the gain in terms of unity in the ontology of Smart comes with risk of loss in terms of comprehensiveness.

There is a middle ground between Smart and Baker which retains the ontological unity in Smart's ontology, and some, or even much, of the intuitiveness of Baker's view. This is Davidson's anomalous monism. Davidson accepts an ontological base similar to that of Smart, but he rejects Smart's reductionism, holding instead that there are no general laws connecting the mental and the physical.[9] I believe that Davidson's argument for anomalous monism of the mental can be extended: There are no general laws connecting the common sense and the physical.[10] A Davidsonian argument for this position can be put in the following way:

(1) The vocabularies of physics and common sense are distinct.
(2) What physical predicates classify homogeneously need not be classified homogeneously in the vocabulary of common sense.
(3) There are no law-like relations between physical predicates and common sense predicates.[11]
(4) Entities fall under laws according to the way they are described.

---

8 We might compare Smart's position to the eliminative materialism of Paul Churchland. Eliminative materialism has a similar base, but the second element is different. Instead of identification we have elimination.

9 See Davidson (1980).

10 A similar argument can be used to argue against the reduction of biology to physiology.

11 This premiss is an empirical claim and is, therefore, in need of some empirical support. Alex Rosenberg (1994) has argued that biology cannot be reduced to physiology or any other physical science. The basic idea behind his argument is as follows: (1) Biological entities are individuated in terms of function while physiological entities are individuated in terms of structure. (2) Entities functioning in similar ways may have different structure and entities that are structured in similar ways may function differently. (3) So, there are no laws connecting biological kinds to physiological kinds. (4) Entities fall under laws according to the way they are described, (C) There are no law-like connections between biology and physiology.

(C) There are no general laws connecting entities described in physical vocabulary with entities described in common sense vocabulary.

If this argument is sound we can hold on to a strongly unified ontology – monism – without thereby having to accept the claim that ordinary objects and properties should be reduced to objects and properties of physics, or else eliminated. If anomalous monism is plausible as a view about the relation between ordinary objects and the constructs of physics, then we need not worry that there is a conflict between the ontology of the physical sciences and ordinary objects. Ordinary objects now turn out to be constructs of basic entities described in common sense terms. But, and this is what would make the position so attractive, ordinary descriptions are autonomous with respect to physicalist descriptions, just as mental descriptions are autonomous with respect to physicalist descriptions in Davidson's anomalous monism of the mental.

By giving up the reductive element of Smart a friend of a Davidsonian view can claim comprehensiveness without having to postulate the possibility of reduction of common sense ontology to the ontology of the physical sciences.

4. Is There a Conflict?

Even if anomalous monism is an option in psychology I am skeptical that it is an option for ordinary objects. The stumbling block is the problem of the many which Peter Unger (1980) has made notorious. The problem of the many is a *reductio* argument which begins with the supposition that there is a single cloud in the sky, and then concludes that there must be either a host of clouds there or else none. As a conclusion about clouds this is not very troubling, but it should be clear that similar arguments can be given with respect to almost any ordinary material object, be it a mountain or a person or what have you. Anomalous monism about the common sense is plausible only if there is a linguistic solution to the problem of the many. And that I doubt. But before I defend that claim, let's look closer at the argument.

Suppose there is a single cloud in the otherwise blue sky, we call it "Puffy". We then begin with the assumption that the following thesis about clouds is true:

Cloud Thesis: A cloud is an aggregate of droplets distributed in a suitable way.[12]

12 What "in a suitable way" means here is just that the droplets are distributed so as to form a cloud; we could say that they are cloudishly arranged.

The word "aggregate" is here borrowed from Unger who uses it to describe the relation between a collection of droplets and a cloud. He (1980, p. 415) offers the following explication of the word:

> [...] the only likely candidates [for being a cloud] will be concrete complexes composed, at least in the main, not merely of some water droplets but of a great many droplets that are "suitably grouped together".

And this seems fair enough even though it falls far short of a definition of aggregation.

Consider now two different aggregates, A1 and A2, that are identified by their constituent droplets and whose sole difference lies in the presence of one particular droplet in A1 and its absence from A2. Then the difference between A1 and A2 is so minute that there should be no reason why A1 was a cloud rather than A2, or vice versa.

Indifference Premiss: A1 and A2 are indifference candidates, i.e., they are both plausible as candidates for being the cloud, but neither is more plausibly the cloud than the other.

From the Indifference Premiss it is tempting to conclude that either both A1 and A2 are clouds or neither is, which would suggest the following thesis:

Equal Status Thesis: If A1 and A2 are indifferent then either both are clouds or neither is.

From the Indifference Premiss and the Equal Status Thesis we get the following conclusion:

Either both A1 and A2 are clouds or neither is.

Since we are assuming that there is exactly one cloud in the sky, neither A1 nor A2 can be the cloud. This alone is not alarming, but since A1 and A2 are as good candidates as any for being the cloud, by parity of reasoning we can generalize the above conclusion:

Conclusion: Either there is a host of clouds in the sky or there is none.

And since we want to say that there is exactly one cloud in the sky, this conclusion is unacceptable. Something must give.

I will discuss two ways in which philosophers have responded to this problem, a linguistic view according to which the problem shows that singular terms and count nouns in our language are vague, and an ontological view according to which the problem shows that there are vague objects.

4.1. A linguistic response

Anomalous monism about ordinary object is an option only if we can give a linguistic solution to the problem of the many. The terms "cloud" and "aggregate of droplets" classify in different ways and, according to linguistic solutions, that is all the problem of the many shows: The terms are applied in different ways without being applicable to different sorts of things.

According to linguistic views the problem of the many shows that names such as "Kilimanjaro" and "Puffy", and general nouns such as "mountain" and "cloud", are vague in the sense that they do not have any determinate reference and extension. One challenge for such views is to give an account of the truth of a sentence such as "There is exactly one cloud in the sky" when, intuitively, there is only one cloud there. If something is to be considered a solution, it must be able to retain what seems, intuitively, to be beyond doubt, for instance, that in certain circumstances it is determinately true that there is exactly one cloud in the sky.

The linguistic view consists of two parts. The first is an explanation of why there are vague words in a language like English, the second is a systematic account of the truth conditions of complex sentences in a language containing vague words. The explanation of why we have vague words in our language is given by a thesis called *semantic indecision*, a thesis that consists of two sorts of claims: a claim about what facts are relevant for meaning determination, and the claim that these facts leave meaning underdetermined. In the case of singular terms semantic indecision will be argued for in terms of which objects are candidate referents of names, and to what extent the meaning determining facts might distinguish among those objects as the actual referents of names.

The second part of the linguistic view is an account of the truth conditions of complex sentences with vague constituents, whether these are common nouns, names or predicates. The sentence "Kilimanjaro is the tallest mountain in Africa" is certainly true but, for friends of the linguistic view, this is not explained by the referent of the name "Kilimanjaro" satisfying the predicate "is the tallest mountain in Africa", as standard semantics has it. They do not believe that there is such a thing as *the* referent of the name "Kilimanjaro" or *the* extension of the common name "mountain".

At this point linguistic views differ among themselves. In what follows, I restrict my discussion to views based on supervaluation. Supervaluation enables friends of the linguistic view to account for the truth values of ordinary English sentences when some constituent expressions are vague. In particular, it allows the sentence "There is exactly one cloud in the sky" to come out true when, intuitively, there is only one cloud there, despite the challenge of the problem of the many; in each precisification there is exactly one object that is a cloud in the sky, and so in each precisification the sentence "There is exactly one cloud in the sky" comes out true.

What the problem of the many shows is that in different acceptable precisifications different objects satisfy the predicate "is a cloud in the sky" and, therefore, different objects will be the referent of the name "Puffy". This leads to the conclusion that it is true that there is exactly one cloud in the sky even though it is not true of any object that it is the cloud in the sky, for different precisifications disagree about which object it is.[13]

Following Smart, the acceptable precisifications might be defined in terms of the basic entities of physics and mereological sums of such entities. The resulting ontology will be strongly unified, although it would resemble the eliminative materialism of Churchland more than the reductive materialism of Smart. However, even if the ontological unity is the result of elimination of ordinary objects, the view does not have the further implications that we ought to give up ordinary discourse and stop using nouns such as "mountain" or "cloud", or names such as "Kilimanjaro" or "Puffy". Ordinary discourse is accepted as a useful fiction. The vagueness of a name such as "Kilimanjaro" consists in it not referring determinately to a single object, namely a certain mountain. That was just a fiction. Instead, the name refers indeterminately to one or another of a host of mereological sums of basic entities. The main virtue of the linguistic view is that it can retain a unified ontological base while keeping at least some of the common sense intuition by being able to account for the intuitive truth values of sentences containing vague expressions.

Our initial question, whether there is conflict between the ontology of the physical sciences and common sense ontology, does not have a simple answer according to the linguistic view. From a purely ontological point of view, there is no conflict since common sense ontology is empty; there are no ordinary objects and what does not exist cannot be in conflict with anything. But does the linguistic view belong to an acceptable ontological stance? The crucial question is whether

---

13 The sentence "There is something that is the cloud in the sky" is determinately true, since it is true in each precisification, and yet it is not true of any thing that it is the cloud in the sky. Here, truth behaves like necessity: the sentence "Someone is the tallest boy in the class" may be necessarily true (at least, if by "tallest boy" we meant "taller or equally tall as any other boy") and yet it is not necessarily true of any boy in the class that he is the tallest one.

the linguistic fiction can do justice to our ordinary intuition, in particular whether it can meet the comprehensiveness requirement, i.e., whether what gets lost by the elimination of common sense ontology can be reclaimed by the linguistic fiction. And here I am inclined to give a negative answer.

Let us first notice that there are certain anomalies which come with the linguistic view spelled out in terms of supervaluation, for instance the sentence:

There is a certain object that is the cloud in the sky,

is going to be true in each precisification, and hence true simpliciter. And yet it is not going to be true of any object that it is the cloud in the sky. Examples of this sort can be multiplied. But there is another and more serious worry which friends of the linguistic view must face. Unger (1980, sect.12, part A) mentions the problem under the heading "the problem of having an object in mind".

> We commonly suppose that, regarding various existing ordinary things, several nearby stones, for example, we can think of each of them, or have each in mind. [...] But if there are millions of "overlapping stones" before me [...] how am I to think of a single one of them, while not then equally thinking of so many others, with each of which "it" might so readily be confused? The presumed *relations* between us, and our minds, and ordinary material complexes look to be in deep trouble.

And a little later Unger (ibid., p. 456) adds:

> I suggest that up until now, at least, not one of us has ever really thought of any existing stone or table or human hand.

Now, what shall one make of this problem? It is not a contradiction and, hence, does not violate the consistency requirement. But it shows that if we accept the linguistic view, we need a major revision of how we think of relations between us and material objects. The difficulty here is that in so far as the linguistic view keeps the intuitiveness of common sense conception of reality, this conception is of a purely linguistic construction. The comprehensiveness requirement is met in the sense that sentences about ordinary objects, such as "There is only one cloud in the sky" and "Kilimanjaro is the tallest mountain in Africa", receive their intuitive truth values. But in another respect the linguistic view fails to meet the comprehensiveness requirement since such sentences do not admit of their intuitive interpretation. My point is not that the linguistic solution must be rejected, only that if one accepts it, one is faced with a conflict between common sense ontology and the ontology of the physical sciences.

4.2. An ontological response

A different response to the problem of the many is to take ontological commitments of ordinary language at face value. Then we insist that there is a single cloud in the sky and, moreover, we take the existential sentence

$\exists x$ (x = the cloud in the sky)

to be true just in case there is a certain object which is the cloud in the sky. This, I maintain, commits us to vague objects.

The view that there are vague objects – the ontological view – requires a rejection of the first premiss of the problem of the many, Cloud Thesis: A cloud is an aggregate of droplets distributed in a suitable way.

This premiss identifies an ordinary object, a cloud, with a construct of some basic entities, in this case an aggregate (or mereological sum) of droplets. But by postulating vague objects such an identification must be rejected. What I suggest in place of Cloud Thesis is a thesis about the constitution of clouds by fusions of droplets:

Cloud Constitution Thesis: A cloud is, at any time, constituted by a fusion of droplets distributed in a suitable way.

I assume that we can help ourselves to a notion of constitution so that Cloud Constitution Thesis has a clear meaning.[14] By making this change in the argument, and corresponding changes elsewhere, we end up with the conclusion that there are either many fusions of droplets that constitute our cloud or else none. But choosing between these disjuncts is not difficult, we opt for the first one: There are many distinct fusions of droplets that are equally a constituter of Puffy. That's what it is for Puffy to be a vague object. Solving the problem of the many in this way is no new idea, Mark Johnston (1997, p. 56) argues for it:

> [...] recognizing that constitution is not ever identity, which involves recognizing a distinction of category between a material object and the matter which constitutes it, is crucial in dealing with the problem of the many.[15]

---

14 Our notion of constitution could be defined along the lines of Judith Jarvis Thomson's notion of constitution in her paper "The Statue and the Clay" (1998).

15 Johnston does not offer any definition of constitution which makes his solution to the problem of the many incomplete. He does, however, make use of supervaluation in cashing out the notion of vagueness that he has in mind. Such a use of supervaluation is, I believe, fundamentally mistaken since supervaluation is exclusively a linguistic device.

The idea that there is a categorical distinction between the matter of which an object is composed and the object itself goes back at least to Aristotle.[16] His main concern in these matters was to give an explanatory account of the difference between constituent matter and a constituted object as can be seen, for instance, at the end of *Metaphysics* Z (1041b 5-9), where Aristotle asks "Why certain materials are a house?", and gives the following answer:

> Because that which was the essence of a house is present. And why is this individual thing, or this body having this form, a man? Therefore what we seek is the cause, i.e. the form, by reason of which the matter is some definite thing; and this is the substance of the thing.

Aristotle also asks whether the essence of an object (say a house) is definable in terms of the materials (say bricks and beams) from which the object is made? To this question Aristotle gives a negative answer. The materials can only account for the material cause of the house, not its formal or final cause. Similar ideas can be seen in contemporary philosophy, in particular in philosophy of biology. Alexander Rosenberg has, for instance, argued extensively that biological entities, such as animals or parts of animals, organs or even genes differ from their constituent materials in such a way that reduction of the biological to the physiological or chemical is impossible.

By accepting constitution as a relation distinct from identity, we must inquire into the ontological status of constituted objects. Ontology with ordinary objects constituted by basic entities has a base in a similar sense in which Smart's ontology has a base. At the ontological base are entities (perhaps the entities of physics) which constitute other entities but which are not constituted by other entities. In this sense the ontological view of vagueness retains what is perhaps the most important intuition behind materialism such as that of Smart. But vague objects violate the outlook of Smart in another respect. His ontology is what we might call finite: Once the ontological base has been specified and mereological sum has been defined, a complete account of the ontology has been given: Everything either belongs to the base or is identical to a fusion of entities belonging to the base. In contrast, common sense ontology such as that of Baker is open ended. What exists isn't exhausted by entities belonging to the base or entities identical to mereological sums of such entities. There are the ordinary objects, tables, chairs, mountains, persons, etc which, although wholly constituted by basic entities, are not

16 Aristotle was more concerned about an explanatory account of the difference between an object and its constituent matter than a descriptive account such as offered by Judith Thomson (1998).

identical to mereological sums of basic entities. And these will most often be vague.

The difference between the finite ontology of Smart and the open ended ontology of Baker becomes particularly clear if we consider what Peter van Inwagen (1990, p. 21) calls "the special composition question":

> [The special composition question] is about the mutual relations that – at least in typical cases – hold among various objects of the same type (among atoms, among planks, among cells, among bricks), relation in virtue of which they are bound together into a whole.

Van Inwagen (ibid., p. 1) holds "that there are no tables or chairs or any other visible objects except living organisms". Accordingly, he offers the following answer to the special composition question:

> $\exists y$ (the *x*s compose *y* if and only if the activities of the *x*s constitutes a life or there is only one of the *x*s).

Smart should also be able to give a decisive answer to the special composition question. His answer will differ markedly from that of van Inwagen. The concepts "life" and "constitution" have no role in Smart's ontology. Instead, he might rely on "mereological sum" and "identity". His answer should be along the following lines:

> $\exists y$ (the *x*s compose *y* if and only if the mereological sum of the *x*s is identical to *y*).

In contrast to these two proposed answers to the special composition question, Baker should decline to give a single answer to it. According to her view, there are all sorts of object. In response to the special composition question she might begin a list in the following way:

> $\exists y$ (the *x*s compose a chair *y* if and only if the activities of the *x*s fulfill the role of a chair).
> $\exists y$ (the *x*s compose a table *y* if and only if the activities of the *x*s fulfill the role of a table).
> $\exists y$ (the *x*s compose a person *y* if and only if the activities of the *x*s fulfill the role of a person).

There is, however, no reason to think that this list has an end nor is there any reason to think that it is possible to get around using sortals on the left hand side of the biconditional and mentioning corresponding roles on the right hand side except for the uninformative:

> ($\exists y$ the *x*s compose an object *y* of kind K) if and only if the activities of the *x*s fulfill the K-role.

Because of the open-ended character of common sense ontology it is bound to seem arbitrary from the point of view of a finite ontology such as that of Smart. But is common sense ontology really arbitrary? I think not, or rather, it is no more arbitrary than is our recognition of ordinary objects. When it comes to ordinary objects there is no question whether we should recognize them. We do recognize them, we may be able to tinker with the exact characterization of such objects, we might be able to disallow a few, but it is not possible to get rid of the class of ordinary objects altogether.

In common sense ontology there is a break between the ontological base and ordinary objects. We give up monism in favor of dualism, or perhaps pluralism. This does not mean that there is a conflict between the ontology of the physical sciences and common sense ontology. An acceptable ontological stance will include both, and each will be autonomous in its own realm. Our ontological stance will be promiscuous, but in the absence of a catholic requirement to the effect that one should be wedded to one and only one kind of objects, such promiscuity is just among the pleasures of life.

Does this mean that all is well? It depends. The promiscuous life is in a danger of logical infidelity. The purity of set theory and logic may seem to force us to accept the monistic catholic requirement. The difficulty arises because of two premisses. The first premiss is the thesis that ordinary objects are vague: The cloud Puffy does not have all its parts determinately. The second premiss is the thesis that sets are precise with respect to their members.[17] Now, these premisses entail that there is no such thing as the set of all and only those droplets that are parts of Puffy at some particular time. And this claim is far from innocent; given some rather weak assumptions, its negation can be proved in classical logic. From the following two premisses:

---

17 Michael Tye is happy to accept vague sets. Having given as an example of a vague set the set of tall men, Tye (1990, p. 536) says: "I classify a set *S* as vague, if, and only if, (*a*) it has borderline members and (*b*) there is not determinate fact of the matter about whether there are objects that are neither members, borderline members, nor non-members". I, on the other hand, believe that sets must be precise with respect to their members. In other words, I believe that set membership is a precise relation, because sets are defined in terms of their members.

(1) There is at least one droplet that is part of Puffy,
(2) There are at most n droplets that are parts of Puffy,

where n is some sufficiently large number, one can derive, using only classical logic and modest principles of set theory, the following proposition:

(3) There is a set which is the set of all and only the objects which are parts of Puffy.

The set-theoretic principles that are needed for the derivation are:

Extensionality: There are no distinct sets with the same members.
Singletons: For any x there is a set with x as its only member.
Pairwise Union: Any two sets have a union.

What this shows is that given set theory and classical logic, an object which has at most n objects of a certain kind as parts (say, a cloud that has at most n droplets as parts) cannot be vague in the sense that its parts do not form a set. Since (3) is provable given the initial assumptions it must be determinately true given the determinate truth of those assumptions. That, however, contradicts my view.

There are three ways one might try to resolve this contradiction. First, one might argue that (3) did not, after all, contradict the claim that there is no set which is the set of all and only those droplets that are parts of Puffy by claiming that, whenever a sentence had an indeterminate constituent, it was subject to some sort of paraphrasing that avoided the contradiction. Second, one might give up set theory. Third, one might give up classical logic.

The first way might be motivated by the observation that the embedded biconditional in (3) has the notion of parthood on the right side, and since parthood can be vague, we need an interpretation of the biconditional which allowed that the whole proposition could be true while a constituent proposition was indeterminate. But I do not see how this could serve my purposes. We can prove (3) given set theory and classical logic which means that, given these assumptions, (3) must be determinately true. So, the truth of (3) given (1) and (2) together with modest set theory and classical logic, is not something an interpretation of the biconditional can call into question; rather, any interpretation of the biconditional must accommodate the truth of (3) given the initial assumptions.

The second way, i.e., giving up set theory, does not look promising either. The problem is that the challenge could be repeated in terms of collections or pluralities. So abandoning sets would not be enough, one would also have to abandon collections and pluralities. And having done that, most of mathematics would have to be abandoned as well, and with that the price has become too steep.

The remaining option is to give up classical logic. Logic does not belong to ontology though it is part of any ontological stance. That, however, is not to say that the logic must be classical in so far as there are alternatives. Now, if we return to the question with which we began, namely:

> Is there a conflict between the ontology of the physical sciences and common sense ontology?

We can answer: If logic must be classical, then there is conflict, otherwise not.

University of Iceland

References

Aristotle (1928), *Metaphysica, The Works of Aristotle Translated into English, Vol. 8*, Oxford, The Clarendon Press.

Baker, L. R. (1999), "What is This Thing Called 'Commonsense Psychology'", *Philosophical Explorations* 2(1), 3-19.

Barrett R. and R. Gibson (1990), *Perspectives on Quine*, Oxford, Blackwell.

Borst, C. V. (ed.) (1970), *The Mind/Brain Identity Theory*, London, MacMillan.

Davidson, D. (1980), "Material Mind", in Davidson (1980a), pp. 245-261.

Davidson, D. (1980a), *Essays on Action and Events*, Oxford, Oxford University Press.

Johnston, M. (1997) [1992], "Constitution is Not Identity", in M. Rea (1997), pp. 44-62.

Kazmi, A. A. (ed.) (1998), *Meaning and Reference*, Calgary, University of Calgary Press.

Maddy, P. (1996), "Ontological Commitment: Between Quine and Duhem", *Philosophical Perspectives* 10, *Metaphysics*, pp. 317-341.

McGee, V. (1998), "'Kilimanjaro'", in A. A. Kazmi (1998), pp. 141-196.

Quine, W. V. O. (1971), "On What There Is", in Quine (1971a), pp. 13-14.

Quine, W. V. O. (1971a), *From a Logical Point of View*, 2nd edn., Cambridge, Mass., Harvard University Press.

Rea, M. (ed.) (1997), *Material Constitution: a Reader*, Lanham, Rowman and Littlefield.

Rosenberg, A. (1994), *Instrumental Biology: Or the Disunity of Science*, Chicago, University of Chicago Press.

Sainsbury, M. (1994), "Why the World Cannot Be Vague", *The Southern Journal of Philosophy* 33, Supplement, 63-81.

Smart, J. J. C. (1970), "Materialism", in C. V. Borst (1970), pp. 159-160.

Strawson, P. F. (1990), "Two Conceptions of Philosophy", in R. Barrett and R. Gibson (1990), pp. 310-320.

Thomson, J. J. (1998) "The Statue and the Clay", *Nous* 32(2), 149-173.

Tye, M. (1990), "Vague Objects", *Mind* 99, 535-557.

Unger, P. (1980), "The Problem of the Many", *Midwest Studies in Philosophy* 5, 411-467.

Van Inwagen, P. (1990), *Material Beings*, Ithaca, NY, Cornell University Press.

Should Philosophy of Science Conform to Scientific Practices?

Harri Sahavirta

Philosophical theories of science are supposed to be (at least partly) descriptive: philosophy of science studies science, which presents itself in scientific procedures. In short, the aim is to give an account of some central features of science: of the structure and practices of science. Therefore, it would be rather natural to require that philosophical theories of science should conform to scientific practices. However, this requirement turns out to be problematic in many respects.

Firstly, scientists do not usually identify themselves (or their own working methods) from proposed philosophical theories. This is perplexing since scientists are supposed to be experts on science – they should know what science is about and what they are doing. This evokes the following question: should philosophical theories conform to scientific practices so scientists can identify themselves. If not, there should be some kind of explanation to why philosophers are more aware of the central features of science than the scientists themselves. Secondly, philosophers have frequently claimed that their theories of science correspond with scientific practices, and they have disapproved of earlier theories, because these are not in harmony with these practices. The result has been often a peculiar kind of chain: successive theories of science claim to describe scientific practices, but are condemned for the fact that they cannot explain said practices. Thus, one may ask why it is so hard to give an accurate description of science. Is there something in the nature of philosophy of science which could explain this? Or should we search for an answer from the nature of science. These above questions may seem simple and innocent but lead to some profound metaphilosophical considerations – to which there are no simple answers.

While searching for answers, one may first notice that philosophers have often proposed theories which are not very intuitive – and sometimes, they are not even supposed to be descriptions of actual scientific practices. Thus, Rudolf Carnap attempted a rational reconstruction of the scientific concepts; he wished to reduce all meaningful concepts to some immediate experiences (from which these concepts could be constituted).[1] Karl Popper opposed such a reduction, because it

---

1 Carnap admitted that his reconstruction in the *Aufbau* did not describe the actual process of science; it was a logical reconstruction *in a rationalizing or schematizing fashion.* Carnap (1969, p. 89).

was not in accordance with the actual scientific practices: scientists do not attempt to reduce all the concept they use to some immediately given data.[2] On the contrary, he claimed that scientists are testing their theories to see where they fail. Carnap admitted that Popper had a point: Popper's theory was consistent with scientific procedures.[3] However, Popper did not accentuate positive evidence and insisted that there are no inductive procedures at all, but instead he declared that research proceeds by conjectures and refutations; scientists attempt to falsify their theories.[4]

Some scientists have agreed with Popper, but many others have found his views odd: normally scientists are trying to find some evidence for their theories. Moreover, in the sixties, Thomas Kuhn challenged Popper's view by noting that it did not comply with scientific practices; the history of science has not proceeded through conjectures and refutations. Likewise, Kuhn disputed the view that science is a logically rigorous enterprise and questioned whether there can be falsifications at all. He stressed that, in fact, scientists are working within some framework and that research is not guided by (the wish to attain) the truth, but by a paradigm and consist of puzzle solving – that is, of attempts to make the observations agree with the paradigm.[5] Kuhn's theory has been very popular. Nevertheless, many scientists (in natural sciences) find this position distorted, because scientists tend to rely on their observations and consider them as facts. In short, scientists are often attracted to (even naive) scientific realism and do not usually regard their results as paradigm-determined. Besides, Popper could claim that Kuhn's theory did not actually comply with scientific practices since there are usually many rival

---

2 Popper (1935; 1992a) claimed that scientific concepts cannot be constructed from some observations sentences because they include universal theories. This was a part of Popper's critique of induction. However, Popper (1994a, p. 48) had earlier pointed out that this does not conform to scientific practices.

3 Popper's early manuscript *Die Grundprobleme* was appraised by Carnap because there was a correspondence between its epistemology and scientific practice. But, in *Logik der Forschung,* a shift had taken a place: Popper defended his demarcation because it clarified and resolved some problems not because scientist would actually use it. Hacohen (2002, p. 247.) Nevertheless, the conformity to actual practices was a crucial point for Popper since the only criterion which could be used to evaluate logically consistent epistemologies was their applicability to factual scientific procedure and he claimed that views of the Vienna Circle did not conform to the scientific practice which used hypothesis and predictions, but his own deductivism was both prescriptive and descriptive. (Ibid., p. 197.)

4 These are Popper's central thesis, See, e.g., Popper (1935) and (1992a).

5 These are very Kuhnian views. See, e.g., Kuhn (1970) and (1979a).

paradigms (and not only one as Kuhn assumed).[6] The final step was taken by Paul Feyerabend who insisted that since science is not a logically rigorous business (as Popper supposed) and one cannot find a clear pattern of the development of science (as Popper and Kuhn had assumed), there are no common patterns or rules for science whatsoever. There is no definite scientific method at all; there are no rules which scientists could not break in the name of science.[7]

Consequently, Popper and Kuhn (and even Feyerabend) reconciled that philosophy of science should conform to actual scientific practices; it should include a valid description of scientific procedures, which is in accordance with the history of science.[8] Nevertheless, both of them proposed theories which have not appeared familiar to scientists. Now, this seems to lead to a predicament: philosophers suggest that scientists do not actually understand the logic of their own methods, nor the sociological and psychological factors, which are included. Plainly, this does not appear to be a very intuitive view, since it would be plausible to claim that scientists are experts in science. However, to contradict the above is not an improvement. On the contrary: it does not sound good to say that philosophers have failed to give descriptions for some evident facts – that is, for something obvious to all scientists. In fact, one may ask why should philosophers even attempt to describe something already distinct (and clear) and how could they fail in this task? To be blunt: it seems, scientists are, on one hand, simple-minded and cannot reflect their own working methods, or, on the other hand, philosophers are simple-minded since they do not recognize evident facts.

Now, one might attempt to solve this predicament by understanding that this kind of situation is all too familiar and present in many reports which attempt to explain human actions. In psychology, scholars often try to explain human behaviour by motives and similar factors, which are not known to the agent. This applies to historical accounts and philosophical studies respectively. Here are some thoughts: firstly, one may be depressed without knowing why. Secondly, it is quite possible that Napoleon was not aware of all of his own motives. And thirdly, Wittgenstein might not identify himself in the philosophical studies written of him. Now this does not in any way mean that the respective studies concerning depression, Napoleon, or Wittgenstein would have failed. In fact, it is clear that self-reflections are not always reliable and sometimes we are not fully aware of what we are doing.

One could say that – in explanations – it is justified to go beyond the level that is known to the object itself. Here, one might even point out that, according to

---

6 Popper (1965) discussed Kuhn's views in the International Colloquium in the Philosophy of Science where Kuhn and Popper met face to face. See, Popper (1979, p. 55).

7 See Feyerabend (1993) and (1979).

8 Kuhn (1979a, pp. 1, 4; 1979b, p. 237) point out this agreement.

Popper, the aim of science is to explain the known by something unknown: research attempts to find some hidden structures which explain the actual phenomena. Consequently, it could be suggested that *descriptions* of science tell us how scientists act and how they understand their own acts, but there are also *explanations* of science and these are descriptions about the hidden structure of these actions. It may even seem that philosophy should supply explanations; study the hidden structures; and not only to describe something that is already evident. But is this the solution to the predicament? There is a point in this but – ultimately – to say (like Kuhn) that scientists are educated into a paradigm and that this makes them blind to see what they really are doing, is to say that the way scientists see science is *fiction*. However, it is problematic to claim that the majority of scientists cannot understand their own work. The point is simply that scientific research is supposed to be a rational and conscious activity – and when we act rationally and consciously, we are supposed to be aware of what we are doing.

To review the above discussion: philosophical accounts of science may reveal factors which are hidden for scientists – but philosophy of science should not drift too far away from scientific practices. However, this does not explain why it is so hard to give an accurate description of scientific practices. Thus, one may still ask why philosophical theories of sciences have not appeared intuitive to scientists and other philosophers. One may search for an answer for this question from two alternative directions: one could suggest that (1) scientific procedures are such that it is hard to analyze them, or that (2) philosophical descriptions are such that it is difficult to apply them to scientific practices. However, it seems that these suggestions just emphasize different aspects of one phenomenon.

Consequently, one might claim that there is no simple, unique, and well-defined practice of science, but that scientific practices are manifold. This may be done by suggesting that different sciences have different methods, or that there are different kinds of research situations which demand different kinds of approaches, or that there are different levels in the development of science. Philosophers usually assume that this is not the case; one theory (or description) is enough to reveal the essential features of science. Hence, Popper insisted that, on one hand, there is a methodological unity: all sciences use the hypothetic-deductive model.[9] On the other hand, Popper also assumed that there is one model, which may be used to describe the development of science. He claimed that science always develops through conjectures and refutations. Kuhn did not declare methodological unity and he insisted that there are different stages in the development of science. Nevertheless, he seems to have been attracted by the insight that there is only one model

9 Popper's methodology for social sciences was based on this idea. See, Popper (1991, p. 131).

for the development. He introduced new dynamics of science: pre-scientific stage, periods of normal science and scientific revolutions.[10]

Therefore, one could say in general that ideally philosophy creates a universal theory of science – that is, a description that applies to all science. It is, in fact, quite typical for philosophers to attempt to give simple and universal accounts – although, there is, in reality, a cluster of practices. This seems to have been Feyerabend's point: scientific practices are so manifold that no definite set of rules or patterns can be applied to all science. Unfortunately, he could not offer anything more than methodological anarchism to solve the problem.[11]

Consequently, it may seem that philosophers offer us too simple and idealized descriptions. However, the problem is not as simple as this. This may be shown by considering the issue from another perspective. One may claim that this state of affairs is due to the nature of descriptions: all descriptions are simplified and idealized. In fact, no description can be perfect – all descriptions are made from some point of view and therefore they are selective. Here Popper and Kuhn agreed: theory formation cannot begin by collecting facts and it cannot proceed by collecting them until a complete description has been reached. This way one would not reach any informative description at all, but just a vast collection of details without any pattern: there would be no difference between relevant and irrelevant details and no boundaries of the description. A description has to have a point of view – some perspective which may be used to select and evaluate the material for the description. Furthermore, one should have some kind of a theory, which provides a pattern for the material. This means that descriptions leave something out, too.[12]

This suggests that there could be many good and relevant descriptions of science – made from different perspectives. Accordingly, one has to ask how we can choose between these descriptions? Now, there is an easy answer: the relevant aspects of science are, of course, those conforming to scientific practices – or which explain these practices. As a result, one has returned back to the beginning: an accurate description of science should describe the actual practices of science, or to reveal the hidden structures of these practices. But, as I have argued previously, it seems that there is neither a simple and unique practice of science, nor a simple and unique hidden structure of science. And even if there was only one hidden structure, it might be difficult to say which one of the rival descriptions has discovered it. As it is, philosophers have often pointed out that a given set of facts may be explained by several consistent theories. It is noteworthy that Popper

10 This is seen in Kuhn (1970).

11 This is to say that Feyerabend (1993) saw the problem but could not solve it.

12 This is the Kuhnian claim that we see the world through a paradigm but also Popper held a very similar view. See, e.g., Popper (1991, pp. 76–77; 1973, p. 261).

and Kuhn used – by far – the very same facts of history of science but explained them differently.

Still, one can argue even further. Popper and Kuhn agreed that we tend to see the world in light of our favourite theories. Popper held that our horizon of expectations (that is, our problems, expectations and theories) form a search-light: we see the world in light of our theories.[13] Kuhn, on the other hand, stressed that the world is always seen through a paradigm. And both of them noticed that human beings – including scientists – tend to be dogmatic. Popper claimed that human beings do not give up their theories unless they have adopted a critical attitude and the theories are refuted. Kuhn insisted that scientists hold on paradigms as long as possible – and sometimes even after that. On this basis, one might claim that Popper wished to give an accurate description of science but – in fact – he selected some facts and saw them through his favourite searchlight. In other words, he imposed his theory of conjectures and refutations upon the history and practices of science. The same holds for Kuhn: he wished to give a description of science but proceeded by creating a paradigm of the scientific process – and thereafter he saw the history of science through this paradigm. Accordingly, both of them created a theory that accounted for the very same facts; and both of them saw these facts in the light of their own theory – and thus the other theory seemed to miss the relevant points. Feyerabend, on the contrary, realized that the proposed searchlights were not enough – and thus he could see no pattern at all.

This may seem to constitute a fine solution the original problem: scientific practices are manifold and descriptions (theories) of science are selections of facts and patterns, which are used to organize our observations of science. Once we have created such a horizon of expectations, or a paradigm, we tend to hold on to it for as long as possible. Unfortunately, this solution confronts a familiar problem: one has elucidated the problems of describing scientific practices and one has given an explanation of Popper's and Kuhn's dispute, but it is not clear that Popper and Kuhn would have accepted the account (or that they would have recognized themselves). So, one has to ask if it is possible that philosophers give rational and conscious argumentations without being aware of what they are actually doing.

On the other hand, all this does not solve the problem of how one can evaluate rival descriptions (theories). It is noteworthy that Popper and Kuhn were very well aware of the fact that all observations are theory-laden and that this evokes problems for comparing theories. Kuhn even insisted that discussion between rival theories is impossible (in some cases); rival paradigms are incommensurable. For this reason, opponents may use the same words but these words have different

13 About the idea of theories as searchlights, see Popper (1995, pp. 118, 127-128; 1992b, ch. 7).

meanings, because their meanings are due to different paradigms.[14] Feyerabend radicalized this thesis of incommensurability: there is no method of science, no hidden structure, no privileged point of view, no difference between science and pseudo-science. Unfortunately, this solution does not conform to the scientific practices: scientists do not identify themselves from it and philosophers do not think that it would describe or explain scientific practices. Popper adopted the contrary approach and claimed that discussion and criticism of rival theories, or frameworks, is always possible – and even quite easy (one just has to try). We can become aware of our frameworks and criticize them and we can replace our frameworks if they do not work. This is done all the time in science.[15] And for this reason, Popper claimed that Kuhn's theory does not conform to the scientific practices.

Consequently, it seems that one can require that philosophical theories of science conform to scientific practices; this is an important evaluative principle for philosophical theories. And to a degree, you may even require that scientists should recognize themselves from these theories; scientists are supposed to be aware of what they are doing. However, philosophy of science also attempts to extract and explain hidden structures included in the scientific procedures. These explanations of the hidden structures are descriptions. And for that reason they are also idealized and selective – while the scientific practices are manifold. Moreover, one tend to see the scientific procedures in the light of some adopted theory; an elevated description. As a result, one might attempt to solve the original problem by saying that, if scientists do not recognize themselves from the theories of philosophers, it is only because scientists and philosophers regard science from different perspectives and conceptualize scientific practices differently. But this, of course, evokes the question: how can we make decisions between rival theories?

University of Helsinki

References

Carnap, R. (1969) [1928], *The Logical Structure of The World*, translated by R. A. George, Berkeley, University of California Press.

---

14 These argumentations are given in Kuhn (1970, pp. 94, 148; 1979b, pp. 266–267). Nevertheless, he wished to avoid (radical) relativism. On this point, see, e.g., Kuhn (1979b, p. 263).

15 Popper formulated these views as criticism of "The Myth of The Framework" and repeated them frequently after the dispute with Kuhn. See, e.g., Popper (1979. p. 56; 1996, p. 52; 1994b, pp. 138–139).

Feyerabend, P. (1979), "Consolations for the Specialist", in I. Lakatos and A. Musgrave (1979).

Feyerabend, P. (1993) [1975], *Against Method*, 3rd rev. ed., London, Verso.

Hacohen, M. H. (2002), *Karl Popper: The Formative Years 1902-1945 – Politics and Philosophy in Interwar Vienna*, New York, Cambridge University Press.

Kuhn, T. S. (1970) [1962], *The Structure of Scientific Revolutions*, 2nd enlarged ed., Chicago, The University of Chicago Press.

Kuhn, T. S. (1979a), "Logic of Discovery or Psychology of Research?", in I. Lakatos and A. Musgrave (1979).

Kuhn, T. S. (1979b). "Reflections on my Critics", in I. Lakatos and A. Musgrave (1979).

Lakatos, I. and A. Musgrave (eds.) (1979), *Criticism and the Growth of Knowledge. Proceedings of the International Colloquium in the Philosophy of Science, London, 1965, volume 4*, Cambridge, Cambridge University Press.

Popper, K. (1935). *Logik der Forschung – Zur Erkenntnistheorie der Modernen Naturwissenschaft*, Wien, Springer.

Popper, K. (1973), *Open Society and its Enemies 2 – The High Tide of Prophecy: Hegel, Marx, and the Aftermath*, London, Routledge.

Popper, K. (1979), "Normal Science and Its Dangers", in I. Lakatos and A. Musgrave (1979).

Popper, K. (1991), *The Poverty of Historicism*, London and New York, Routledge.

Popper, K. (1992a), *The Logic of Scientific Discovery*, London and New York, Routledge.

Popper, K. (1992b), *Objective Knowledge – An Evolutionary Approach*, Oxford, Clarendon.

Popper, K. (1994a), *Die beiden Grundprobleme der Erkenntnistheorie*, edited by T. Eggers Hansen, 2nd ed., Tübingen, Mohr.

Popper, K. (1994b), *Knowledge and the Body-Mind Problem*, edited by M. A. Notturno, London and New York, Routledge.

Popper, K.(1995), *Conjectures and Refutations – The Growth of Scientific Knowledge*, 5th ed., London and New York, Routledge.

Popper, K. (1996), *The Myth of The Framework – In Defence of Science and Rationality*, edited by M. A. Notturno, London and New York, Routledge.

Science and Philosophy – a Map of the Field

Arto Siitonen

1. Introduction

The words "science" and "philosophy" require reflection and critical analysis, when a comparison between the indicated areas of research is suggested. Otherwise the envisaged comparison easily develops into a conflict, fuelled by prejudices, mistrust and authority claims. Prejudices are not a privilege of lay persons. Scientists may have overt or concealed contempt for philosophy; philosophers may belittle the significance of research which is as it were "confined within the limits of a special science".

Scientific and philosophical inquiries share the origin in ancient Greek culture and are cultivated abreast in universities and other research institutes. There are various connecting links between them. The following themes play a key role in the shared area between science and philosophy: (1) the relations between inquiry and common sense; (2) problems, methods, hypotheses, argumentation, theories, reliability, validation; (3) foundational issues; (4) codification of research: How to allocate heuristic strategies and presentation of results in a proper way?; (5) the ethics of science: the desirable conduct of research professionals, and the moral challenges presented to research and its applications; and (6) social, political and economic issues affecting research.

The concept of science can be interpreted in a wide sense, in which it contains the social and the human as well as the natural sciences. Science itself can be an object of research, as is the case in *science studies* and in *philosophy of science*. Many valuable contributions to philosophy have been written by professional scientists.

In the present study it will first be charted which various routes from philosophy to science and from science to philosophy have been established in the literature concerning their relations. After that the theme of scientific philosophy, or of philosophy as a science, will be addressed. Finally, some reflections are presented.

2. Various Approaches

Science may seem to pose a *challenge to* philosophy, concerning issues like fruitfulness, progress, and objectivity. This means that scientific research is taken as a kind of model to which philosophy is compared. It was in this spirit that Immanuel Kant compared the achievements and methods of the new natural sciences to those of philosophy. He required in the preface to the second edition of his *Critique of Pure Reason* (1787) that philosophy should reach "the secure path of a science" ("den sicheren Gang einer Wissenschaft"; B XXIII). In the introduction to that work, he raised the question "How is metaphysics, as a science, possible?" (B 22); here one may substitute the word "philosophy" for the word "metaphysics".

The idea of *philosophy as a science* forms the very name of the *Festschrift* dedicated to Paul Weingartner, edited by Edgar Morscher and others (1981). Bernard Bolzano's *Wissenschaftslehre* (1847) does not actually distinguish philosophy from (other) disciplines of science. In Bolzano's view, philosophy is one *among the sciences*. Johannes Rehmke (1929) studies philosophy *as basic science*. Edmund Husserl (1981; originally 1911) and Heinrich Scholz (1961) consider philosophy *as rigorous science*.

A further possibility is to postulate a *scientific philosophy*. In Germany, there appeared from 1877 to 1916 a journal called *Vierteljahrschrift fuer wissenschaftliche Philosophie*. It was founded by Rickhard Avenarius together with Karl Göring, Max Heinze and Wilhelm Wundt. Its contributors included, among others, Ernst Mach, Friedrich Jodl, Friedrich Paulsen, Otto Liebmann, Eduard Zeller, Edmund Husserl, Gottlob Frege, and Moritz Schlick.

Charles S. Peirce wanted to place philosophy on a scientific basis throughout his career. The first part of Franz Brentano's book (1970; originally1903) concerns the contrast between scientific philosophy and philosophy of prejudices. Hans Reichenbach compared the new, scientific philosophy to the old, traditional philosophy in his book from 1931. In his book of 1951, he traced the historical roots of these two philosophies and analyzed the main problem areas of scientific philosophy. Bertrand Russell wrote an article on *scientific method in philosophy* (1954; originally 1918).

Gerhard Frey (1970), Derek Gjertsen (1989) and Oswald Schwemmer (1990) have each of them written a book on the theme *science and philosophy*. Cf. also the article by F. Paulsen (1877) and the book by Wolfgang Stegmüller (1969). The book edited by Yehuda Elkana (1974) traces the *interaction between* science and philosophy. Paul Dupont (1920) studies the *scientific connections* of philosophical problems. Karl R. Popper's article (1952) traces the roots of philosophical problems *in science*. Joseph Agassi (1964) has a contrary approach to that of Popper in his article on the roots of scientific problems *in philosophy*.

Closely connected to the question of roots is that of *modeling* philosophical problems through science; cf. Plato's dialogue *Meno* 84d–85b. Rudolf Carnap (1976; originally 1928) claims to have *dissolved* two philosophical problems by way of confronting them with science. Eino Kaila (1928) stresses that philosophy has, in respect to science, not only an *analytic* task of clarifying its presuppositions but also that of building up a *synthesis* on the basis of the results of science.

The idea of scientific philosophy was severely put in question by Ludwig Wittgenstein (1958). There are presumably many *critics of philosophy* among scientists; cf. S. Weinberg (1992), especially chapter VII: "Against Philosophy". In turn, some philosophers have *criticized science*; cf. Martin Heidegger (1954, p. 4), according to whom "science does not think". Compare also the book by John Passmore (1978) on critics of science. Science as a model of rationality has been put in question by Paul Feyerabend (1979).

3. On the Idea of Scientific Philosophy

Rickhard Avenarius (1877) raised the question "How is scientific philosophy, or philosophy as science, possible?" In order to be that, according to Avenarius, philosophy must fulfill the *formal condition* of conceptual organization of its contents, and the *material condition* that it is empirically founded. Scientific research needs, in order to be completed, theoretical abstraction, and thus the concepts used in various sciences become philosophical in their character. Accordingly, science itself is possible only through philosophy. This leads to a unity of science, into which philosophy and all other disciplines of research are integrated.

Avenarius had a unified conception of the enterprises of science and philosophy. He contrasted scientific philosophy with speculative philosophy and thought that the latter had been flourishing especially in Hegel's system. It would just disappear from the scene due to the development of scientific philosophy. This was the message of the journal which he founded and to the first volume of which he wrote the introduction. One may see in the logical empiricism of the following century a further development of both his formal and his material condition of scientific philosophy.

In his article "Philosophy and the Sciences: a Classification", Charles S. Peirce put forward the following thesis: "Philosophy is *positive science*, in the sense of discovering what really is true [...]" (1955, p. 60). The relation between philosophy and other sciences is characterized by him as follows: "If philosophy glances now and then at the results of special sciences, it is only as a sort of condiment to excite its own proper observation" (ibid., p. 66). From observation of sciences, philosophy goes on to imitation:

> Philosophy ought to imitate the successful sciences in its methods, so far as to proceed only from tangible premises which can be subjected to careful scrutiny, and to trust rather to the multitude and variety of its arguments than to the conclusiveness of any one (ibid., p. 229; in the article "Some consequences of four incapacities").

Edmund Husserl (1981, esp. pp. 8, 11, 16, 56) sets the conditions of *philosophy as a rigorous science*, or of *scientific philosophy*, as follows:

> Secure foundations, conceptually demarcated and thoroughly clarified, goal-setting problems, correct methods, insightful theories, a certain logical harmony.

It takes an obscurantist to not accept these features as signs of sound philosophy. It seems that philosophers from widely diverging schools can underwrite this list of desiderata, whatever their differences in thinking and articulation. Among these philosophers can be many who do not even accept the characterization "scientific" for their philosophy or for philosophy in general, but who have no difficulty in claiming that they are in line with Husserl's criteria. A basis is needed to sustain any architecture of ideas. One has to know what the problems at issue are, and there must be reliable ways of treating them. Without a view of the whole enterprise and its significance, no edifice of thoughts can be intelligible. Finally, consistence is required when the statements which belong to a philosophical theory are presented.

However, there may be differences in the ways in which the terms contained in Husserl's claim are interpreted. A sign of such differences can be perceived already in the text of the article. Husserl is complaining of the poor state of logic (ibid., p. 51) at a time when logic was in fact developing more intensively than ever before (Peirce, Boole, Schröder, Frege, Peano, Russell, etc.). Obviously his view on the concept of logical harmony was dramatically differing from the prevailing consensus among the *avant-garde* of logicians. Moreover, one may wonder: which methods are correct, then? Which are the leading problems? What kind of basis is firm enough? When is a theory insightful? These questions were controversial in 1911, and they still are.

One may note, furthermore, that Husserl's conditions can easily be accepted by scientists to characterize their disciplines, too – although among them the interpretations of the key terms in his list would presumably vary at least as much as among the philosophers.

Husserl contrasts scientific philosophy to the kind of philosophy which addresses to itself the question on meaning and value of cosmos and life, *Weltanschauungsphilosophie*. Its typical representative is, according to him, Wilhelm Dilthey, whose historicism he criticizes. But he also puts in question naturalism and includes into it the psycho-physiological psychology, to which he contrasts his phenomenological psychology. The latter, according to him, deserves the qualification "fully scientific psychology" (ibid., p. 39). This claim, to say the least, is controversial.

According to Bertrand Russell, there are two kinds of motives behind philosophical research: those which come from religion and ethics, and those which come from science. The philosophy which is inspired by science, may lay stress either on the results or on the methods of science. Russell's (1954, p. 96) aim is to consider "applying to philosophical problems certain broad principles of method which have been found successful in the study of scientific questions". What distinguishes the provinces of philosophy and the special sciences, is that philosophical propositions are general and *a priori*. Scientific philosophy merges with logic, and its approach to the problems will be piecemeal and tentative, like that of other sciences, its results will be "partial and probably not wholly correct" (ibid., p. 109). By inventing fruitful hypotheses, scientific philosophy is able to approximate truth, like other sciences.

Hans Reichenbach in his early years introduced the idea of scientific philosophy in his 1931 booklet on modern philosophy of nature:

> A new scientific philosophy, arising from the results of scientific research and dedicated to rigorous standards, is beginning to take shape. It has offered unexpected solutions to a number of old problems and has raised several novel questions whose existence had been overlooked at earlier times [...] The new philosophy is convinced that it can solve its problems only in close connection with scientific and mathematical research. It has elected as its method the analysis of actual scientific knowledge rather than the analysis of reason. (1958, p. 79.)

By the expression "analysis of reason", Reichenbach is directing his criticism to Kant, to whose philosophy he had been committed in his dissertation in 1916. He contrasts the new philosophy to the claims of those "who see in philosophy a superscience that creates its own knowledge out of pure reason, intuition, or similar pretentious sources of truth, those who, owing to their faith in such superscience, look down upon the 'merely partial' results of the special sciences [...]" (ibid., p. 80). He opposes philosophers who are "deprecating the achievements of science" (ibid.)

As one can see in the opening words of the booklet, Reichenbach stresses the problem solving process as the basic common factor of science and philosophy. Against the obsolete idea of the autonomy of philosophy, he argues in favour of the autonomy of problems, which is "basic in science and epistemology" (ibid., p. 83).

> It has always turned out that an unprejudiced analysis of problems, whether carried through from a philosophical, or mathematical, or physical point of view, furnishes the same results. (ibid.)

Except for problems, also methods are shared between science and philosophy: "It should be clear that this philosophy treats genuine philosophic problems by modern scientific methods" (ibid., p. 104). Due to its analytic approach to science, philosophy will face "a transition from the intuitive to the scientific method, from isolated speculation to scientific co-operation. Philosophy changes from being a sweeping world view to becoming a progressive science" (ibid., p. 83).

Especially the last sentence in the above quotation sounds utterly optimistic. Reichenbach is well aware that the new, scientific philosophy advocated by him means a "radical break with the philosophic method of the past" (ibid.) His radicalism is reinforced by the claim that the "modern task can only be performed without consideration of tradition, in close contact with the problems that scientific discoveries pose for the philosopher" (ibid., p. 84).

In his book *The Rise of Scientific Philosophy*, Reichenbach claims that philosophy and science should be compatible. If this is not the case, and if philosophy appears incomprehensible, it is the philosophers who are responsible (cf. 1951, p. viii).

For Reichenbach, correspondingly as for Avenarius, the contrast to scientific philosophy is speculative philosophy. It is curious that Reichenbach does not mention the journal which was founded by Avenarius, or Husserl's 1911 article – or Russell's 1918 article. Neither does he tell where he had picked up the very term "scientific philosophy" when applying it for the first time in his 1931 article. Did he deliberately hide his traces?

Gerhard Frey (1970) presents an analysis of the concept of scientific philosophy and its limitations. The essential points of his study are captured by the following paraphrases (cf. esp. p. 97, 108, 133f.):

> The problem of a scientific philosophy is only solvable if we consider all possible scientific methods as legitimate. One may not draw a sharp line between scientific and metaphysical philosophy. There is an area of philosophy – epistemology and philosophy of science – that widely allows a scientific treatment. Occasionally it then also appears that certain classical problems of philosophy are not decidable from the scientific point of view. For instance, the general question whether there are or are not evident statements, is not scientifically decidable. Moreover, there are inside special sciences metaphysical questions, and science, as well as epistemology and philosophy of science, have also metaphysical implications.
>
> Treating the problems of scientific philosophy does not require that philosophy should be subordinated under special sciences, but these problems are genuine philosophical problems. For instance, the problem of universals and the problem of existence can be treated by the method of logical analysis. This method can be applied also to hermeneutics and to history of philosophy.
>
> The limits of scientific philosophy are not exactly determinable. It is not easy to decide whether there are metaphysical questions that would allow a scientific

> treatment and solution. However, the task of scientific philosophy will be to widen the limits of possible scientific treatment of problems as much as possible.

In comparison to Reichenbach, Frey does not consider scientific philosophy as an overall transformation of the philosophical heritage that would substitute the speculative approach. For him, scientific philosophy is a certain approach to problems that has its limitations. On the other hand, he uses the concept "scientific" in a wide sense, including not only logical and empirical methods but also interpretive methods.

Wittgenstein did not accept the idea of a scientific philosophy. In *Tractatus*, he claimed that philosophy is not one of the natural sciences (§ 4.111) but it "sets limits to the much disputed sphere of natural science" (§ 4.113). Natural scientific hypotheses, such as Darwin's theory, have nothing to do with philosophy (cf. § 4.1122). Psychology is as far away from philosophy as other natural sciences (§ 4.1121). The difference between saying and showing distinguishes science from philosophy (cf. § 6.53).

In his lectures in Cambridge during the term 1933–34, Wittgenstein (1987, p. 18) rejected the idea that science could function as a kind of model for philosophy:

> Philosophers constantly see the method of science before their eyes, and are irresistibly tempted to ask and answer questions in the way science does. This tendency is the real source of metaphysics, and leads the philosopher into complete darkness.

One may think that asking and answering questions is something that tightly connects science and philosophy with each other. It may be; only the way is false, if Wittgenstein is right. There is also something else that is unconventional here. According to such advocates of scientific philosophy as Reichenbach, philosophy is by no means "the real source of metaphysics" but rather an antidote against it. What connects Reichenbach and Wittgenstein, is their anti-metaphysical orientation, but they proceed from it into diametrically opposite conclusions concerning the significance of science to philosophy.

What about Wittgenstein's argument? It consists of two sentences: "I want to say here that it can never be our job to reduce anything to anything, or to explain anything. Philosophy really *is* 'purely descriptive'" (ibid., p. 18). The context tells that by the method of science, Wittgenstein understands the search for laws and generalizations. Russell in his 1918 article characterized philosophical propositions as "general" (cf. above), as being even more general than any propositions of special sciences. As Wittgenstein's argument indicates, his approach was exactly opposite to Russell's. In itself, Wittgenstein's argument is rather meager – one would like to know more about why scientific philosophy is apt to lead into darkness. Concerning the theme of explanation, is it not the philosopher's task to make things understandable and not only describe them? If philosophy is exclusively descriptive, on the other hand, is it then not closely related to the so-called *idiographic* sciences, such as history (cf. Wilhelm Windelband's distinction between the nomothetic and the idiographic sciences)?

3. Reflections

There is a rich literature concerning the relations between science and philosophy. Today, the expression "scientific philosophy" is seldom used. It has been practically replaced by the term "philosophy of science". This does not exclude continuation as to the themes and the moral and intellectual commitments.

Many of the representatives of the idea of scientific philosophy – from Kant on – thought that scientific research can be advisory or even paradigmatic for philosophy. In philosophy of science, the accentuation is slightly different. From its perspective, science may be said to offer a challenge to philosophy in the sense that certain conceptual issues require clarification. What makes science science? How to account for and justify the procedures of observation, hypothetical reasoning, inductive reasoning, and interpretation? Questions and problems abound; the art of raising them and pondering on them is certainly one of the strongest threads connecting various forms of inquiry together.

University of Helsinki

References

Agassi, J. (1975), "The Nature of Scientific Problems and Their Roots in Metaphysics", in Agassi (1975a), pp. 208–239.

Agassi, J. (1975a), *Science in Flux*, Dordrecht, Reidel.

Avenarius, R. (1877), "Zur Einführung", *Vierteljahrschrift für wissenschaftliche Philosophie* 1, 1–14.

Bolzano, B. (1981) [1837], *Wissenschaftslehre*, Aalen, Scientia.

Brentano, F. (1970), "Wissenschaftliche Philosophie und Philosophie der Vorurteile", in Brentano (1970a), pp. 4–45.

Brentano, F. (1970a), *Versuch über die Erkenntnis*, A. Kastil (ed.), Hamburg, Meiner.

Carnap, R. (1976) [1928], *Scheinprobleme in der Philosophie*, Frankfurt a.M., Suhrkamp.

Dupont, P. (1920), *Les Problèmes des la Philosophie et leur enchainement scientifique*, Paris, Alcan.

Elkana, Y. (ed.) (1974), *The Interaction between Science and Philosophy*, Atlantic Highlinds, N. J., Humanities Press.

Feyerabned, P. (1979), *Science in a Free Society*, London, New Left Books.

Frey, G. (1977), *Philosophie und Wissenschaft*. Stuttgart, Kohlhammer.

Gjertsen, D. (1989), *Science and Philosophy*, London, Penguin Books.

Heidegger, M. (1971) [1954], *Was heisst Denken?*, Tübingen, Niemeyer.

Husserl, E. (1981) [1910–11], *Philosophie als strenge Wissenschaft*, Frankfurt a.M., Klostermann.

Kaila, E. (1928), *Beiträge zu einer synthetischen Philosophie*, Turku, University of Turku..

Kant, I. (1965) [1781/1787], *Critique of Pure Reason*, Translated by.Norman Kemp Smith, New York, St Martin's Press.

Morscher, E., Neumaier, O. and G. Zecha (eds.) (1981), *Philosophy as a Science. Essays in Scientific Philosophy*, Bad Reichenhall, Comes.

Passmore, J. (1978), *Science and its Critics*, London, Duckworth.

Paulsen, F. (1877), "Über das Verhältnis der Philosophie zur Wissenschaft. Eine geschichtliche Betrachtung", *Vierteljahrschrift für wissenschaftliche Philosophie* 1, 15–50.

Peirce, C. S. (1955), *Philosophical Writings of Peirce*, edited by J. Buchler, New York, Dover.

Rehmke, J. (1929), *Philosophie als Grundwissenschaft*, 2nd.ed., Leipzig, Meiner.

Reichenbach, H. (1958), "Aims and Methods of Modern Philosophy of Nature", in M. Reichenbach (1958), pp. 79–108). (Published originally in German as a separate booklet in 1931.)

Reichenbach, H. (1951), *The Rise of Scientific Philosophy*, Berkeley / Los Angeles, University of California Press.

Reichenbach, M. (ed.) (1958), *Modern Philosophy of Science. Selected Essays by Hans Reichenbach*, Westport, Conn., Greenwood Press.

Russell, B. (1954) [1918], "On Scientific Method in Philosophy", in Russell (1954a), pp. 95–119.

Russell, B. (1954a), *Mysticism and Logic and other Essays*, Harmondsworth, Penguin Books.

Schwemmer, O. (1990), *Die Philosophie und die Wissenschaften*. Frankfurt a.M., Suhrkamp.

Scholz, H. (1961), *Mathesis Universalis. Abhandlungen zur Philosophie als strenge Wissenschaft*, ed. by H. Hermes, F. Kambartel and J. Ritter, Basel / Stuttgart, Schwabe.

Stegmüller, W. (1969), *Metaphysik, Skepsis, Wissenschaft*, Berlin / Heidelberg / New York, Springer.

Weinberg, S. (1992), *Dreams of a Final Theory*, New York, Pantheon Books.

Wittgenstein, L. (1963) [1921], *Tractatus Logico-Philosophicus*, translated by D. Pears & B. McGuinness, London / New York, Routledge & Kegan Paul.

Wittgenstein, L. (1987), *The Blue and Brown Books*, Oxford, Blackwell.

Still Room for Philosophy of Mind?

Gunnar Svensson

1. The Challenge from Neuroscience

Reflecting on the nature of consciousness has traditionally been regarded as one of the central tasks of philosophy. However, since today neuroscientists are able to conduct detailed empirical investigations of the micro structures and the neuronal activity in the human brain, many scientists are convinced that now the time has come for a more *scientific* study of consciousness. In the words of a well-known neuroscientist: "We are now able to study, in a scientifically respectable way, what philosophers previously have only been able to speculate about".[1] Of course, this recent development within neuroscience poses a serious challenge to philosophy. Has philosophy of mind throughout the centuries been confined to mere speculations? Are past philosophical studies of the mind simply pre-scientific studies of what today can be empirically studied by neuroscience? In short, is philosophy of mind simply bad, uninformed science? In this paper, I look closer into this matter and ask whether philosophers ought to re-educate themselves into neuroscientists, or whether there still is, even in the age of neuroscience, room for what used to be called philosophy of mind.

2. Philosophy of Mind

Philosophy of mind has been at the center of philosophy at least since the dawn of modern philosophy. According to the standard view, Descartes set the stage with his dualism, according to which the mind is something fundamentally different from, yet somehow correlated with, the workings of the human brain. The exact nature of this correlation was the topic of century-long discussions between philosophers of different camps. The debate aroused by this problematic relation between mind and brain, and in a wider sense between mind and matter, led to a number of fundamentally different philosophical positions in ontology and epistemology. At one extreme were the idealists. According to them, the mind is all

1 V. S. Ramachandran from University of California at San Diego. The quotation is from the Millenium Conference "Man and Mind" at Karolinska Institute, Stockholm, Sweden, in June 2000.

encompassing and nature essentially mental in character. At the other extreme were the materialists. According to them, there is nothing but matter. They regarded mind as a mere chimera and human beings as nothing but intricate biological machines. It was hard to see how any empirical, or scientific, investigation could settle the matter. The problem was clearly a metaphysical and not an empirical one – and as such, of course, the victim of much speculative thought.

During the late 19th century, a number of attempts were made to make philosophical studies of the mind more scientifically respectable. In 1874, Franz Brentano published his *Psychologie vom empirischen Standpunkt*, in which he advocated a descriptive study of the mind, and, in 1890, William James, one of the early experimentalists in psychology, published his impressive two-volume work *The Principles of Psychology*. However, "empirical" and "scientific" must here be understood in ways different from the ones common today. The mind and the mental phenomena were regarded as essentially "inner", and, thus, the method to study them closer, "scientifically" or "empirically", was through introspection.

In 1925, C. D. Broad published his *The Mind and Its Place in Nature*, an extended version of the Tarner Lectures that he gave in 1923. In this book, Broad tried to give a comprehensive treatment of the received view of the nature of mind and mental phenomena. In this work, as well as in other contemporary but less ambitious projects, the mind was unhesitatingly regarded as a natural phenomenon. The philosophical challenge was to describe the character and the workings of the mind in a manner similar to the one in which we describe the character and the workings of other natural phenomena, i.e., to describe their place in nature. In the early 20th century, in this sense investigating and describing the nature of the mind was still regarded as one of the most central tasks of philosophy.

3. Philosophy of Mind during the 20th Century

During the 80 years that have past since Broad gave his Tarner Lectures, the scene has changed dramatically. Mental phenomena have by philosophers been regarded as epiphenomena, as nonentities, as conceptual or linguistic illusions, as myths, as identical to electro-chemical neuronal processes, as mere functional properties of the body, as emergent quasi-properties, and even as totally fictitious. Common to most of the philosophical "theories" is that mental phenomena are explained away as not being in any strict sense natural phenomena. Although we cannot help referring to them in our ordinary ways of speaking, they are regarded as something that in reality are just states, aspects, or properties of what do exist in the strict sense, viz. physical phenomena. According to these present-day views, since there are no really existing mental entities, there are strictly speaking no really existing minds either. Thus, the old metaphysical problem about the nature of the mind and its relation to the brain is dissolved – and, along with it, the problem about the mind's place in nature.

What are we to say about this development? Has the recent philosophical discussion reached a conclusion similar to the one reached by the neuroscientists referred to above? Many philosophers tend to agree that this is really the case. But, before passing the verdict, let us look a little closer at the development within philosophy of mind during the 20th century.

3.1. Epiphenomenalism

The classical "easy-way out" for a materialist is epiphenomenalism. It is not really a 20th century theory, but it has had its recent proponents. According to epiphenomenalism, mind and mental phenomena are merely by-products of the physical processes in the body and the brain; impotent by-products just mirroring the physical processes – effects lacking the power of acting as causes. However, even when seen as epiphenomena mental phenomena do have *some* kind of existence. Thus, the mere expulsion of them from the physical world does not save the epiphenomenalist from accounting for their existence. What, more exactly, does it mean for the mind to be an epiphenomenon, caused and acted upon by the physical system it mirrors, yet itself totally lacking causal power? What kind of existence is that? This pressing question is usually evaded and never answered by the epiphenomenalists. But the question remains to be answered if epiphenomenalism is to be even considered as an option for a serious position within philosophy of mind.

3.2. Logical Behaviorism

The anti-metaphysical movement during the 1930s and the 1940s held that the question about the ontological status of mental events is a pseudo-question. A number of attempts were made to show that what *prima facie* seem like immediate and pressing questions are really chimeras, nonsensical statements generated by conceptual confusions. The focus of these attempts was on the mental concepts and on the psychological language in which we describe mental phenomena. The most elaborated attempt of this kind was Gilbert Ryle's *The Concept of Mind* from 1949. Ryle concentrated on the linguistic expressions used for describing various aspects of our mental life. He found that they were all essentially expressing behavior, dispositions, tendencies, etc, i.e., in intricate ways describing bodily states and movements. "Logical behaviorism" was the label coined for this position.

According to logical behaviorism, mental phenomena are, in a fundamental sense, nonentities. After analyzing the logic of the relevant psychological concepts, or the corresponding linguistic expressions, it seems to logical behaviorists clear that what we talk about, when we talk about mental phenomena, are nothing but bodies and their complex behavior. Thus, the mental is not something hidden and only partly disclosed or expressed in behavior – it is simply *nothing* apart from behavior. In the opening chapter of his *The Concept of Mind*, Ryle tried to capture what he believed to be the source of our tendency to postulate something inner, something only partially disclosed in speech and behavior. He called it Descartes' myth, and claimed to have disclosed the mythical character of the mind by the arguments in the book. However, an attentive reader will be confounded by the last section in chapter VI, entitled "The Systematic Elusiveness of the I". From what is said in that section, which has not received as much attention as it deserves, it seems clear, at least to me, that the whole project of logical behaviorism fails. There is indeed a kind of subjective character of experience, the existence of which cannot be eliminated by the intricacies of Ryle's logical analyses of the use of our linguistic expressions. Thus, there seems in the end to be something that eludes the reduction to behavior. Not until given an account of the existence of this elusive subjective character of experience, can we take seriously the logical behaviorist position within philosophy of mind. Nothing in that direction is offered by Ryle, nor, as far as I know, by any other proponent of logical behaviorism.

3.3. Identity Theory

The 1950s saw a revival of the old view that questions about the status of the mind and the mental are real questions, not pseudo-questions. But, although mental phenomena were regarded as real phenomena, they were not seen as having any kind of separate existence. According to the new kind of theory, often called the identity theory, mental states and processes are simply *identical* to electro-chemical states and processes in the brain. The proponents spent much effort on arguing that the identification was not different in kind from other identifications made in the course of science, e.g., the identification of temperature with mean kinetic energy, or the identification of lightning with electric discharge. It was claimed that experience is just the inner aspect of physical processes in the brain. Thus, when we use our psychological vocabulary, what we talk about is something, the very same thing, that we can also talk about by using our physical vocabulary. This kind of attempt at identifying the mental and the physical does in a sense solve the old problem of causation and the correlation between mind and brain: Since there is only one kind events, there is simply no question about the one causing the other.

However, appealing as it may seem at first, this view raises more questions than it answers.

Firstly, why do only some physical events have an inner conscious aspect? Secondly, our conscious experience is in a sense unified; we do not experience a set of disassociate events, in the sense in which the physical processes in the brain are disassociate. Where does the unification take place? And, if there really is a place of unification, is it only the brain-activity in that place that has a conscious inner aspect? What is it that makes certain physical processes conscious and not others? Questions like these can be multiplied, and so far none of them has been answered in any satisfactory way.

Since no empirical study of the brain has shown any trace of a unifying process, some philosophers have drawn the conclusion that conscious experience is not *strictly* identical to physical processes in the brain, but instead an emergent holistic property dependent on the whole neural activity in the brain. According to this view, the emergent property is caused by, and for its existence dependent on, the neural processes – somehow independent of the underlying neural activity, yet with a causal power to interact with it. But is this not just an *ad hoc* solution to the problem? And does it not generate even more puzzling questions about emergence, existence, and causation?

Leaving all these difficult questions aside, I believe that the crucial difficulty for all forms of identity theories, along with the research programs associated with them, is that they appear to be, strictly speaking, impossible to fully understand. The reason is that they all rest on a metaphor about an inner and an outer aspect of the same event – a metaphor that we, in this special context, cannot make any sense of. Let me spell this out in little more detail.

When an identity theorist claims that mental phenomena are identical to neural processes, he does not mean that mental phenomena are illusory, but rather that the real event "looks" different from the inside than from the outside, i.e., different from the perspective of one who empirically observes the neural activity of a brain than from the perspective of the person whose brain is observed. The real event is in this case said to have an "inner" as well as an "outer" aspect. The outer aspect is what we describe in a neurological report; the inner aspect is what the person experiences "in his mind". But "inner" as well as "outer" are words primarily with a physical reference; they refer to different locations in space. Even the word "aspect" has primarily a spatial reference; to see something from different aspects is to view it from different angles, or from different directions. If we take these expressions literally, then the conscious experience must somehow be located inside the brain, even inside the neurons. Of course, that is absurd. To regard the mind as inside the scull may be quite informative, but when we push the metaphor further, and talk about the inner aspect of the brain activity, then the metaphor simply breaks down. We are left with a spatial picture that is to be understood in a non-spatial way. How can we possibly do that? All we are left with is that "inner" must be understood as "not outer", i.e., in this case as "non-physical". But, if "inner" here means nothing but "not outer", how shall we understand the identification between the inner and the outer? To claim that something essentially non-physical is identical to something essentially physical seems like an outright logical contradiction! To my mind, the metaphor simply cannot be understood, and thus, neither can the identity theory in any of its different forms.

3.4. Functionalism

A quite different approach to the question of the nature of the mind became popular in the 1960s and the 1970s; the approach was called "functionalism". This was at the dawn of the computer era. Since computers could simulate human thought and reach results that hitherto only humans had been able to reach, the question arose in what sense machine intelligence and machine thinking differ from ordinary intelligence and thinking. The functionalists claimed at first that concerning cognition there was no essential difference between man and machine. The words we use when describing human cognition could also be used to describe machines; the role of these words is to refer to functional properties. Later, this view was extended to include our whole mental vocabulary. According to functionalism, mental properties are all functional properties. We can use our ordinary mental vocabulary to describe the functions of complex machines just as well as we can use it to describe the functions of human beings. Thus, according to functionalism, mental states and processes are not exclusively human, nor exclusively biological; any system functionally equivalent with a human being can be said to have mental states and go through mental processes.

Now, the shortcomings of functionalism are fairly evident. It resembles logical behaviorism in that it concentrates on the role played by the words in our language in describing the sayings and doings of others, and just like logical behaviorism it encounters problems when it comes to describing one's own mental life. Many of our mental concepts are of course to a large extent functional, but others, the phenomenal, seem essentially different. Thus, many functionalists spent much effort on arguing that, contrary to first sight phenomenal properties are after all essentially functional in character. Be that as it may, the fact remains that to oneself, the mental realm is not just mapped by mental concepts, it actually does exist. Being conscious, experiencing, is not just something revealed in the functioning of the body, but something with a character of its own. A functionally equivalent system may or may not be conscious in the sense in which I am conscious. Thus, there is an essential fact that is left unexplained by the functionalists, viz. that my own mental states and processes are conscious in a way that is exclusively revealed to me.

I am well aware of the fact that many philosophers with a functionalist bent have argued that reactions like mine are unfounded, or based on misleading intuitions or mistaken views. However, my argument is more basic than that. I am pointing to the fact that something *essential* is lacking in the functionalistic framework: an explanation of the fact that there is an inner awareness of even the functional properties of the processes in the brain. Since no such explanation is given, neither any room for such explanations is left, I can only draw the conclusion that functionalism, regardless of its other merits, simply fails as an attempt to explain the nature of the mental.

3.5. Eliminativism

Even more radical was the view of the eliminativists of the 1970s and the 1980s. They argued that our mental concepts are systematically misleading, and that the illusory philosophical problems about the mind would disappear if we could only eliminate our ordinary concepts and replace them with concepts more in harmony with the actual facts. Our ordinary mental vocabulary was seen as expressing an old and out-dated theory of the mental, often referred to as "folk-psychology" – a theory not well suited for describing the results of today's detailed and systematic studies of the mental realm. The eliminativists argued, that the words we use to describe our mental life, and with them our ordinary mental concepts, ought to be replaced by words better suited for reflecting the real facts in the world. Thus, according to the eliminativists, our ordinary psychological vocabulary ought to be replaced by a vocabulary correctly categorizing the electro-chemical states and processes in the neurons making up the human brain. This elimination of our psychological vocabulary may seem hard to swallow, but its defenders argued forcefully that such conceptual changes do occur within the normal course of science, and they gave several examples from the history of science of radical theory changes involving simultaneous changes in conceptual structure. The examples were aimed at showing that our reluctance to even seriously considering this radical view of the mental is due to our linguistic and theoretical prejudices.

Now, we seem to be back where we started. Philosophers join with scientists in their claim that neuroscientists today are able to study in a scientifically respectable way what philosophers have only been able to speculate about. According to the eliminativists, we have better give up our preconceptions and begin with an open mind to study neural processes in the brain. Their conviction is that in that way our old philosophical puzzles will gradually disappear, and eventually an account of the mental will be reached fully in line with our accounts of other natural phenomena.

Again, I believe that something essential is overlooked here. Changing concepts and the conceptual structure may of course eliminate a number of problems having to do with the conceptual structure itself, but such a change would not alter the underlying facts. The fact that mental phenomena are conscious in a way that no physical phenomenon is, cannot be erased, eliminated, or explained away by any change in our conceptual apparatus. Not even the most passionate constructivist could seriously hold that the inner awareness in itself is a social or logical construction. Thus, to me it is evident that even eliminativism fails in its attempt to dissolve the most crucial problem about awareness, and, consequently, it also fails as a serious position concerning the nature of consciousness.

3.6. The Verdict

What are we, then, to say about the development within philosophy of mind during the last century? From my scrutiny of the major positions, it seems as if they all fail – and they seem to fail on the *very same* crucial issue, viz. failing to account for the very existence of the inner awareness, or the "stream of consciousness" as William James called it.

To claim that there simply *is* an inner awareness, or to claim that *future* research within neuroscience will reveal its neurological character, does not seem satisfactory. Neither does it seem satisfactory to claim that the existence of this inner awareness is illusory. Since we are all familiar with our own awareness, with our own stream of consciousness, to claim that there *is* no such awareness is of course self-refuting. Unless a purported theory of the mental can outline a direction in which we, at least in principle, could begin to approach the problematic issue of awareness, the theory is simply inapt to deal with the most crucial problems within philosophy of mind. Advocating closer empirical investigations of the brain activity is of course advocating further brain studies, but why believe that closer investigations of the physical processes would ever reveal something of an essentially different character?

Thus, we can now, after our scrutiny of the development within philosophy of mind during the last century, conclude that the philosophical discussion during the 20th century has *not* reached a conclusion similar to the one phrased by some of today's neuroscientists. None of the many proposals, or "theories" put forward by recent philosophers, seems able to deal with the most crucial issues about consciousness. Thus, the old metaphysical problems about the nature of mind and the relation between mind and brain are still very much alive. But the question remains, whether today's scientists are correct in their conviction that philosophers are just speculating while neuroscientists are able to really investigate the matter in a scientifically respectable way. Therefore, let us now turn to questions concerning the scientific approach to mind and consciousness.

4. The Scientific Study of Mental Phenomena

Any study, in order to be regarded as scientific, must be characterized by being to a certain extent systematic. The reactions of Brentano (1874), James (1890), and others, towards the end of the 19th century were reactions against speculations and intuitive arguments. What they advocated was a more systematic, i.e., a more scientific study of the mental realm. In Brentano's case it was a question of finding a systematic description of mental acts. In James's case it was a question of describing mental phenomena on the basis of systematic empirical studies.

However, what James and his contemporaries in the early days of psychology focused on were the states and processes revealed in introspection. And, even if introspective studies were conducted in a detailed and systematic way, they could not meet any ordinary standards of objectivity. There are no ways in which to independently check the adequacy of reports about a subject's own experiences, and there are no ways to compare the results from different introspective investigations independently of what is reported by the subjects themselves. None of the ordinary scientific means of eliminating subjective factors distorting empirical observations can be used in introspective psychology. Thus, introspection is no longer regarded as an acceptable source of empirical data. In order to meet the standards of today's science, any study of the mind and the mental realm must be based on repeatable empirical investigations and on evidence intersubjectively attainable. Of course, the problem of attaining such evidence in psychology is very different from the case in other natural sciences.

A common method in science to come to grips with phenomena that are hard to study directly is to find some kind of empirical criteria for these phenomena. Since mental phenomena are to a large extent revealed, or expressed, in what we do or say, it is not difficult to find behavioral criteria for many, or most, of our mental states and processes. Thus, the step from introspective psychology to methodological behaviorism is a very natural one. The behaviorists claimed that they were indirectly studying states and processes in the mental realm by directly studying the relevant behavioral criteria. Of course, any given set of criteria can be criticized for not correctly revealing the mental phenomena that they are said to be the criteria for. Thus, the behaviorists spent much effort on finding criteria so intimately connected to the phenomena themselves that they could be regarded as almost infallible. Once found, empirical criteria that infallibly reveal the presence of a certain phenomenon are easily regarded as revealing the phenomenon itself. In such cases, the step from regarding them as empirical criteria to regarding them as part of the definition of the phenomenon itself is not a great one.

However, at this point one must be extremely careful. Empirical criteria for psychological phenomena are commonly used simply because we have no direct means for studying the minds of other people. But, when we ordinarily use psychological words in our language, what we are talking about is our own and other people's mental phenomena, not the behavior by which we come to know about the mental states of others. If we come to regard behavioral, or empirical, criteria as similar to definitions of mental phenomena, then the scene is dramatically changed. What we then would be talking about, when using psychological words in our language, would be our own and other people's behavior, not the mental phenomena that we can be introspectively aware of. The step from behavioral criteria to definitions is in this case equivalent to the step from methodological to logical behaviorism. And, if we accept that step, then it follows, either that there is no inner awareness, a claim that we have already seen to be self-refuting, or that we no longer have any means in our language to speak about our own inner awareness. Neither of these two alternative conclusions seems acceptable. Thus, it seems of utmost importance in this case not to take the step from empirical criteria to definitions.

In many cases in science, where "operational" criteria have been introduced for concepts that refer to phenomena that are hard to investigate, like behavioral criteria in psychology, there is an open question whether there are any real phenomena referred to by the ordinary concepts or not. Thus, there is an impetus in science to look closer, in more detail, often with the help of instruments of different kinds. What one is searching for is often a kind of mechanism behind the phenomena studied on the cruder level. In the psychological case, the search is for a mechanism behind the outer behavior, i.e., a bodily mechanism that reveals itself in the behavior of the body. According to the empirical approach of today's science, it is of course evident that the mechanisms behind our human behavior are essentially tied to the neurological processes in the brain. And, the rapid development within science and technology during the last few decades has given rise to a number of new methods for studying the neural activity in the brain in detail. Thus, it should come as no surprise to find that many scientists today advocate neurological instead of behavioral criteria for mental states and processes. Neither should it come as a surprise to find that some philosophers argue that these neurological criteria should be taken as definitions. The move is essentially the same as the move from methodological to logical behaviorism some decades ago.

However, here, as in the case of behaviorism, one must be extremely careful. The step from empirical criteria to definitions has the same consequences in the neuroscience case as in the behaviorist one. Either one has to accept that there is no inner awareness, or one has to accept that we have no means for speaking about this inner awareness. Again, the first alternative seems self-refuting and the second one at least very problematic – especially for a scientific approach to consciousness.

In this section, I have pointed to some structural features in empirical science. These features are not in any sense dependent on the kind of instruments used in the actual empirical observations, nor on the scale or level of the empirical processes in focus. The scientific way of approaching questions concerning the world is to look closer and closer, and to find even finer mechanisms behind the crude phenomena previously studied. But, however fruitful it has proven to be in other areas of research, this kind of approach seems simply inadequate for answering the crucial questions concerning consciousness.

5. The Conceptual Issue

There is yet another problem for the neuroscience approach to the mind – a problem not primarily about the empirical aspect but about the kind of concepts used. The scientific way of empirically looking closer and closer in search of a phenomenon, or the mechanism behind the phenomenon, seems to require that there is something to look for at a specific location at a specific time. However, in the psychological case this prerequisite can be questioned.

Neuroscientists often unhesitatingly refer to the brain activity in some specific parts of the brain by using psychological words from our ordinary language. The person whose brain is observed is said to hear, see, think, feel, smell, remember, etc. Behind this use of these ordinary words lies a conviction that a mental phenomenon must be located at some specific place, or in some specific region. By looking closer at the activity in the brain, the neuroscientist is trying to locate that place or region. But why believe that there is such a location? Many words in our language do of course refer to different kinds of objects in the world; many nouns do – but not all of them. Why believe that words referring to mental phenomena do refer in a way similar to words referring to physical phenomena?

If we concentrate on the use of our ordinary psychological vocabulary, it is clear that it in no way concerns the neural activity in the brain. We use these words to talk about how people act, feel, see, hear, think, dream, and speak. What other people do, say, and show, are what I go by, when ascribing mental attributes to others. And likewise, what I do, say, and show, are what other people go by, when ascribing mental attributes to me. The character of the brain activity is obviously not in question in the ordinary use of such words.

If we consider how we learn to use our psychological vocabulary, it is also clear that its primary use is for describing other people. I learn to use the psychological vocabulary by noticing how other people use these words about themselves and others – and also about me. I learn to recognize the complex criteria for the correct use of these words by being reared into the linguistic community. In this process, I learn what counts as mental states of different kinds. When they concern other people, these criteria are about their outer behavior, about what they do, show, and say. When they concern myself, these criteria are instead about my own inner awareness. However, since I learn to distinguish different kinds of states within my own awareness by learning to use words in our common language, other people's ascriptions of these states to me are fundamental for my own ascriptions. Thus, since other people go by my outer behavior, the structuring of my inner awareness is in a fundamental way dependent on my own outer behavior. One can say that the use of our psychological vocabulary has two-fold criteria: one in the first-person case, and the other in the third-person case. But, it is extremely important to notice that these two kinds of criteria are not separate: the first-person case, the inner criteria, is logically dependent on the third-person case, the outer criteria.

From these very sketchy remarks, it seems to follow that our complex outer behavior is fundamental for the use of our psychological vocabulary. Thus, our outer behavior seems to be fundamental for the meaning of the words describing our mental life. Why believe that these words in *any* sense refer to neural activities in certain parts of the brain? At least at first sight, the criteria for the use of these words have *nothing at all* to do with what happens in the brain. Why believe that a closer study of the body, or the brain, would reveal anything of interest to what is referred to by these words?

What *is* of importance for the use, and thus for the meaning of the words in our psychological vocabulary, seems rather to be the complex behavior of human beings in our complex natural and human surroundings. Thus, if we are really interested in what we talk about when using our psychological vocabulary, we ought to engage in investigations very different from the ones that neuroscientists are engaged in. We should rather engage in investigations more in line with the kind of logico-linguistic ones that the philosopher Ludwig Wittgenstein was struggling with towards the end of his life some fifty years ago.

6. Conclusion

The crucial problem that I have focused on in this paper, viz. how to account for the fact that there is an inner awareness, turns out to be a problem that for structural reasons, empirical as well as logico-linguistic ones, seems forever beyond the reach of science. Thus, the claim that neuroscientists today in a scientifically respectable way are able to study what philosophers previously have only been able to speculate about seems simply false. The claim turns out to be based on a misunderstanding of the fundamental problems involved. From the arguments sketched above, it seems to follow that no neuroscience study of the activity in the brain can even touch the fundamental problems that have been at the center of philosophy since the days of Descartes.

When correctly understood, the subjective character of mental phenomena reveals that there is a fundamental gap between the mental and the physical realm – a gap impossible to bridge by ordinary scientific means. This conclusion may seem comforting for a philosopher, but I am not the one to advocate a disregard for the findings of neuroscience. On the contrary, I believe that there is much for philosophers to learn from the scientific study of neural processes in the brain. Of course, it is a fatal mistake to be blinded by the spectacular empirical findings of today's neuroscience into misunderstanding the character of the fundamental questions about the mental realm. But, it is just as fatal to disregard the findings of today's neuroscience, since that is to leave the floor for neuroscience to sweep away the deeper philosophical questions about the essentially mental character of mental phenomena.

Thus, to conclude, it is my firm conviction that there *is* still today a very *essential* role to play for philosophy of mind – if only correctly understood.

Stockholm University

References

Brentano, F. (1874), *Psychologie vom empirischen Standpunkt*, Leipzig, Duncker & Humblot.

Broad, C. D. (1925), *The Mind and Its Place in Nature*, London, Routledge and Kegan Paul.

James, W. (1890), *The Principles of Psychology*, New York, Henry Holt & Co.

Ryle, G. (1949), *The Concept of Mind*, London, Hutchinson.

Why the Argument from Causal Closure Against the Existence of Immaterial Things is Bad

Daniel von Wachter

Some philosophers, whom I call immaterialists, believe that there are besides material things also immaterial things, such as souls, ghosts, or a God. Others, whom I call materialists, believe that there are no such things but that there are only material things. Immaterialists can defend their view by providing evidence (e.g. by giving a teleological argument for the existence of God) or by reporting perceptual experiences of the things in question. Materialists can defend their view by criticising the arguments of the immaterialist one by one, by arguing that the immaterialist's perceptual experiences are unreliable, and by giving arguments against the existence of the immaterial things in question, such as the argument from evil against the existence of God. However, some materialists want to take a short cut by using what I call *Sweeping Naturalistic Arguments* (SNA). They try to defend materialism by putting forward metaphysical principles that support immaterialism because they exclude that a physical event may have a non-physical cause. In this article I criticise the SNA that claims that science has shown that the physical domain is causally closed. To pave the way I first describe what it would be for a material event to have an immaterial cause.

1. What Would It Be for a Material Event to Have an Immaterial Cause?

There are two ways in which a material event might have an immaterial cause: it might have an immaterial *event* cause or an immaterial *agent* cause.

If a material event, *M*, has an immaterial event cause, then the complete cause of the event at a certain time is either wholly or partially immaterial. (By saying that *A* caused *B*, or that *A* was a cause of *B*, I mean that *A* was the complete cause of *B* at a certain time or a part of it. An event usually has only one complete cause at a time but of course different complete causes at different times.) The material event, *M*, has an event, *C*, as complete cause. *C* takes place earlier than *M*, and there is a causal process of which *M* and *C* are stages. *C* might be a wholly immaterial event, that is, it consists in an immaterial thing, e.g. a human soul, being in a certain way (i.e. having certain properties) at a certain time. Or *C* might be partially immaterial and partially material, that is, there is a material event and an immaterial event, at the same time, whose conjunct is the complete cause of *M* at that time. In that case immaterial and material things co-operate in causing.

If a material event, *M*, has an immaterial *agent* cause, then *M* has no preceding cause, it has no event cause at all. It does not occur by chance or by accident either. Rather, its occurrence is due to an agent. Its occurrence is an agent's choice.[1] Some authors mean other things by an agent cause, but of the conceptions available only those along these lines claim that there is a way in which an event may come about besides event causation, and therefore only these require extra consideration here.

2. Causal Closure

Now let us address the argument from Causal Closure. Jaegwon Kim's (1998, p. 40) way of stating the principle of Causal Closure (CC) is this: "If you pick any physical event and trace out its causal ancestry or posterity, that will never take you outside the physical domain. That is, no causal chain will ever cross the boundary between the physical and the nonphysical". In other words, using "physical" and "material" synonymously, no material event, e.g. a brain event, has an immaterial cause. There is a weaker version of the principle which claims that *if* a material event has a cause, then that cause is wholly material. Further, one can formulate CC as a modal principle claiming that it is *impossible* that a material event has an immaterial cause. Let us call this the Modal Principle of Causal Closure. Unless indicated otherwise I shall mean by CC the non-modal principle that no material event has an immaterial cause.[2] However, my criticism should apply to any version of the principle that supports materialism.[3]

In either version is CC incompatible with immaterialism. (It is compatible with the existence of immaterial things that do not causally interact with material things, and in some versions it is also compatible with a material event having an immaterial cause besides its material cause (as defended by Lowe 2003), but here I want to consider only immaterialism that assumes that there are material events that have an immaterial cause and no full material cause.) So any support for CC is support for materialism. One way to defend CC would be to criticise the immaterialist's arguments one by one. If there are no immaterial objects, then there is

1 I have spelled this out in more detail in Wachter (2003).

2 For different versions of CC, see Lowe (2000, pp. 27–31) and Montero (2003).

3 For defences of physicalism with appeal to CC see Papineau (1993, ch. 1) and Melynk (2003).

nothing immaterial that could cause material events and hence CC is true. But some materialists use CC for a Sweeping Naturalistic Argument. They want to defeat immaterialism quickly by defending CC without criticising the immaterialist's arguments one by one. They say, for example, that CC is a principle of rationality or that it is supported by the success of science and that therefore we should reject immaterialism. If there are good such reasons for assuming CC then the Sweeping Naturalistic Argument from Causal Closure is successful. But I shall now argue that there are no such reasons.

CC is usually defended with reference to science. This can be done in two ways. First, one may argue that CC is a principle of rationality or of science. Secondly, one may argue that CC is supported by the success of science.

Kim (1998, p. 40) points out, that "if you reject this principle [CC], you are ipso facto rejecting the in-principle completability of physics – that is, the possibility of a complete and comprehensive physical theory of all physical phenomena". Therefore, "no serious physicalist could accept such a prospect". However, regardless of whether you would like to be a "serious physicalist", the question is just why one should assume CC. It is true that if CC is false, then (assuming that immaterial events are not governed by laws of nature) there are events in the universe for which there is no complete scientific explanation, i.e., no explanation in terms of the laws of physics. If God moved away the stone from Jesus' grave, then that movement of that stone has no complete scientific explanation. Do we have a reason to rule out this possibility? We need to look at the two arguments I just mentioned in more detail. The first argument claims that science or rationality presupposes CC, the second argument claims that CC is a result of science or is supported through the success of science. In order to be able to tackle these arguments we need to discuss first briefly whether a material event's having an immaterial cause would constitute a "violation" of the laws of nature.

3. Material Events with Immaterial Causes and the Laws of Nature

One way in which science might support CC would be through discovering laws of nature that entail CC. So what do laws of nature say about CC and about potential counter-instances to CC? There are three available answers:

(1) Counter-instances to CC, i.e. cases where a material event has an immaterial cause, would be "violations" of the laws in question and would show that some formula *L*, which was thought to be a law is not a law. Counter-instances to CC would falsify the laws. All evidence for the laws of nature is therefore evidence for CC. This would have been

David Hume's (1999, pp. 10, 12) answer, as we can see from his treatment of miracles.

(2) Counter-instances to CC would be "violations" of the laws of nature, but if they are non-repeatable counter-instances they would not falsify any laws.

(3) Counter-instances to CC would not be "violations" of the laws of nature because laws do not entail regularities of succession, but they say which forces there are in what kind of situation, and also in cases that are counter-instances to CC there are the forces that the laws of nature describe.

I have formed answer (2) in accordance with Richard Swinburne's (1970) treatment of miracles as violations of the laws of nature. The reasoning behind it is this. Assume you have a lot of evidence for some formula *L* being a law of nature. *L* explains many experimental data, is simple, and fits with the other laws. But then you have evidence that some event *C* occurred that would be a counter-instance to *L*. If you try to formulate a law that predicts *C* it becomes much more complicated or it yields many wrong predictions. In that case, Swinburne argues, *L* should still be taken to be a law and *C* should be regarded as a non-repeatable counter-instance to *L*. Giving up *L* would be against the evidence that supports *L*.

Answer (3) is the one I endorse (and shall defend in more detail elsewhere). It is widely assumed that laws entail regularities of the type "events of type *A* are always (or so and so often) followed by events of type *B*", which we may call Humean regularities. Against this, I think J. S. Mill got it right already in 1872 when he wrote "All laws of causation, in consequence of their liability to be counteracted, require to be stated in words affirmative of tendencies only, and not of actual results" (Mill 1973, p. 445). Consider, for example, the law of gravity $F = G\, m_1 m_2 / d^2$. This law does not say that bodies if situated in a certain way will move in a certain way. Rather it says that in a situation of a certain kind there is a force of a certain kind. It does entail a regularity; but not a Humean regularity but a regularity of the type "In a situation of type *S* there is *always* a force of type *T*". Concerning the actual movement of bodies the law of gravity says only that a body will accelerate in a certain way *if nothing else is acting on it*. Thus it says something about the movement of bodies only in some of the cases where it applies. In many cases the law applies and claims that there is a certain force but it does not make a claim about how the bodies will actually move. The class of cases in which the law makes a claim about what kind of event will be followed by what kind of event is a subclass of the cases where the law applies. Many object to the Humean regularity theory of laws now that laws are *more* than regularities, still accepting the Humean assumption that laws entail (Humean) regularities.[4] I, however, hold that laws do not entail Humean regularities at all. They entail regularities about what happens in all cases where nothing else than the objects referred to is acting but not regularities about what happens in all cases where the law applies.

It may be objected that although single laws on their own do not entail regularities, all the laws together do. We can, for example, make predictions about the movement of a certain body taking into account not only gravitational forces but also electro-magnetic forces and others. But as for every prediction based on the law of gravity the phrase "If nothing else is acting on the things in question" needs to be added, so this phrase has to be added to any prediction based on laws of nature. You may have good reasons for assuming that there are no further material things that will be acting on the things in question or for assuming that no God will act on the things in question, but the evidence that supports the laws of nature does not itself warrant this assumption. You need independent evidence for this, for example, evidence that there are no souls that would act on the things in question, or that there is no God, or that God would not act on the things in question because it would be bad. Because laws do not say what things there are not, also the totality of all laws of nature do not entail Humean regularities. For that reason Mill was right when he said that all laws "require to be stated in words affirmative of tendencies only, and not of actual results". I take forces to be a kind of tendency. For our purposes here it is enough to note that the laws of motion require to be stated in words affirmative of forces only, and not of actual movements. I leave it to another time to extend this to laws that are not concerned with forces and hence with movement of bodies.[5]

Imagine God moved the stone away from Jesus' grave. Just like that, without using any material object in order to push it. Would this have been a violation of the laws of nature? The standard view says yes, assuming that the laws of motion entail regularities of which this movement would be an exception. But what do the laws of motion say about this case? They say that there are certain forces, e.g. a gravitational force acting on the stone. There is no reason to assume that God, in order to move the stone, abolished any of these forces. They were all there, God just overrode them. Therefore there was no violation of the laws of nature.

If (1) were correct then the evidence for the laws of nature we believe in would support CC. If answer (2) or answer (3) are correct, then answer (1) is false because counter-instances to CC would not falsify any laws. I have suggested that answer (3) is correct. However, what I say in what follows could easily be adapted to the view that answer (2) is correct. Let us now consider the possible arguments for CC.

---

4 E.g. Armstrong (1983, p. 85).

5 A good discussion of laws of conservation of energy in this vein is Larmer (1988, ch. 5).

4. Is Causal Closure a Principle of Science?

Now consider the idea that CC is a principle of science or of rationality. Thus John Heil (1998, p. 23) writes for example: "Modern science is premised on the assumption that the material world is a causally closed system". The idea is that science assumes that every material event has a material cause and that science could not proceed without this assumption, perhaps comparable to how it could not proceed without using induction.

To this I reply first that science is only to a limited extent concerned with producing scientific explanations of single events. Perhaps history is concerned with what happened to the stone at Jesus' grave and why it happened, and cosmology is looking for the true explanation of a certain supernova. But often science is concerned with more general questions, such as how a body accelerates if it falls down, how photosynthesis works, or how quantum theory is to be squared with the theory of general relativity. For such more general questions it is irrelevant whether some immaterial object once upon a time caused some material event. All you need is the assumption that in the relevant experiments no immaterial objects interfered. If there are ghosts messing around in your experiments, then that will impede your scientific work, just as it will impede with your work if you do not manage to protect your experiments from other things acting in it. But that God once moved a stone or that souls cause some material events in human brains would not prevent us from discovering the laws governing stones and elementary particles.

What about the cases where science does look for explanations of particular events? Well, why should the fact that some events have immaterial causes and hence no law-based explanation prevent science from *looking for* law-based explanations? If there are counter-instances to CC, then science will not *find* true law-based explanations for these events because they have no such explanation, but nothing prevents science from looking for them. Furthermore, science may be able to find for some counter-instances to CC, for example for brain events that have immaterial causes, as claimed by Eccles (1994, ch. 6), that they have no complete material cause and thereby find evidence for the existence of certain immaterial objects. That would be an interesting discovery too.

Are there any other ways in which CC might be a principle of science? Is the assumption of CC necessary for the search for laws of nature? No, the existence of some counter-instances to CC would not impede the search for laws as long as we can conduct experiments with the things in question in which no such counter-instances occur. And even if there were so many counter-instances to CC of a certain kind that we could not discover a certain law, they would prevent us from finding this law, but they would not prevent us from looking for this law.

So I see no reason here for assuming that CC is a principle of science.

5. Is Causal Closure Supported by the Success of Science?

Sometimes it is suggested that CC is supported by the success of science. Science has found material causes for many material events. And science can investigate what the cause of a certain material event was which is a candidate for being a counter-instance to CC. But this will not yield much support for CC because there are so many particular events in the past and in the future where science will never look for, and find, evidence that they had material causes. More promising is the idea that CC is supported by the fact that science has made many successful predictions and is likely to do so in future, and by the fact that science has discovered laws of nature and is likely to do so in future.

If a material event had an immaterial cause then its occurrence could not be predicted through knowledge of laws and of material events that took place earlier than the immaterial cause (except when there are laws that govern situations that are partly immaterial, but I ignore this possibility here). That a scientific prediction, for example of an eclipse or of a spring tide, was successful, shows, therefore, that no immaterial things intervened in the causal processes that led to the predicted event. That many scientific predictions were successful shows that often no immaterial things intervened in causal processes. That many scientific predictions were made and few were unsuccessful shows that not often did immaterial things intervene in causal processes where clear predictions were calculable. Similarly, if it is likely that many scientific predictions in the future will be successful, then it is likely that not often will immaterial things intervene in causal processes where clear prediction are calculable. If, however, there are also many events for which thus far we cannot calculate clear predictions based on the laws of physics, like some weather phenomena or some movements of animal bodies, then the high number of successful scientific predictions and the low number of unsuccessful predictions does not support the view that there are no material events with immaterial causes. It only shows for the sorts of events where we can calculate clear law-based predictions that immaterial things do not often intervene in the causal processes leading to them.

That seems to me to be our situation. We can make many law-based predictions of eclipses, spring floods, etc., and most of them are successful. But there are not only many events in the past whose causes we have not investigated and will never be able to investigate, there are also many events in circumstances where we are thus far unable to make predictions. For example, many events involving animals we cannot predict, such as that the sparrow in front of my window just moved its wing as it did, that the cow ate this yellow flower rather than that one, or that I just whistled an *A* flat rather than an *F*. This is of course not in itself evidence for there being immaterial objects at work, but it means that for all these areas the success of science does not make it improbable that there are immaterial objects. The success of science is compatible with there being immaterial objects acting in these areas. Whether and in which immaterial objects we should believe depends on what evidence there is for them. If there is no positive evidence for them, then we should not believe in them. But if there is evidence, the success of science does not count against it. That is the answer to the question whether the success of science counts against the existence of immaterial objects in areas where we cannot make predictions.

Moreover, the success of science does not even make it very improbable that sometimes there are immaterial things acting in circumstances where on other occasions in circumstances of a *similar kind* we can make successful predictions. If God moved away the stone from Jesus' grave, how should that impede science today or how should it be in conflict with the results and the success of science? It does not conflict with science, and the reason for this is that it is what Swinburne calls a "non-repeatable" event. All we need in order to be able to discover the laws of nature are experiments where we rightly assume that no immaterial objects interfere. Knowledge of laws thus acquired allows us to know about certain events in certain circumstances that they would not occur unless some immaterial thing caused it. Historical evidence for the occurrence of such an event, e.g. that a certain man rose from the dead, is then evidence for the existence of an immaterial object that could cause such an event.

Melynk (2003, pp. 160-162) defends the SNA from CC arguing that since physics has found "sufficient physical causes for physical effects of very many kinds" we should conclude by *induction* that CC is true. I reply that this conclusion is not justified by induction. We are justified by induction when we conclude from the fact that there are certain forces acting in a certain case, that probably such forces are acting in *all* situations of the same kind. But Melynk's reasoning is not of this kind, it is not a proper induction. As we are not justified in assuming that because *often* two things accelerate towards each other (because of gravitation), two things *always* accelerate towards each other, so we are not justified in assuming that because *often* material events have a full material cause, material events *always* have a full material cause. Swinburne (1997, p. 237) makes a similar point: "Only if there was some well-confirmed formula which states after how much work scientists would achieve successful predictions in any field could we reasonably infer (if it follows from the formula) that they will obtain successful predictions in all fields if they work hard enough. There is no such formula. In its absence we must affirm that increasing success does not indicate future total success".

If you do not agree with my conception of laws of nature, consider this argument against Melynk. It is true that science has found full material causes for many material events of many kinds. But that is in conflict neither with the assumption that certain events in the past had immaterial causes (e.g. the stone's moving at Jesus' grave), nor with the assumption that in certain areas (e.g. human brains) some events have immaterial causes. Let me explain.

It would perhaps be in conflict with the assumption that certain events in the past had immaterial causes if science had searched for, and found, full material causes of the vast majority of events, but of course science has not done that. And that science has found material causes for events of *many kinds,* does not support Melynk's point because the fact that events of a certain kind *sometimes* have material causes does not show that they *always* have material causes. To illustrate, the fact that the movement of pieces of iron is sometimes caused through gravitation does not show that it is always caused through gravitation; sometimes it is caused through magnetic forces. Events of the same kind can have causes of different kinds at different occasions. That is why there is no inductive support for CC here.

The fact that science has found full material causes for many material events of many kinds is not in conflict with the assumption that in certain areas some events have immaterial causes. For example, the assumption that some events in human brains are caused by human souls is not in conflict with science having discovered material causes of many sorts of material events outside human brains. This assumption would only become improbable if science discovered full material causes of so many brain events that there were no room for immaterial causes. That is why there is no inductive support for CC here. Science has discovered for many kinds of material events what kind of material event they cause if nothing else is acting in the situation, but it has not discovered that material events never have immaterial causes.[6]

What people who say that science has shown that there are no material events that have immaterial causes might have in mind is that science has found material causes of some kinds of material events that some men formerly believed to have immaterial causes. Imagine, for example, that some men believed that northern light is caused by the god Thor when he is jealous. The discovery that northern lights are caused by interactions of energetic particles from outside the atmosphere with atoms of the upper atmosphere does something to show that this is false: it shows that a northern light has no immaterial causes later than the ones identified by science. However, this does little to show that there are no counter-instances to CC at all. The evidence for the hypothesis that Thor causes northern lights was very weak to start with, because there is little evidence that Thor exists and that he has the power and a reason to cause northern lights, and there may be other counter-instances to CC for whose existence there is better evidence. What needs

---

6 For a further argument against the claim that the success of science supports CC, see Montero (2003).

to be investigated is how strong the evidence for, say the existence of human souls or of God is, and whether we can think of material causes of the alleged counter-instances to CC and whether the explanations of the alleged counter-instances to CC in terms of immaterial causes are *better* than the ones in terms of material causes.

6. Conclusion

I have argued that it is neither a principle nor a result of science that there are no counter-instances to the principle of Causal Closure (CC). The Sweeping Naturalistic Argument (SNA) from Causal Closure based on science is unsuccessful.

There is another popular SNA, claiming that, because immaterial things are so different from each other, it is impossible that there is causal interaction between them. John Heil (1998, p. 23), for example, writes "if minds and bodies are distinct kinds of substance, it is hard to see how such causal interaction could occur". My view on this, which I cannot defend here, is that the only way to show that there are no such interactions is by showing that there are no immaterial things. I suggest therefore that the materialist, like his immaterialist opponent, has to use non-sweeping arguments. He has to criticize the arguments for the existence of God (see e.g. Swinburne 2004), souls (see e.g. Foster 1991; Swinburne 1997), etc., one by one, and he can further support his claim by giving positive arguments against the existence of such things, e.g. the argument from evil against the existence of God (see. e.g. Howard-Snyder 1996). But there is no short cut for materialists.

University of Munich

Acknowledgements

I wish to express my gratitude to the Free State of Bavaria for making this work possible through the Bayerischer Habilitationsförderpreis.

References

Armstrong, D. M. (1983), *What Is a Law of Nature?* Cambridge, Cambridge University Press.

Eccles, J. C. (1994), *How the Self Controls its Brain,* Heidelberg, Springer.

Foster, J. (1991), *The Immaterial Self: A Defence of the Cartesian Dualist Conception of the Mind*, London, Routledge.

Heil, J. (1998), *Philosophy of Mind: A Contemporary Introduction*, London, Routledge.

Howard-Snyder, D. (ed.) (1996), *The Evidential Argument from Evil*, Indianapolis, Indiana University Press.
Hume, D. (1999) [1748], *An Enquiry Concerning Human Understanding*, Oxford, Clarendon Press.
Kim, J. (1998), *Mind in a Physical World: An Essay on the Mind-Body Problem and Mental Causation*, Cambridge, Mass., MIT Press.
Larmer, R. A. (1988), *Water into Wine? An Investigation of the Concept of a Miracle*, Montreal/Kingston, McGill-Queen's University Press.
Lowe, E. J. (2000), *An Introduction to the Philosophy of Mind*, Cambridge, Cambridge University Press.
Lowe, E. J. (2003), "Physical Causal Closure and the Invisibility of Mental Causation", in S. Walter and H.-D. Heckmann (2003), pp. 137-154.
Melnyk, A. (2003), "Some Evidence for Physicalism", in S. Walter and H.-D. Heckmann, pp. 155-172.
Mill, J. S. (1973) [1872], "A System of Logic", in *Collected Works of John Stuart Mill*, edited by J. M. Robson, Toronto, University of Toronto Press.
Montero, B. (2003), "Varieties of Causal Closure", in S. Walter and H.-D. Hackmann, pp. 173-187.
Papineau, D. (1993), *Philosophical Naturalism*, Oxford, Blackwell.
Petrus, K. (ed.) (2003), *On Human Persons*, Frankfurt/ Lancaster, Ontos Verlag.
Swinburne, R. (1970), "Violation of a Law of Nature", in Swinburne (1989), pp. 75-84.
Swinburne, R. (ed.) (1989), *Miracles*, New York, Macmillan.
Swinburne, R. (1997), *The Evolution of the Soul (Revised Edition)*, Oxford, Clarendon.
Swinburne, R. (2004), *The Existence of God (Second Edition)*, Oxford, Clarendon.
Wachter, D. von (2003), "Free Agents as Cause", in K. Petrus (2003), pp. 183-194.
Walter, S. and H.-D. Heckmann (eds.) (2003), *Physicalism and Mental Causation: The Metaphysics of Mind and Action*, Exeter, Imprint Academic.

Science and the Philosophers

Kenneth R. Westphal

1. Introduction

The advent of distinctively Modern European philosophy at the turn of the seventeenth century was occasioned by two major developments: the painful recognition after thirty years of religious war that principles of public conduct must be justified independently of sectarian religious dogma; and the growth of natural science, especially discoveries in astronomy that linked terrestrial and celestial physics in a newly mathematicized, explanatory mechanics founded by Galileo and dramatically extended by Newton. The roles of reason and empirical evidence in inquiry, and their superiority to custom and tradition for knowledge of nature were undeniable, though their respective roles and proper epistemological accounting were far from obvious. I shall review briefly some key points in the advent of natural science in order show that some fundamental philosophical predilections have obscured the proper roles of reason and evidence in the philosophical accounting of scientific knowledge. Though these predilections are more readily apparent among seventeenth and eighteenth century philosophers, they are no less prevalent in mainstream analytical philosophy of science. Their diagnosis augurs a fundamental philosophical reorientation.

2. The Advent of Modern Science

Galileo directly disputed authority as a criterion of truth in scientific matters. Unlike Bacon, he did not think that sensory evidence could serve as this criterion, for he was well aware of the relativity of motion and the illusions and appearances that can infect observation. He relied on rigorous mathematical formulability as a criterion. This was not a Pythagorean or Platonic vision; mathematical formulae alone were not enough. On the contrary, he held that mathematical formulation of laws of nature demonstrate genuine regularities in natural phenomena. This requires that mathematical formulae be fitted to careful observation, and the joint satisfaction of these two demands must also be rationally intelligible.

Galileo formulated a genuinely scientific methodology. His 'resoluto-compositive' method involves making an hypothesis, *e.g.*, that the speed attained by a ball falling a certain distance is constant, regardless of the slope it descends. If this hypothesis is true, then in certain circumstances that ought to exemplify or instantiate the hypothesis, certain consequences follow from the hypothesis, and those consequences ought to be observed in those circumstances. If observation accords, within tolerable limits, with those deductive consequences, this is taken to confirm the hypothesis.

Galileo eschewed the investigation of ultimate causes of motion in favor of investigating the actual properties of motions:

> The present does not seem to be an opportune time to enter into the investigation of the cause of the acceleration of natural motion, concerning which various philosophers have produced various opinions .... Such fantasies, and others like them, would have to be examined, but with little gain. For the present it suffices our Author that we understand him to want us to investigate and demonstrate some the attributes [*passiones*] of a motion so accelerated (whatever be the cause of its acceleration) .... (Galilei, *Opere* 8:202, cf. 7:260-261; 1974, pp. 158-159, cf. 1967, pp. 234-235)

This is to say, physics can be pursued independently of metaphysics – and ultimately physics constrains the metaphysics of nature. Descartes was incensed by this, but Galileo showed himself to be a scientist more than a metaphysician in this regard, and practicing scientists rightly followed suit.

Newton's gravitational theory integrated Kepler's celestial physics with Galileo's terrestrial physics, producing a general mechanics that applied not only to our earth and solar system, but to the universe as a whole. Newton insisted that his laws of motion be 'deduced from the phenomena'. In this, he improved on Galileo's method. Newton called his method 'analysis and synthesis'. Basic principles are to be derived by a careful quantitative analysis of particular phenomena, in this case, observed motions. Once successful principles are determined for those particular motions, the principles are generalized to see if they can account for other cases of motions:

> ... whatever is not deduced from the phenomena must be called an hypothesis; and hypotheses, whether metaphysical or physical, or based on occult qualities, or mechanical, have no place in experimental philosophy. In this experimental philosophy, propositions are deduced from the phenomena and made general by induction. The impenetrability, the mobility, and the impetus of bodies, and the laws of motion and the law of gravity have been found by this method. And it is enough that gravity really exists and acts according to the laws which we have set forth and is sufficient to explain all the motions of the heavenly bodies and of our sea. (General Scholium; Newton 1999, p. 943)

The synthetic aspect of Newton's method lies in this final phrase, showing that the principles reached by analysis of particular phenomena in fact serve to explain accurately a wide range of related phenomena. Newton knows that this method is fallible, but also that there's none better (Query 31; Newton 1953, pp. 404–405).

The fallibility of Newton's method is easily over-estimated because of the contemporary fixation on prediction and retrodiction. By requiring that natural laws be 'deduced from the phenomena', Newton rejected Galileo's appeal, in part, to purely *a priori* demonstrations based on geometry and thought experiments, and on Galileo's glib appeal, in effect, to what are now called *ceteris paribus* clauses.[1] Newton's method requires accounting for the discrepancies between his idealized basic gravitational model and actual natural phenomena. Through repeated use of the same explanatory resources, Newton sought to account for whatever associated events and circumstances produced such discrepancies. Repeatedly using the explanatory resources of his theory of gravity to account for the exact measurements of the phenomena in question provides for convergent, increasingly accurate, ever less idealized measurement of causal parameters that explain those phenomena. The success of such repeated approximations, progressively reducing the idealizations of the scientific model and explanation, proves that there *is* a genuine causal phenomenon by measuring that phenomenon ever more exactly. This is a vastly richer, more demanding and far more illuminating account of empirical success of a scientific theory than mere prediction and retrodiction (Harper 2002a, esp. pp. 95, 100, 102; 2002b; Smith 2002).

Newton was notoriously reticent about the status of gravity. Newtonian mechanics is committed to its existence, while explaining neither its causes nor its manner of acting. The problem was that gravity appears to involve action at a distance, and – according to the reigning mechanical philosophy, which still followed Aristotle by recognizing only action by contact – that was preposterous, utterly impossible. Despite Newton's reticence on this point, his mechanics revised Boyle's claim that matter is inert. Though this redefinition has some precedents in Gilbert's work on magnets and in Kepler's speculations about solar force, Newton's theory gave unprecedented support to the claim that matter has active properties. Newton knew he was transcending the bounds of 'mechanical', that is, corpuscular, philosophy:

> [The force of gravity] acts not in proportion to the quantity of the *surfaces* of the particles upon which it acts (as mechanical causes are wont to do [*sic*]), but in proportion to the quantity of *solid* matter .... (General Scholium; Newton 1999, p. 943)

---

1 Galilei, *Opere*, 7:171–173; (1967), 145–148; letter to Pietro Carcavy, 5 June, 1637 (*Opere*, 17:90–91). I thank Cinzia Ferrini for discussion of this point.

To the charge that he lapsed into discredited Aristotelian occult qualities, Newton replied that he did not postulate a single alleged cause of a particular effect, but rather a general force, formulable in a quantified law of nature, which explained a huge range of otherwise diverse phenomena, previously thought to be unrelated:

> To tell us that every species of things is endow'd with an occult specifick Quality by which it acts and produces manifest Effects, is to tell us nothing: But to derive two or three general Principles of Motion from Phænomena, and afterwards to tell us how the Properties and Actions of all corporeal Things follow from those manifest Principles, would be a very great step in philosophy, though the Causes of those Principles were not yet discover'd: And therefore I scruple not to propose the Principles of Motion above-mentioned [*sc.*, gravity and inertia], they being of very general Extent, and leave their Causes to be found out. (Query 31; Newton 1952, pp. 401–402)

Newton allows that there may well be a cause of gravity; he didn't know how to explain gravity itself, and his later work speculated about various possible forms of mechanical aether that might explain gravity. Despite the controversy about even apparent action at a distance, Newton was quite right that it sufficed for his explanatory, scientific purposes to show that bodies do behave in accord with the law of gravity he formulated, which provided excellent grounds to treat gravity as a cause or force, even if perhaps not an ultimate cause or force. François De Gandt (1995, pp. 265–272) notes that Newton's mathematical theory of orbital motion forged an important kind of theoretical independence from metaphysical and physical questions about the ultimate nature of space, time, or gravity. Newton is thus entitled to his reticence about the status of gravity.

In his chemical researches, Robert Boyle relied on 'transdiction', inferring from observed phenomena some properties of subobservable components of observable bodies, such as corpuscles (Mandelbaum 1965, ch. 2). Boyle argued essentially by analogy, and by refusing to take merely logical possibilities as objections to the analogy. Newton's third rule of reasoning in philosophy does the same:

> Those qualities of bodies that cannot be intended and remitted, and that belong to all bodies on which experiments can be made should be taken as qualities of all bodies universally. (Newton 1999, p. 795)[2]

This rule may appear to concern only gross bodies that happen to be out of sight, say, in outer space. However, it concerns 'all bodies universally'. Unrestricted in this way, this rule covers sub-observable bodies, too, as Newton clearly stated in

2 By 'intended and remitted', Newton means 'intensified or decreased'.

related passages (Newton 1999, pp. 409, 795–796). Newton's laws of motion rely on mass, and mass must be attributable to any and every part of a body; otherwise the laws of motion cannot be successfully quantified. This is why Newton needs 'transdiction', and the theoretical success of Newtonian Mechanics justifies attributing to any sub-observable parts of bodies the same kinds of properties his theory ascribes to gross bodies: volume, mass, mobility, rigidity, inertia, gravity. Newton's quantitative analysis of molar phenomena provides enormously strong grounds for attributing specific properties to matter, even to unobservably small bits of it.

I have deliberately stressed Galileo's and Newton's methods. Bacon recognized that good science requires the joint use of sensory observation and rational analysis, but he did not appreciate the extent to which both scientific observation and rational analysis are driven by the effort to provide an accurate quantitative, mathematical treatment of physical phenomena. The Modern exact sciences are built on this effort to bring all three components – observation, reason, and mathematics – together.

In discussing the relevance of the almost purely mathematical framework of *Principia* Books I and II to Newton's "System of the World" (Book III), De Gandt (1995, p. 267) remarks:

> The solidity of the inductive fabric is due to its mathematical framework, which makes it possible to establish an extremely tight network in which observation and theory advance on and regulate each other.

Borrowing terminology from logical empiricism, this might suggest that Newton's mathematics forms the 'correspondence rules' between his theoretical and observational language. This suggestion is too glib. Newton's mathematical framework plays a constitutive role in his theoretical postulates and, as De Gandt notes, for the mutual regulation of theory and observation (cf. Smith 2002).

A third point about the development of natural science must be noted briefly. The Age of Newton was also the age of the founding of national and international SCIENTIFIC SOCIETIES. Among other aims, these societies served, developed and highlighted the importance of two key features of natural science: They served to disseminate scientific information and techniques, and they initiated the development and use of the peer review process. The replication of key observations and experiments to check reported observational or theoretical results rightly became a *sine qua non* of scientific integrity and the legitimacy, the very justification of scientific data and theory.

3. Modern Philosophical Reaction

One fascinating point about these elementary features of mechanics is how recalcitrant they have been to philosophical understanding, right from their inception. Modern philosophers largely agreed that sensation and conception formed poles on a continuum, though rationalists and empiricists disagreed about which pole was clear and trustworthy, and which was a dim, confused, vague intimation of the other pole. In accounting philosophically for scientific knowledge, empiricists emphasized observation, stressing Bacon's ideal of unprejudiced, 'immaculate' perception. Rationalists emphasized rational reflection, to which they gave higher priority than mathematics, though none of them denied the relevance of observation.

Descartes (to Mersenne, 29 June 1638) upbraided Galileo for philosophizing without a foundation, because Galileo didn't worry about ultimate causes of motion, but only about specific laws governing specific kinematic phenomena. Galileo acknowledged that his new physics required a new metaphysics (and theology); to this extent he agreed that Descartes's issues were legitimate. However, Descartes gave priority to epistemology over metaphysics, and priority to metaphysics over natural science. Descartes sought first to analyze reason, then to analyze sensation, and then combine these in an account of empirical knowledge. He purported that we have a set of basic ideas, or 'common notions' of 'eternal truths', which are given to us innately by God, and which are known to be true by their intuitive self-evidence, through the 'light of nature'. Descartes held that developing reliable natural-scientific knowledge requires determining which clear and distinct ideas known by reason are instantiated by the objects of perception, and how the objects we perceive produce their effects (both on one another, and on us) in virtue of their instantiating those ideas of reason. Empirical knowledge is not deductively derived from clear and distinct basic knowledge because objects instantiating any of several different arrangements of simple natures might produce any given empirical phenomenon. The role of experiment in Descartes' theory of science is to ascertain the parameters of empirical phenomena as precisely as possible so as to eliminate, so far as possible, alternative arrangements of simple natures instantiated in physical objects, so as to leave us with one clearly superior explanation (*Principles*, letter from the Author).

The problem with Descartes's proposal is that his set of alleged 'common notions' of 'simple natures' is derived by *a priori* reflection and alleged intuitive self-evidence. Descartes granted that his explanations might be false; God may have arranged the world to function differently than Descartes' explanations propose (*Principles* IV, §204). This is especially true of explanations which appeal to unobservable physical micro-structures. (This is the issue of transdiction, again.) Descartes sought to guarantee that he at least had the correct set of 'common notions' with his proof of God's existence and veracity, and that God implants in us ideas of the simple natures on the basis of which God created the universe.

Descartes chose this high rationalist road because of his fascination with the ideal of *scientia*, namely the axiomatic-deductive model of logical and mathematical knowledge, and because of his abhorrence of skepticism. Because custom and tradition are social and historical phenomena, because they have so seriously misled us about the nature of nature, and because the route to genuine knowledge of nature was discovered by brilliant, scientifically minded individuals, it seemed obvious, undeniable, that genuine knowledge is an individual phenomenon, and that only an individualist epistemology can defend realism about the objects of knowledge. Any social or historical account of empirical knowledge can only land us in error, relativism, or worst of all skepticism. Descartes' concern to avoid or refute skepticism embedded infallibilist notions of epistemic justification deeply into the core of epistemology, along with the axiomatic-deductive model of rational justification. The individualism and infallibilism of Descartes' epistemology meet in Descartes' foundationalism, the thesis that there are some basic elements of knowledge that are each known by each of us individually, while all other items of knowledge are derived from them. Descartes' foundationalism was rationalist; the basic items of knowledge concern elementary truths that are self-evident to reason.

Descartes's physics, based on these methods, was far from adequate. Descartes's fascination with geometry led him to focus on extension as the sole essential attribute of material substance, to the exclusion of mass or force (*Principles* II, §64). As a result, Descartes repeated the error that Galileo had exposed, that acceleration is proportional to distance (extension of a motion) rather than to time (To Mersenne, 13 Nov. & 18 Dec. 1629). Descartes, too, sought to resolve phenomena into their most basic elements (Rule 2). Descartes differed from Galileo about this, however, because Descartes thought he could know in advance, by *a priori* reflection, the full set of basic elements; whereas Galileo derived his basic elements piecemeal solely through resoluto-compositive analysis of the natural phenomenon itself.

Nevertheless, Descartes' epistemological package has dominated epistemology down to the present day, even among many philosophers, including empiricists, who regarded themselves as Descartes' philosophical antipodes. Descartes' epistemological package consists in these four theses:

(1) Only individualist epistemology can uphold realism about the objects of knowledge;
(2) Human knowledge has foundations;
(3) The structure of empirical knowledge ultimately conforms to the axiomatic-deductive model of *scientia*;
(4) Empirical justification is infallibilist.

This package has dominated epistemology, in part because it has provided part of the framework within which philosophers have thought, so that these particular views are revealed as key implicit assumptions or enthememes in their reasoning, and only rarely as explicit premises, much less conclusions to philosophical demonstrations.

The relations of Hume's philosophy to Descartes' is a rich, fascinating, important, and still largely unwritten chapter in the history of philosophy. Though Hume rejected Descartes' rationalism and dualism, he also recommended Descartes' works as propaedeutic to his own. Seizing on the alleged clear and unequivocal nature of sensory impressions, and on the empirical basis of natural science, Hume dispensed with Descartes' *cogito* and modeled his philosophy on Newton's (*En* §§1, 8, 9). Accordingly, Hume identified a set of basic objects and a set of basic laws governing those objects, which he claimed were the minimum sufficient basis for an entire philosophical account of human nature, including human knowledge. Hume's basic objects are impressions and ideas; his basic principles are concept- and verification-empiricism, and his basic laws are that impressions and ideas are associated solely by their resemblance, spatial or temporal contiguity, or constant conjunction, including the special case of cause and effect. On this basis, Hume officially defended total empirical skepticism as the inevitable consequence of Cartesian infallibilism, deductivism, and individualism. Though he recognized that such a dire conclusion reflected directly back onto his basic assumptions and principles, he claimed to find no fault with them, and pled the privilege of the skeptic to be as befuddled by his own skepticism as he was by his official epistemology (*T* 400).

One striking feature of Hume's philosophy is the astonishing though rarely noticed way in which it reveals the great extent to which Modern empiricists latched onto the empirical basis of natural science, while disregarding the fundamental role of mathematics in physical mechanics: Hume's infamous psychological laws of association, which are central to his claim to follow the Newtonian model, are entirely qualitative. Hume does not even suggest how to quantify them mathematically; he does not even recognize that this is a question worth asking, because he didn't understand that mathematical quantification is a *sine qua non* for (candidate) causal laws. Hume's attempt to 'introduce the experimental method' (*T* 1) into moral (as contrasted with natural) philosophy starts on the wrong foot.

Kant finally broke the Modern presumption that sensation and conception are poles of a continuum. Instead, they are distinct, jointly necessary components of human knowledge. Kant also broke another key bone of contention between Modern rationalists and empiricists, namely that rationalists believed that our having *a priori* concepts sufficed to legitimate metaphysics, while empiricists believed that banishing metaphysics required rejecting *a priori* concepts. Kant argued that we do indeed possess *a priori* concepts, though through his very sophisticated theory of cognitive reference, Kant shows that we can only use *a priori* concepts in legitimate cognitive judgments about spatiotemporal objects and events. Singular cognitive reference requires singular sensory presentation, and this singularity of reference requires spatiotemporal specification. Definite descriptions alone cannot do the job, because no purported definite description determines or indicates whether it is empty, definite, or ambiguous. (This is one lesson of Kant's criticism of Leibniz's principle of the identity of indiscernables.) Kant also broke with Cartesian foundationalism, by rejecting internalism about epistemic justification, and by rejecting the foundationalist assumption, still prevalent today, that our inner experience has epistemic priority over our outer experience.

Kant's transcendental idealism tried to justify and to explain the applicability of mathematics to nature, and thus also our use of mathematics in natural science. In his *Metaphysical Foundations of Natural Science*, Kant refuted the philosophical objections to action at a distance, he showed how to use reference frames to establish any relevantly 'absolute' space within which to understand the motions of bodies, and he sought to show how gravity can be understood as essential to, and so as inherent in, matter as such. Kant's work is indeed a watershed in the history of philosophy.

Kant's *Critique of Pure Reason* sought to lay the philosophical foundation for a deeply rationalist account of natural science. Like Descartes, Kant held that epistemology has priority over metaphysics, which in turn has priority over physics; Newtonian mechanics cannot be properly founded without an *a priori* explication and defense of its most basic concepts and principles. Kant's notion of 'properly founding' natural science retained the traditional model of *scientia*, and sought to justify Newton's three Laws as corrolaries constructed *a priori* on the basis of Kant's principles of causal judgment defended in the *Critique of Pure Reason*. Unfortunately for Kant's project, even if his arguments for transcendental idealism were valid, which they are not, his idealism fails to provide an adequate basis for Kant's model of *scientia*: his metaphysical constructions of the concept of matter fail to show that gravity is inherent in matter, and they also fail to justify *a priori* Newton's Laws of Motion, or even Kant's variants of Newton's laws (Westphal 2004, chapters 5, 6).

Modern philosophers, including Kant, failed to develop an adequate account of natural scientific knowledge. Their repeated failures to understand natural science may appear to support Quine's dictum that two kinds of people are drawn to philosophy: those interested in philosophy, and those interested in its history. A closer look, however, belies this patronizing dismissal of our philosophical predecessors.

4. Contemporary Philosophical Reaction

Because Carnap's reductionist program in the *Aufbau* failed, Quine rejected the traditional epistemological aim of reconstructing the whole of empirical knowledge based on one's own sensory stimulations. Instead he recommends 'naturalized epistemology', which can appeal to psychology in order to understand how we come to believe what we do about the world. The only genuine form of empirical knowledge, and the only genuine form of epistemic justification, is natural-scientific. The normative aspirations of epistemology now reduce to analyzing and distinguishing more from less effective methods of generating empirical knowledge. Unlike either Descartes or Hume, Quine (1969, p. 84) wants to 'let consciousness fall where it may'. Quine (1953, pp. 17–18, 44; 1969, p. 83) insists that 'physical objects' are a posit we make to provide the simplest account of our sense stimuli. Consequently, unobserved 'theoretical' entities postulated by scientific theories are no more, and no less, legitimate than ordinary physical objects. And once we reject reductionism, there is nothing illegitimate about affirming the existence of physical objects, along with many of our beliefs about them.

This all sounds sober and up-to-date, but did Quine really jettison Descartes' epistemological package? Note first, 'assigning' is an activity, it is something we do; to 'assign' sense stimuli to an individual object requires that we are self-conscious. Either Quine is not entitled to 'let consciousness fall where it may', or he is not entitled to his account of physical objects as a simplifying 'posit'. The only legitimate sense of 'hypothesis' is a thesis one proposes, or posits, in order to account for some range of manifest phenomena. Hypotheses are the province of self-conscious intelligent inquirers. Quine cannot both 'let consciousness fall where it may' and treat physical objects as an explanatory, simplifying 'hypothesis' to account for our sensory stimuli. To this extent, at least, the Modern epistemological tradition continues in and through Quine. This is no accident.

At its inception, analytic philosophy rebelled against metaphysics generally and especially 'hegelianism' (however understood), and rooted itself deeply in the eighteenth century. This is apparent in the prevalence of sense-datum theories and the like in early analytic epistemology, which was driven mostly by reductionist semantics and metaphysics, not by epistemological concerns with justification or truth. The reactionary character of early analytic philosophy was expressed plainly by Russell, who in 1922 proclaimed:

> I should take 'back to the 18th century' as a battle-cry, if I could entertain any hope that others would rally to it. (Russell 1994, 9:39)

The eighteenth century Russell advocated was epitomized by Hume, not Kant. More importantly, Hume's views were identified with his *Enquiry Concerning Human Understanding*, not his much more profound *Treatise of Human Nature*. Hume's associationist psychology was to be replaced by powerful new logical techniques that afforded, among other advances, logical constructions of physical objects out of sense data. Certainly this was a project worth trying!

Russell's deep allegiance to Hume was seconded by Quine, who held:

> On the doctrinal side, I do not see that we are farther along today than where Hume left us. The Humean predicament is the human predicament. (Quine 1969, p. 72; cf. pp. 74, 76)

Central to the Humean predicament is Hume's fork, according to which the only propositions we can know *a priori* are analytic ('relations of ideas'), whilst we can only know synthetic propositions ('matters of fact') *a posteriori*. Famously, Quine rejects the analytic/synthetic distinction and on the basis of his semantic holism rejects the distinction in kind between the *a priori* and the *a posteriori* – though whether his criticisms are sound remains controversial (Creath 1991; Hanna 2001, pp. 171–180, 281–285). None of these shifts, however, affect the most basic Cartesian orientation of Quine's philosophy: Quine, too, believes that any genuine form of meaning can only come from sensation (1969, p. 75; *cf.* 1974, p. 1), and Quine, too, believes unquestioningly in the priority of inner experience over outer experience. This is evident in his declaration:

> Save the surface [of the sentient body] and you save all. (Quine 1969, p. 155; cf. pp. 75, 82–83, 158; 1960, p. 22; 1974, p. 3; 1975, p. 68; 1990, p. *xii*)

Quine's focus on sensory stimulation appears also in his account of radical translation:

> In experimentally equating the uses of 'Gavagai' and 'Rabbit' it is stimulations that must be made to match, not animals. (Quine 1960, p. 31)

Because Quine is committed to 'the Humean predicament', matching stimulations can only be a matter of any observer matching various of his or her own sensory stimulations – whatever that may mean, and however that may be done. Once admitted, the egocentric predicament swallows all. At its core, Quine's semantics is incoherent. Quine's rejection of intensions is part of his rejection of any Cartesian *cogito* and any form of 'mentalism'. This rejection requires adopting a behavioristic approach to 'meaning'. Behaviorism of any variety, however, requires studying the organism's response to environmental stimuli and its action in and on its environment. Unfortunately, Quine's infamous theses of the inscrutability of reference and the indeterminacy of translation preclude any unambiguous identification of any organism's environment, its environmental stimuli, or even of the observed organism itself! What kind of 'matching' of sense stimulations can this predicament possibly afford? If radical translators are human, they too must start in this ego-centric predicament, trying to 'assign' their sense-stimuli to physical objects and events. Indeed, they must not merely try to do this, they must succeed at such assignments, *if* they are to study the vocables uttered by any presumed human subject within his or her native environment. Yet exactly what such alleged success could consist in, and how any such success could be ascertained, are quite mysterious, for in their own all-too-human case, all any purported radical translator can do nothing more than try to match up his or her own sense-stimuli in ways that afford a coherent web of belief about the world, including those components of the world that consist in the (purported) target subject, and his or her environment and vocables. Quine's behavioristic program simply cannot get started at all. It is thus little wonder that in 1998, when asked what progress philosophy had made in recent decades, Quine literally had nothing to say (Boxer 1998, cited by Hanna 2001, p. 284n). By the nature of his case, that's all he could say. For his honesty he deserves genuine credit.

To his credit, Quine read Hume's *Treatise*. One supremely important section of Hume's *Treatise* has no parallel in the *Enquiry*, namely Hume's brilliant and profound analysis in 'Of Scepticism with regard to the senses'. Quine cites precisely this section, indeed in direct connection with, and apparently as his basis for, asserting that physical objects are simply a 'posit'. This is Quine's basis for contending that 'on the doctrinal side', we have not advanced beyond Hume. As an historically rooted philosopher, I'm pleased that Quine read and singled out this crucial section of Hume's *Treatise*. Quine cannot, however, be credited with a very deep understanding of Hume's analysis.

In this section, Hume realized that his three laws of psychological association cannot account even for our concept of physical object (what Hume called 'body'). The most Hume can do is develop occasioning causes of our belief in physical objects, in terms of three (non-quantified) psychological propensities to respond to certain patterns of impressions with certain beliefs that pertain to the persistence and re-identifiability of alleged physical objects. Hume espouses what we would now call an 'error' theory of perception (Westphal 1998a, §4); Quine does likewise, apparently for the same reasons. Unfortunately for Hume, the beliefs with which we respond to those patterns cannot be defined in terms of his official concept empiricism. (If they could be so defined, he would not have needed to introduce special psychological propensities, which significantly transcend the principles of his official empiricism.) In fact, Hume's propensities smuggle into his account concepts that can only count as *a priori*. But like Hume, who called them a 'fiction', Quine prefers to call physical objects a 'myth', rather than even consider whether any of our nonlogical concepts are in fact *a priori* – quite aside from the issue of whether our use of them is cognitively legitimate.

However, the plausibility of there being *a priori* concepts is greatly bolstered by Hume himself. His analyses of abstract ideas, and of the ideas of space and of time, provide a devastating *critique* of his own official 'copy theory' of impressions and ideas. One key point is this. Because impressions are particulars, any impression or idea is a concrete, complex object. As such, any impression has a variety of aspects. Yet Hume himself contends that, through our use of language, rooted in 'distinctions of the understanding' and 'judgement', we can identify and distinguish these many aspects of any one impression or idea (Westphal, 2005).

It must be said plainly: Empiricist accounts of conceptual content or linguistic meaning simply have never deserved the philosophical prominence they have enjoyed. It must be said, too, that part of why empiricist accounts of meaning have been so durable is that philosophers have not studied the history of philosophy carefully or seriously enough to discover how problematic such theories are, and were already shown to be – whether implicitly by their advocates, or explicitly by their critics – throughout the Modern period (cf. Turnbull 1959; Westphal 1989, ch. 4; 1998a; 2000; 2002b). *Pace* Quine, those who are truly interested in philosophy simply *must* study the history of philosophy, and must do so philosophically, historically, and textually. Such study aims not simply to understand the views of other philosophers, but by confronting a wide variety of divergent philosophical views, schools, and orientations, such study enables us, in ways nothing else can, to reflect on, identify, and assess our own philosophical inheritance, orientation, and predilections.

Now if Quine insisted on holding incoherent views, that's his affair. But as in the case of Rorty, shame on all those philosophers who willingly jumped on Quine's band-wagon without examining his views carefully enough to recognize that his new philosophical clothes were transparently ill-conceived. The history of empiricism is a philosophical goldmine, precisely because it reveals just how fundamentally inadequate empiricism is.

In *The Empirical Stance*, Bas van Fraassen complains of the poor understanding of science current even among self-styled philosophers of science, noting that these misunderstandings were learned from their teachers, in particular from Quine (*ES* 11, cf. 29). One of these misunderstandings (omitted by van Fraassen) is especially glaring: no natural scientist begins from the egocentric Cartesian-Humean predicament; natural scientists begin with observing natural phenomena, natural events in nature. 'The positing of bodies', simply is *not* 'rudimentary physical science' (*pace* Quine 1975, pp. 67–68). Quine can suppose otherwise only by conflating or shifting among several distinct, even incompatible kinds of 'naturalism' (Haack 1993, pp. 118–138). Quine's attempt to refashion natural science to accord with that predicament is just one more attempt to assimilate science to unquestioned philosophical predilections, and in this regard is flatly inconsistent with Quine's professed scientism, that science alone is the legitimate model and instance of genuine knowledge. One key source of naturalism, according to Quine (1981, 72), is 'unregenerate realism'. Yet this key tenet is one to which Quine's semantics and epistemology do not and cannot entitle him. Quine's logical acumen is rightly renowned, but his epistemological views in fact demonstrate by *reductio ad absurdum* the insufficiency of 'the logical point of view' for understanding human empirical knowledge.

Van Fraassen praises the astonishing 'critical armamentarium' developed by analytic philosophy (*ES* 18, 30).[3] I certainly agree that analytic philosophy has developed indispensable critical resources, yet I contend that it is far from complete or sufficient, especially regarding issues concerning human knowledge: there are vast critical resources that have yet to be assimilated from traditions and stances typically neglected, if not outright rejected, by analytic philosophers.

Van Fraassen rightly stresses how widely variable our understanding of 'empiricism' has been throughout philosophy (*ES* 32–34, 201–225), and he endeavors to stake out a new version of empiricism. His new version of empiricism attempts to evade standard philosophical objections to empiricism by distinguishing between philosophical theses and philosophical stances. The standard objections to empiricism all bear on empiricist theses. Van Fraassen proposes to scuttle these objections by identifying empiricism with a certain kind of stance. He argues that empiricism cannot consist in express empiricist theses, because so far no sound principle of empiricism has been found. All key theses of empiricism have been refuted, indeed by empiricists themselves, and nothing can foreclose on future proposals for key empiricist theses suffering the same fate. If empiricism continues, despite the critical demise of every key empiricist thesis to date, then empiricism must consist of something other than, certainly in addition to, its key theses (*ES* 46–47). This non-propositional core of, *e.g.*, empiricism is a philosophical *stance*. Most generally, philosophical stances are 'philosophical positions that cannot be captured in dogmas' (*ES* 47). Thus van Fraassen insists that there is more to a philosophical position than its propositions, and this 'more' is essential to a philosophical position being a philosophical *stance*; it's an attitude rather than a belief or set of beliefs.

Van Fraassen's empirical stance is this:

> Empiricists reject demands for explanation, especially explanation by postulate, and any forms of abduction. They hold some version of the naive principle empiricism,

---

3 This may seem surprising in view of his early training in phenomenological philosophy of science. However, at the time he studied, only early Husserl was on the 'continental' agenda; the later Husserl and Heidegger were yet to be discovered.

> that information only comes through the senses. They have certain kinds of ideals of epistemic rationality, of significance (or meaning). They admire natural science, and their idea of rationality does not bar disagreement. (*ES* 47)

Important to van Fraassen is that none of these factors is itself a belief, though any of them may be associated with various empiricist beliefs. By adopting the empiricist stance, philosophers are free to espouse their key Principle of Naive Empiricism, that 'experience is the only source of information' (*ES* 45, 47), without fear of theoretical refutation. Van Fraassen deliberately leaves this principle undeveloped, for his main concern is with identifying and calling our attention to philosophical stances.

Van Fraassen's express concern with philosophical stances performs a genuine philosophical service: It draws philosophical attention to the basic context within which philosophers develop, present and defend their express views or theses. The question, 'What else besides propositions belongs to a philosophical view?' has been too long suppressed, especially in analytical philosophy. Simply raising this issue calls for richer forms of philosophical reflection than have been common in the analytic tradition, in which philosophers have followed Mill's reduction of 'reflection' to 'introspection', and then banished introspection as a disreputable philosophical method (Scharff 1995). As a result, their methods remain profoundly pre-Kantian (Westphal 2004, ch. 1). It should be noted that stance theory entered feminist philosophy forcefully with Nacy Hartsock's (1983) work, which developed in part through reconsidering Marx's reflections on how class membership conditions what one knows, believes, and pays attention to. 'Stances' are news to today's heirs of logical empricism because they have resolutely neglected feminist and marxist epistemology, and have forgotten their own positivist roots within a particular cultural circle (see below).

Paul Teller (2001, pp. 123–125) rightly notes that van Fraassen is very much a voluntarist; he's in the business of advocating and recommending his views and stance, more than arguing for or 'proving' them. How compelling a recommendation can van Fraassen provide for his empirical stance? Consider two significant details. First, van Fraassen condemns the revival of essentially Seventeenth Century metaphysics within the analytic tradition (*ES xviii*, 1–30), without noticing the great extent to which analytic philosophy began by reviving essentially Eighteenth Century Humean epistemology, as attested by Russell and Quine. To reject 'demands for explanation' and 'any forms of abduction' is to reject natural scientific knowledge almost entirely. If perception alone can provide sufficient basis for belief, whilst the entire non-observational theoretical structure of a science can at best be a candidate only for acceptance (*ES* 89–90; 2001, pp. 151–153), then we are only entitled to accept, but not to believe, that even sub-observable particles or quantitities of matter have mass. On van Fraassen's empiricist principles, 'transdiction' can never warrant belief. Newton's methods show otherwise (Harper 1991 to 2002b).

Second, van Fraassen rejects explanation and abduction on the basis of antecedent, non-scientific philosophical predilections, on the basis of his presumed philosophical 'stance', despite the fact that not even a coherent (not to say 'plausible' or 'illuminating') formulation of its key Principle of Naive Empiricism, that only sensation provides information about the world, has been given in, well, how many centuries? Van Fraaseen condemns metaphysics in these terms:

> .... I see metaphysical concoctions not as underpinnings but as the canopies of baroque four-poster beds. ... Metaphysical theories purport to interpret what we already understand to be the case. But to interpret is *to interpret into* something, something granted as already understood. (*ES* 3)

Van Fraassen contends that this metaphysical strategy attempts to explain the understood in terms of the hopelessly obscure and unfounded (*ES* 3). Yet van Fraassen regularly deploys the epistemological version of this same strategy by providing 'an empiricist view' of whatever topics he writes on. Like the other philosophers discussed above, van Fraassen, too, is engaged in cutting science to fit his antecedent epistemological presumptions, in his case 'empiricist' presumptions. However, van Fraassen's strategy is even worse than that of the metaphysicians he condemns: Until and unless a coherent version of his Principle of Naive Empiricism can be stated, restricting the phenomena of human knowledge, especially scientific knowledge, to what fits his empiricist view is to interpret knowledge into the antecedently incoherent and thus demonstrably false! This strategy cannot be given any legitimate voluntarist recommendation, unless van Fraassen wishes to join the ranks of philosophical irrationalists, who offend against reason far worse than the post-German idealists he thinks fell prey to the illusions of reason (*ES* 2).

The term 'empiricism' has become a philosophical mantra, like 'naturalism', which is recited almost automatically by analytic epistemologists or philosophers of science to proclaim their philosophical sobriety.[4] The real point of the term 'empirical' is that empirical evidence is crucial for justifying, evaluating, formulating, or even understanding empirical claims, claims about specific features of the world. Empiricists, however, have no monopoly on this core sense of 'empirical'; rationalists, pragmatists, phenomenologists and others acknowledge, indeed insist on this point, too. The real issue is how best to understand the nature and role of empirical evidence within human experience and empirical knowledge. Accordingly, in the interest of honesty in advertising, van Fraassen's position must be re-named 'the empiri*cist* stance'.

---

4 Regarding 'naturalism', see Rouse (2002).

It is important to note that van Fraassen continues the individualist tradition in epistemology; this is built into his Bayesian understanding of rational justification. The individualism required by the Bayesian approach to rational justification is expressed concisely by Paul Teller, in comments on van Fraassen's book. In recommending a moderate contextualism, Teller (2001, p. 125) states:

> The pursuit of knowledge is a deeply social enterprise; and many of us accept, in the spirit of reliabilism, that we are better off, if, on the whole, we do not brush aside the contrary opinions of those we acknowledge as worthy interlocutors. But to go this far is already to accord some weight to both the opinions and the values of many of our peers even though nothing rationally compels us to do so.

On Teller's as on van Fraassen's view, nothing rationally compels us to accord any weight to the opinions or values of our peers, because rational justification is an individual affair. Is this individualism tenable, even in its own epistemological terms? Does van Fraassen's empiricist stance offer any real basis for understanding how and why science is 'a deeply social enterprise'? Consider the following case.

One philosopher has taken the naive principle of empiricism, that all information reaches us only through sensation, very seriously and developed a genuine epistemology based squarely on it: Fred Dretske. On his view, mere receipt of information does not suffice for knowledge, because information yields belief or knowledgely only by decoding. Dretske tries to preserve the essentially empiricist model of his epistemology by defining knowledge recursively, positing an elementary level of information acquisition at which no background information is required to decode this information. Perhaps there is such a primitive level at which there is no distinction between receiving information and decoding it. However, if background information isn't relevant at this level, certainly background interests of the organism are, such as interests in safety, food, shelter, mates, or the like. More importantly, this basic level is at best proto-cognitive. If decoded information has a proposition-like structure (in Dretske's terms, it has a digital rather than an analog form), no decoded information at this level can have an express *propositional* structure, because such structure is linguistic.

Surprisingly, Dretske says nothing about the role of language either in information decoding or in our information channels themselves. Dretske attempts to account for conceptual content solely in terms of referential opacity. Lacking from his account of conceptual content is thus the key role of inferential articulation in both identifying and using conceptual content (Sellars 1981). The fundamental importance of inferential articulation for specifying conceptual content reinforces the central role of language in information decoding and processing. These brief observations lend contemporary support to Leibniz' (*New Essays* 2:1.2) dictum: nothing is in the intellect that was not first in the senses – except for the intellect itself! Leibniz' closing qualifier is the Achilles' heel of empiricism, including contemporary empiricism (cf. Turnbull 1959; Westphal 2005).

Yet the crucial need to account epistemologically for the structure and functioning of the mind is only one lesson of Dretske's information theory. His theory also puts paid to the egocentric predicament, including Quine's version of it. 'Saving the surface' does *not* 'save all'! This is because what information a subject can receive from a signal is a function of the state of the environment and the causal laws governing it, as well as the sensitivities of an organism's receptors and the organism's decoding capacities and abilities. Only within a determinate kind of environment can any organism receive – much less decode – any genuine information, and to know what information an organism decodes we must determine what of its information *it* saves, not what information might be saved (by some omniscient – and hence non-natural – radical translator?) from its sensory input. More significantly, Dretske's own account and examples of simple concept acquisition betray some of the essentially linguistic and social dimensions of human knowledge. Dretske's examples of simple concepts are 'red' and 'robin'; his discussion of acquiring these simple concepts expressly involves a *teacher*, a second person who helps instill the correct use of the terms 'red' and 'robin' in the nascently lingual child, obviously, in the presence of relevant instances of those concepts (Westphal 2003, §27).

One striking feature of the Cartesian epistemological package that has remained *the* epistemological package right into the present day, expressly found in Quine's, Van Fraassen's, and Dretske's views, is epistemic individualism. This is one key reason why philosophical accounts of science have failed to account philosophically for natural-scientific knowledge. Whatever may have been their explicit epistemological or methodological theories or commitments, in their scientific practice, eighteenth century scientists exhibited clear recognition of some of the key social dimensions of epistemic justification, which lie in part in constructive mutual critical assessment. What compels us rationally to attend to the critical assessment by our peers of our own views is recognition of our own human fallibility, a feature recurrently manifest in our attempts to formulate or to justify any worth-while empirical claim or theory.[5]

Precisely because they have used 'empiricism' as a membership badge, empiricists have refused to engage the more thorough critiques of empiricism developed by philosophers with different orientations or 'stances'. Highlighting philosophical stances almost unwittingly (cf. *ES* 49) raises again a very important though almost entirely neglected issue remaining from early logical positivism. Early logical positivists recognized that their views were shared within a particular

---

5 On the social dimensions of legitimate consensus, see Longino (1994); on the social dimensions of evidence, see Haack (2003), pp. 57–91.

'cultural circle', and they delimited the relevant range of protocol sentences to those uttered by 'scientists of our cultural circle'. This is how Hempel defended Carnap against the charge that Carnap had offered nothing but a coherence theory of truth that obviously failed to distinguish the truth from any elaborately detailed, coherent piece of fiction (Westphal 1989, pp. 56–57). Van Fraassen recognizes that highlighting philosophical stances invites the objection that stances lie beyond rational assessment, because they are stances rather than propositions. He insists that stances can be evaluated, and that it does in fact happen – 'welcome to the real world' (*ES* 61–62). This casual observation misses the point of the objection. Here van Fraassen is in a bind: the individualist, Bayesian resources central to his view of epistemic rationality and justification are incapable of addressing this problem (*ES* 67), simply because his Bayesian principles concern estimating the evidence bearing on propositions, they do not at all bear on the merit of expressly *non*-propositional stances! The closest Carnap came to resolving this problem lay in his pragmatic approach to choosing among linguistic frameworks. Unfortunately, Carnap's pragmatic approach this such choices is inconsistent with his own cardinal distinction between 'internal' and 'external' questions. Carnap's semantics simply can't handle the issue of assessing or recommending linguistic frameworks (Westphal 1989, ch. 4).

Like Carnap, van Fraassen proudly hails the (alleged) fact that the most telling criticisms of empiricism have come from empiricists themselves (*ES* 38). This pride is symptomatic of a key problem. In fact, the most telling criticisms of empiricism that were *recognized* or acknowledged by empiricists all came from empiricists, because those criticisms were all within the empiricist camp, and were cast in terms of whether various empiricist principles sufficed for meeting various empiricist goals. The devastating criticisms of empiricism offered by non- or anti-empiricists – Kant or Hegel, or more recently those of Blanchard, Weinberg, Tuschling & Rischmüller, or Frederick Will – were neglected, because they weren't developed by philosophers of 'our [logical empiricist] cultural circle'. This persistent refusal by empiricists to reflect on their own philosophical stance is very much a manifestation of the voluntarist underpinnings of the empiricist stance, as Teller (2001, pp. 123–125) rightly emphasizes. Recall, *e.g.*, Quine's (1969, p. 134) 'preference' for ontological desert landscapes. Such 'preferences' entered analytic philosophy as ungrounded Carnapian decisions to adopt a linguistic framework. Yet preferences didn't then, don't now, and simply cannot provide philosophical reasons. Instead, such 'preferences', like the adoption of a wholly non-propositional 'stance', really are expressions of allegiance to a certain philosophical cultural circle; historical perspective reveals that the empiricist stance is highly partisan. Once again, this is why philosophers, to *philosophize* adequately, *must* attend to philosophical history, within its historical context. It is almost the only way for us to gain a critical vantage point on our own present perspective.

Van Fraassen hopes to show that stances can be assessed in part by identifying assumptions, commitments or principles shared by competing philosophical stances (or so he said in response to questions at the Pasadena meeting of the American Philosophical Association, March 2004). Surely these are important, but *ex hypothesi* they cannot suffice for assessing competing stances. Unless at least one stance is just flat inconsistent with those shared views, the shared basis does not suffice to determine the controversial claims or substances at issue between competing stances.

The underlying problem here, not yet recognized in van Fraassen's book, though he once nearly verges on it (*ES* 40), is Sextus Empiricus' Dilemma of the Criterion. Competing stances disagree about relevant standards and methods of justification. Most briefly, the Dilemma of the Criterion poses the problem, How can basic criteria of justification be established, when those very criteria themselves are subject to fundamental dispute, without vicious circularity, dogmatism, infinite regress, or arbitrary (and hence disputable) assumption? The currency of this ancient problem was renewed in the Modern period by Bayle; Descartes sought to answer it, though unsuccessfully (Westphal 1987–88). This Dilemma was included in Nagel and Brandt's (1965) classic anthology, *Meaning and Knowledge*. It then largely vanished from sight until Moser and vander Nat (1995) again anthologized it. Pyrrhonian scepticism was given a hearing by Robert Fogelin (1994), though oddly he omitted the Dilemma and greatly softened the Pyrrhonist challenge to knowledge by restricting it to issues of certainty.[6] Although some analytic epistemologists have tried to solve the problem, I have argued elsewhere that none of them have succeeded (Westphal 1989; 1998b).

Sextus' Dilemma of the Criterion puts paid to the foundationalist model of deriving justification from some ultimate starting point, regardless of whether it consists in principles or data. It likewise puts paid to the individualist model of rationality. The trope of circularity can be resolved, along with the Dilemma of the Criterion, only by recognizing the centrality of constructive self- and mutual criticism in rational justification and by rejecting justificatory infallibilism. Rational justification consists in assessing the merits of any claim in view of its adequacy for its intended domain, its superiority to prior and contemporaneous alternatives, and its continued adequacy for its domain in view of renewed occasions of its use, often in changed circumstances. Fallibilism is a fact of our empirical inquiry – and of our '*a priori*' philosophical inquiries, too! Trying to eliminate human fallibility by forcing our empirical knowledge into the axiomatic-

---

6 Pyrrhonian equipollence arguments do not merely undermine certainty; they undermine knowledge altogether by (purportedly) showing that no view is any more justified than its alternatives. Fogelin, who urges us to take Pyrrhonian scepticism seriously, does not himself take Pyrrhonian scepticism seriously enough.

deductive model of *scientia* has failed every time (Westphal 1987–88; 2004). The proper response to human cognitive fallibility is constructive self- and mutual criticism. Mutual constructive criticism has, of course, been a mainstay of empiricist (and, more broadly, analytic) *practice*, which empiricist philosophers fail properly to appreciate due to their individualist views of rational justification. Epistemic justification is in part social because any moderately interesting claim has implications that reach far beyond what any one of us could ever work out, much less evaluate. Others who use our apparently justified statements, whether in similar or quite different circumstances, can generate and assess information that can bear on the formulation, use, or justificatory status of our original statement. Only rarely have analytic philosophers noted the importance of self-criticism; none have inquired into its possibility.

Whoever believes that nothing rationally compels us to attend to informed criticism from others is committed, far too deeply committed, to one side of an old and long-since discredited Enlightenment dichotomy, central to the Cartesian epistemological package, according to which reason simply must be an individual phenomenon, if it is to have any critical purchase on tradition, which manifestly is a social phenomenon (Westphal 2002a). This dichotomy either denies or ignores the crucial importance of mutual critical assessment in rational justification, and is deeply at odds with rational practice within the natural sciences. Accepting this dichotomy, whether implicitly or explicitly, whether in theory or in practice, is not a proper stance for philosophers who claim to admire science as the very epitome (if not sole instance) of empirical knowledge.

5. Hegel's Transcendental & Pragmatic Realism in Epistemology

Surprisingly, the sound approach to these issues comes from the Kantian tradition rejected by Russell and Moore, and by analytic philosophers ever after (analytic Kant studies not withstanding), including van Fraassen (*ES* 2). Indeed, the sound approach to addressing these issues was developed by Kant's greatest student, Hegel. Right in the middle of the Introduction to *The Phenomenology of Spirit*, Hegel paraphrased Sextus' Dilemma. Unrecognized by friends and foes alike, this key to Hegel's epistemology lay dormant, and Hegel's acute epistemology lay unrecognized, until very recently. In his Introduction, Hegel provides a subtle and powerful analysis of the self-critical structure of our cognitive awareness of ourselves and of objects apparent to us in our surroundings. While dealing with a plethora of other issues, Hegel devotes the whole of the *Phenomenology* to resolving Sextus' Dilemma and the epistemological conundrums recounted above (Westphal 2003).

Hegel was the first philosopher to recognize that Kant's own accounts of causal judgments, and of the requisite degree of regularity and variety among the contents of our sensory manifold, in fact provide genuinely transcendental proofs of what we now call mental content externalism, of anti-individualism in Burge's (1992, pp. 46–47) sense. Moreover, Hegel recognized that these transcendental proofs refute Kant's own transcendental idealism, that they also show the main fallacy in Kant's direct arguments for his idealism, and that these criticisms are based squarely on Kant's own analyses and principles in the Transcendental Analytic! Demonstrably, the Cartesian/Humean predicament is *not* the human predicament (Westphal 2003; 2004).

Hegel expressly developed the pragmatic and social dimensions of epistemic justification, and he recognized that these historical, social and pragmatic dimensions of human empirical knowledge are consistent with, indeed they require, realism about the objects of our empirical knowledge. While defending our possession and legitimate use of certain *a priori* concepts, Hegel recognized the key insight of Kant's cognitive semantics, that these concepts can only be used in legitimate cognitive judgments about spatio-temporal phenomena. Hegel's infamous 'idealism' is in fact a realist form of causal holism.[7] More surprising yet, Hegel's philosophy is rooted in empirical scientific enquiry:

> Not only must philosophy accord with the experience nature gives rise to; in its *formation* and in its *development,* philosophic science presupposes and is conditioned by empirical physics. (*Enz.* §246 Remark)

Hegel's metaphysics was derided by analytic philosophers who failed to recognize Hegel's acute epistemology and his equally acute cognitive semantics. This twin failure resulted directly from analytic philosophers' anachronistic return to the eighteenth century metaphysics *and* epistemology of Hume's first *Enquiry.* Hegel is, of course, the original and past master of philosophical stances and their internal critical assessment: this is the central point of his phenomenological examination of forms of consciousness in the *Phenomenology of Spirit.* Finally, Hegel, building on Kant, is the fountainhead of pragmatic realism, the one tradition in epistemology that put paid to Descartes' epistemological predicament (above, §3).

---

7 Two key principles of Hegel's causal holism are that the causal characteristics of things are central to their identity conditions; because causal dispositions are relational, the identity conditions of things are mutually inter-defined.

6. Concluding Unscientistic Postscript

I have argued that concern with the *empirical* basis of commonsense and natural-scientific knowledge is no monopoly of empiricists. It behooves epistemologists and philosophers of science to become much more careful and articulate about what counts as 'empirical' and what counts as 'empiricist'. It behooves epistemologists and philosophers of science to become much more reflective and sophisticated about their own philosophical stances, as well as those of others. It behooves empiricists to consider much more carefully the wide range of presuppositions required even to have, to recognize, or to consider their favorite forms of 'empirical evidence'. 'Empiricism' in philosophy of science simply can*not* be assimilated to 'empiricism' in epistemology! More generally, philosophical understanding, appreciation, and above all critical assessment of natural science and its findings cannot be gained by assimilating natural-scientific knowledge to pre-conceived philosophical models of natural science. Instead, a genuinely 'naturalized' philosophy of science must examine scientific knowledge as it is developed and embodied in the natural sciences themselves, both historically and currently. In these regards, careful and discerning history of science is crucial for the philosophy of science – at the very least, because scientists have shown a far greater understanding and mastery of scientific methods and scientific knowledge than have philosophers, whose track record in this domain is often far from impressive.[8]

For examining these issues philosophically, the history of philosophy is utterly invaluable, especially that chapter of the history of philosophy occupied by the reputed prince of philosophical darkness, Hegel. The siren song of science has been too long distorted by philosophers with tin Cartesian ears. Excessive specialization in the field has only exacerbated these problems. We philosophers are long overdue to lift a methodological page from Nietzsche, who observed:

> ... to see differently in this [vedantic or Kantian] way for once, *to want* to see differently, is no small discipline and preparation of the intellect for its future 'objectivity' – the latter understood not as 'disinterested contemplation' (which is a non-concept and absurdity), but rather as the capacity to have one's pro and contra *in one's power*, and to shift them in and out: so that one knows how to make precisely the *difference* in perspectives and affective interpretations useful for knowledge. ... There is *only* a perspectival seeing, *only* a perspectival 'knowing'; and *the more* affects we allow to speak about a matter, *the more* eyes, different eyes, we know how to bring to bear on one and the same matter, that much more

8 Though certainly not always: see, *e.g.*, the exemplary research of Stein (2002; cf. Shimony 2002a), Shimony (2002b), and Harper (1991 to 2002b).

> complete will our 'concept' of this matter, our 'objectivity' be. (*GM* 3 §12, cf. *FW* §295, *EH* I §§1, 9)

To recognize that philosophers take stances is an important first step towards understanding them and their roles in the development and assessment of express philosophical views and theses. But we must avoid the error all too frequently found among philosophical schools in the twentieth century, including logical positivism, of 'acknowledging' as worthy interlocutors only those who subscribe to the same philosophical stance. Above all, this thwarts the identification and assessment of one's underlying philosophical stance. Descartes, at least, recognized that his epistemological inquiry was only possible under the special conditions of being able to set aside all of his everyday, moral, and religious concerns, involvements and beliefs. Descartes' epistemological package, most of which became *the* epistemological stance, failed to understand even Galilean or Newtonian mechanics. Due to its atomistic tendencies, it has fared even worse with quantum theory, which is deeply holistic (Teller 1986; 1989).

I close with a note of emphatic agreement with van Fraassen, who observes:

> Philosophy itself is a value and attitude-driven enterprise; philosophy is in false consciousness when it sees itself otherwise. To me, philosophy is of overriding importance, to our culture, to our civilization, to us individually. For it is the enterprise in which we, in every century, interpret ourselves anew. But unless it so understands itself, it degenerates into an arid play of mere forms. (*ES* 17)

Amen! Yet if we are to get our philosophical stances in order, if we are to achieve genuine philosophical self-understanding, and if we are to account philosophically for natural scientific knowledge, we must fundamentally transform our contemporary philosophical stances. Many of the requisite methods and materials lie waiting in the unfamiliar and still largely uncharted scope and depths of the Kantian-Hegelian tradition. Routes into this philosophical gold mine have been cleared; now the time is ripe for a fresh look at their achievements.[9]

University of East Anglia

---

9 I am grateful to the organizers of the XV Internordic Philosophy Symposium for their kind invitation, which allowed me to develop these thoughts, and for discussions at the Symposium and comments from Susan Haack on the penultimate draft, all of which prompted many improvements. Remaining flaws remain solely my responsibility.

References

Boxer, S. (1998), "At the End of a Century of Philosophizing, the Answer is Don't Ask", *New York Times* 15 Aug., Late Edition - Final, §B, p. 9, column 1.

Burge, T. (1992), "Philosophy of Language and Mind: 1950-1990", *The Philosophical Review* 101.1:3-51.

Carnap, R. (1928), *Der logische Aufbau der Welt*, Berlin, Welkreis; R. A. George, tr., *The Logical Structure of the World*, Berkeley, University of California Press, 1969.

Carnap, R. (1950), "Empiricism, Semantics, and Ontology", *Revue International de Philosophie* 4, revised form published in *Meaning and Necessity*, Chicago, University of Chicago Press, 1956, pp. 205-221.

Carnap, R. (1963), "Replies and Systematic Expositions", in P. A. Schilpp (ed.), *The Philosophy of Rudolf Carnap* (The Library of Living Philosophers), pp. 859-1013.

Creath, R. (1991), "Every Dogma has its Day", *Erkenntnis* 35(1-3), 347-389.

DeGandt, 1995, *Force and Geometry in Newton's Principia*, C. Wilson, tr., Princeton, Princeton University Press.

Descartes, R. (1964-), C. Adam and P. Tannery, eds., *Oevres de Descartes*, 13 vols., revised edition, Paris, Vrin/C.N.R.S..

Descartes, R. (1984, 1985, 1991), J. Cottingham, R. Stoothoff, and D. Murdoch, trs., *The Philosophical Writings of Descartes*, 3 vols, Cambridge, Cambridge University Press.

Dretske, F. I. (1981), *Knowledge and the Flow of Information*, Cambridge, MIT/Bradford Press.

Galilei, G. (1890-1910), *Le Opere di Galileo Galilei*, A. Favaro, ed., Florence, Edizione Nazionale, designated '*Opere*'.

Galilei, G. (1974), *Two New Sciences*, S. Drake, tr., Madison, University of Wisconsin Press.

Haack, S. (1993), *Evidence and Inquiry*, Oxford, Blackwell's.

Haack, S. (2003), *Defending Science – within Reason*, Amherst, NY, Prometheus.

Hahn, L. E., and P. A. Schilpp (eds.) (1986), *The Philosophy of W. V. Quine*, LaSalle, Ill., Open Court.

Hanna, R. (2001), *Kant and the Foundations of Analytic Philosophy*, Oxford, The Clarendon Press.

Harper, W. L. (1991), "Newton's Classic Deductions from Phenomena", *PSA 1990* 2:183-196.

Harper, W. L. (1997), "Isaac Newton on Empirical Success and Scientific Method", in L. Earman and J. D. Norton (eds.), *The Cosmos of Science*, Pittsburgh, University of Pittsburgh Press, pp. 55-86.

Harper, W. L. (1998), "Measurement and Approximation: Newton's Inferences from Phenomena versus Glymour's Bootstrap Confirmation", in G. Weingartner, G. Schurz and G. Dorn (eds.), *The Role of Pragmatics in Contemporary Philosophy*, Vienna, Hölder-Picher-Tempsky, pp. 265-287.

Harper, W. L. (1999), "The first six propositions in Newton's argument for Universal Gravitation", *The St. John's Review* 45.2:74-93.

Harper, W. L. (2002a), "Howard Stein on Isaac Newton: Beyond Hypotheses?", in Malament (2002), pp. 71-111.
Harper, W. L. (2002b), "Newton's Argument for Universal Gravitation", in I. B. Cohen & G. Smith (eds.), *The Cambridge Companion to Newton*, Cambridge, Cambridge University Press, pp. 174-201.
Harper, W. L., B. H. Bennett and S. R. Valluri (1994), "Unification and Support: Harmonic Law Ratios Measure the Mass of the Sun", in D. Prawitz and D. Westerståhl (eds.), *Logic and Philosophy of Science in Uppsala*, Dordrecht, Kluwer Academic Publishers, pp. 131-146.
Hartsock, N. (1983), *Money, Sex, and Power: Toward a Feminist Historical Materialism*, New York, Longman.
Hegel, G. W. F. (1968-), H. Buchner & O. Pöggeler, eds., *Gesammelte Werke*, Hamburg, Meiner.
Hegel, G. W. F. (1980), *Phänomenologie des Geistes*, in *Gesammelte Werke*, vol. 9.
Hegel, G. W. F. (1992), *Encyklopädie der philosophischen Wissenschaften* (3d ed., 1830), in *Gesammelte Werke*, vol. 20, designated '*Enz.*'.
Hume, D. (1975), *An Enquiry Concerning Human Understanding*, in P. H. Nidditch (ed.), *Enquiries Concerning Human Understanding and Concerning the Principles of Morals*, 3d ed., New York, Oxford University Press, designated '*En*'.
Hume, D. (2000), D. F. Norton and M. J. Norton (eds.), *A Treatise of Human Nature*, Oxford, Oxford University Press, designated '*T*'.
Kant, I. (1902-), Könniglich Preussische (now Deutsche) Akademie der Wissenschaften, ed., *Kants Gesammelte Schriften*, 29 vols, Berlin, G. Reimer, now De Gruyter.
Leibniz, G. W. (1875-90), C. I. Gerhardt, ed., *Die philosophische Schriften von Gottfried Wilhelm Leibniz*, Berlin, Weidmann.
Leibniz, G. W. (1981), *New Essays on Human Understanding*, P. Remnant & J. Bennett, trs., Cambridge, Cambridge University Press.
Ladyman, J. (2004), "Empiricism versus Metaphysics", *Philosophical Studies* 121.2:133-145.
Lipton, P. (2004), "Epistemic Options", *Philosophical Studies* 121.2:147-158.
Longino, H. E. (1994), "The Fate of Knowledge in Social Theories of Science", in F. Schmitt (ed.), *Socializing Epistemology*, Lanham, Md., Rowman & Littlefield, pp. 135-57.
Malament, D. (ed.) (2002), *Reading Natural Philosophy: Essays in the History and Philosophy of Science and Mathematics*, LaSalle, Ill., Open Court.
Mandelbaum, M. (1964), *Philosophy, Science, and Sense Perception*, Baltimore, Johns Hopkins University Press.
Moser, P. & A. vander Nat (eds.) (1995), *Human Knowledge: Classical and Contemporary Approaches*, New York, Oxford University Press.
Nagel, T. & R. Brandt (eds.) (1965), *Meaning and Knowledge*, New York, Harcourt, Brace & Jovanovich.
Newton, I. (1952), *Opticks*, New York, Dover.

Newton, I. (1999), *The Principia: Mathematical Principles of Natural Philosophy*, I. B. Cohen, A. Whitman, assisted by J. Budenz, trs., Berkeley, University of California Press.
Nietzsche, F. (1967-), G. Colli & M. Montinari (eds.), *Kritische Gesammtausgabe*, Berlin, De Gruyter.
Nietzsche, F. (1967), *Ecce Homo*, W. Kaufmann, tr., New York, Vintage, designated '*EH*'.
Nietzsche, F. (1974), *The Gay Science*, W. Kaufmann, tr., New York, Vintage, designated '*GS*'.
Nietzsche, F. (1998), *On the Genealogy of Morality*, M. Clark & A. Swensen, trs., Cambridge, Mass., Hackett Publishing Co., designated '*GM*'.
Quine, W. V. O. (1953), *From a Logical Point of View*, New York, Harper & Row.
Quine, W. V. O. (1960), *Word and Object*, Cambridge, M.I.T. Press.
Quine, W. V. O. (1969), *Ontological Relativity and Other Essays*, New York, Columbia University Press.
Quine, W. V. O. (1974), *The Roots of Reference*, LaSalle, Ill., Open Court.
Quine, W. V. O. (1975), "The Nature of Natural Knowledge", in S. Guttenplan (ed.), *Mind and Language*, Oxford, The Clarendon Press, pp. 67-82.
Quine, W. V. O. (1981), *Theories and Things*, Cambridge, Mass., Harvard University Press.
Quine, W. V. O. (1990), *The Pursuit of Truth*, Cambridge, Mass., Harvard University Press.
Quine, W. V. O. (1995), *From Stimulus to Science*, Cambridge, Mass., Harvard University Press.
Rouse, J. (2002), *How Scientific Practices Matter*, Chicago, University of Chicago Press.
Russell, B. (1994), J. Passmore, general ed., *The Collected Papers of Bertrand Russell*, London, Routledge.
Scharff, R. (1995), *Compte After Positivism*, Cambridge, Cambridge University Press.
Sellars, W. (1981), "Mental Events", *Philosophical Studies* 39:325-345.
Sextus Empiricus (1934), *Outlines of Pyrrhonism*, Rev. R. G. Bury, tr., in *Works*, 4 vols., vol. 1, Cambridge, Mass., Harvard University Press.
Shimony, A. (2002a), "Introduction: The Character of Howard Stein's Work in Philosophy and History of Physics", in Malament (2002), pp. 1-8.
Shimony, A. (2002b), "Some Intellectual Obligations of Epistemological Naturalism", in Malament (2002), pp. 297-313.
Smith, G. E. (2002), "The Methodology of the *Principia*", in I. B. Cohen & G. Smith (eds.), *The Cambridge Companion to Newton*, Cambridge, Cambridge University Press, pp. 138-173.
Stein, H. (2002), "Bibliography of Howard Stein", in Malament (2002), pp. 407-410.
Teller, P. (1986), "Relational Holism and Quantum Mechanics", *The British Journal for Philosophy of Science*, 37:71-81.
Teller, P. (1989), "Relativity, Relational Holism, and the Bell Inequalities", in J. Cushing and E. McMullin (eds.), *Philosophical Lessons from Quantum Theory*, South Bend, Ind., Notre Dame University Press, pp. 208-223.
Teller, P. (2001), "Whither Constructive Empiricism?" *Philosophical Studies* 106:123-150.
Teller, P. (2004), "What is a Stance?", *Philosophical Studies* 121.2:159-170.

Turnbull, R. (1959), "Empirical and A Priori Elements in Broad's Theory of Knowledge", in P. A. Schilpp (ed.), *The Philosophy of C. D. Broad*, New York, Tudor, pp. 197–231.
Van Fraassen, B. (2001), "Constructive Empiricism Now", *Philosophical Studeis* 106:151–171.
Van Fraassen, B. (2002), *The Empirical Stance*, New Haven, Yale University Press, designated '*ES*'.
Van Fraassen, B. (2004a), "Précis of *The Empirical Stance*", *Philosophical Studies* 121.2:127–132.
Van Fraassen, B. (2004b), "Replies to Discussion on *The Empirical Stance*", *Philosophical Studies* 121.2:171–192.
Westphal, K. R. (1984a), "Was Nietzsche a Cognitivist?", *The Journal of the History of Philosophy* 26.3:343–363.
Westphal, K. R. (1984b), "Nietzsche's Sting and the Possibility of Good Philology", *International Studies in Philosophy* 16.2:71–90.
Westphal, K. R. (1987–88), "Sextus Empiricus *Contra* René Descartes", *Philosophy Research Archives* 13:91–128.
Westphal, K. R. (1989), *Hegel's Epistemological Realism*, Philosophical Studies Series, vol. 43, Dordrecht, Kluwer.
Westphal, K. R. (1998a), *Hegel, Hume und die Identität wahrnehmbarer Dinge*, Frankfurt/Main, Klostermann.
Westphal, K. R. (1998b), "Hegel's Solution to the Dilemma of the Criterion", revised version in J. Stewart (ed.), *The Phenomenology of Spirit Reader: A Collection of Critical and Interpretive Essays*, Albany, SUNY Press, pp. 76–91.
Westphal, K. R. (2000), "Hegel's Internal Critique of Naïve Realism", *Journal of Philosophical Research* 25:173–229.
Westphal, K. R. (2002a), "Rationality and Relativism: The Historical and Contemporary Significance of Hegel's Response to Sextus Empiricus", *Esercizi Filosofici* (Trieste) 6:22–33; Italian translation (by Dr. C. Ferrini): "Razionalità e relativismo: Il significato storico e contemporaneo della risposta hegeliana a Sesto Empirico", *Etica e Politica* 4.1 (2002), http://www.units.it/~etica/2002_1/index.html
Westphal, K. R. (2002b), "'Sense Certainty', or Why Russell had no 'Knowledge by Acquaintance'", *The Bulletin of the Hegel Society of Great Britain* 45/46 (2002):110–123.
Westphal, K. R. (2003), *Hegel's Epistemology: A Philosophical Introduction to the Phenomenology of Spirit*, Cambridge, Mass., Hackett Publishing Co.
Westphal, K. R. (2004), *Kant's Transcendental Proof of Realism*, Cambridge, Cambridge University Press.
Westphal, K. R. (2005), "Hume, Hegel and Abstract General Ideas", *Bulletin of the Hegel Society of Great Britain* 51/52.

II

Perspectives on Naturalism

Quine and Wittgenstein on the Dissolution of Philosophical Problems

Martin Gustafsson

1. Introduction

It is a remarkable and thought-provoking fact that the two thinkers who have had the most profound influence on post-war analytic philosophy seem to have such radically different views on the nature of philosophical inquiry. According to Quine, good philosophy is continuous with empirical science. Philosophers, like physicists or biologists, are scientists in search of an organized conception of reality. Methodologically they may be more oriented towards language than most other scientists, but herein, says Quine, lies no principled distinction. Reflection on verbal discourse is important in many other theoretical sciences as well, and does not make the philosopher's results independent of empirical data and experimentation: "if the theoretical scientist in his remote way is bound to save the eventual connections with non-verbal stimulation, the philosopher in his remoter way is bound to save them too." (Quine 1960, p. 276)

Wittgenstein, by contrast, thinks the achievement of philosophical clarity requires a kind of investigation very different from scientific research. According to Wittgenstein's diagnosis, the source of philosophical confusion is not that philosophers aren't scientific enough. On the contrary, he claims that disorder in the thinking of philosophers arises precisely because of their tendency to "constantly see the method of science before their eyes" and be "irresistibly tempted to ask and answer questions in the way science does. This tendency is the real source of metaphysics, and leads the philosopher into complete darkness." (Wittgenstein 1958, p. 18) While Quine thinks progress in philosophy requires that we eradicate the alleged distinction between conceptual and factual inquiry, Wittgenstein thinks it is precisely such obliteration which makes us go astray in the first place: "Philosophical investigations: conceptual investigations. The essential thing about metaphysics: it obliterates the distinction between factual and conceptual investigations." (Wittgenstein 1974b, §458)

What, exactly, does this difference between Quine and Wittgenstein amount to? A closer look at their ways of reasoning makes it clear that the relation between their views is more complicated than what the above juxtaposition may be taken to suggest. One possible misunderstanding needs to be discarded at once. In separating philosophy from science, Wittgenstein is not claiming that scientists never do philosophy, or that conceptual clarification is totally disconnected from scientific endeavors. In fact, Wittgenstein lists Boltzmann and Hertz among the figures who have had the deepest influence on his thinking. Hertz's introduction to *The Principles of Mechanics* was of particular importance to him. Throughout his life, Wittgenstein kept on referring to this piece of text as a paradigm of philosophical inquiry and urged his students to read it carefully. So, the point of his making a distinction between philosophy and science is not to draw boundaries between academic disciplines, but to capture a methodological difference of deeper significance.

It should also be noted that Quine is perfectly willing to admit that there are significant differences between philosophy and the other sciences. His claim is that these differences are differences in degree rather than in kind. He is not denying that the connection between philosophical investigations and empirical data is extremely remote and indirect.

Another complication is that the conception of philosophy against which Quine tends to contrast his own outlook is not one to which Wittgenstein would want to subscribe. Quine's primary target is the traditional idea of a first philosophy, which he finds in the writings of Descartes and onward, all the way up to Carnap's *Aufbau*. "[M]y position," he says, "is a naturalistic one: I see philosophy not as an *a priori* propaedeutic or groundwork for science, but as continuous with science." (Quine 1969, p. 126) This shows that the view of philosophy from which Quine first and foremost wants to distance himself is a view that Wittgenstein would also want to reject. Wittgenstein's way of distinguishing philosophy from science does not involve any claim to the effect that philosophy can, in some non-empirical, *a priori* fashion, provide a foundation for empirical research. This is clearly so with respect to the later Wittgenstein; arguably, it is equally true of the author of the *Tractatus*.

If we look at Quine from Wittgenstein's point of view, he seems to have a much too limited conception of the alternatives available for thinking about the relation between philosophy and science. Quine tends to argue as if *either* you accept that philosophy is continuous with science, *or* you will have to subscribe to the traditional idea of a *prima philosophia*. Whereas what Wittgenstein wants to achieve is something different from both these alternatives: according to him, the task of philosophy is neither to give us new factual knowledge, nor to constitute the pillar upon which the special sciences must rest, but to engage in purely clarificatory investigations aimed at dissolving conceptual confusions.

From Quine's point of view, the idea that there is such a third, clearly distinguishable alternative, is mistaken. Quine seems to think that, in the end, the very idea of a sharp distinction between empirical and conceptual investigations must presuppose some version of the idea of a first philosophy. Even if Wittgenstein tries to steer clear of this traditional conception, Quine would probably say that his ideas about philosophy's distinctive character reveals that he is still held captive by a traditional and misconceived picture of the relation between philosophy and science.

Of course, I cannot within the limited space of the present paper give a full account of the differences between Quine's and Wittgenstein's conceptions of philosophy. My aim here is much more restricted. I will focus on one aspect of the difference between Quine's and Wittgenstein's conceptions, an aspect which I take to be of central importance. More precisely, I will look at one of the few places in Quine's corpus where he relates his own view of philosophy to what he takes to be Wittgenstein's conception, namely, sections 53 and 54 of *Word and Object*. Interestingly enough, Quine here claims that there is a similarity between his own views and Wittgenstein's. Comparing Quine's discussion with an example from Wittgenstein, I will argue that this alleged similarity is an illusion. Indeed, I hope my comparison will give a clearer sense of how different the philosophical outlooks of these two thinkers really are.

2. Quine on Explication as a Way of Dissolving Philosophical Problems

Sections 53 and 54 of *Word and Object* constitute Quine's most sustained discussion of *explication* as a philosophical method. Quine's paradigm example is the set-theoretical explications of the notion of ordered pair offered by mathematicians such as Wiener and Kuratowski. According to Quine, ordered pairs might seem indispensible in science. But at the same time they can be troublesome, due to the fact that they do not obey the principle of extensionality: the ordered pair $\langle x,y \rangle$ is not the same as the ordered pair $\langle y,x \rangle$. The way to dissolve these troubles, Quine argues, is to show that they proceed from scientifically superfluous usages. Here is how the dissolution is supposed to work. First, we should note that the scientifically relevant employments of the notion of ordered pair are captured by a single, very simple postulate: If $\langle x,y \rangle = \langle z,w \rangle$, then $x=z$ and $y=w$. This postulate encapsulates the scientific work done by the notion of ordered pair. Wiener's achievement, Quine argues, was to show that the same work can be done by a set-theoretical, extensional substitute. He proposed that we identify $\langle x,y \rangle$ with the set $\{\{x\},\{y,\varnothing\}\}$ (where '$\varnothing$' signifies the empty set). Identical sets have the same members; hence, if $\{\{x\},\{y,\varnothing\}\} = \{\{z\},\{w,\varnothing\}\}$, then $x=z$ and $y=w$. According to Quine, the possibility of such an extensional construal shows that any worry we might have about the non-extensional character of ordered pairs can simply be left aside as scientifically immaterial. Wiener dissolves our philosophical qualms by showing that, as scientists, we do not have to care about them.

Wiener is not offering a conceptual analysis in any traditional sense of the word, for he makes no synonymy claim. According to Quine, the point is just that nothing of scientific importance is lost if we reinterpret talk about ordered pairs as talk about sets. On the contrary, such reinterpretation involves scientific gains. We are able to handle observational data in a more economical manner, since the old notion of ordered pair is replaced by set-theoretical notions that are already in use elsewhere.

One aspect of this benefit is that Wiener's explication involves the elimination of what Quine calls ontological commitments. The explication takes us from a theory which postulates ordered pairs in addition to sets, to a theory which identifies ordered pairs with sets. Our earlier worries were connected with the view that ordered pairs have something which distinguishes them from sets: a perplexing, non-extensional character which seemed to require philosophical explanation. Wiener's explication should make us realize that this distinctive property is best seen as an illusion.

Quine argues that a similar strategy of dissolution applies to other, philosophically more significant perplexities. Consider the many traditional worries associated with the dualist conception of mental states and events as being distinct from physiological states and events. According to Quine, the scientific purpose of positing such distinct mental states and events can only be to develop a systematic theory which explains and predicts the behavior of human organisms – just like the positing of atoms and molecules helps us predict and explain the behavior of physical objects. "But," Quine remarks, "if a certain organization of theory is achieved by thus positing distinctive mental states and events behind physical behavior, surely as much organization could be achieved by positing merely certain correlative physiological states and events instead. [...] The bodily states exist anyway; why add the others?" (Quine 1960, p. 264)

The point here is exactly analogous to the one made about ordered pairs. Quine isolates what he takes to be the scientific work done by the notion of mental states and events. He then points out that that work may just as well be done by physiological substitutes. Without any scientific loss, we can reinterpret talk of distinct mental entities as talk about physiological entities, and thereby get rid of the traditional worries about the mental – for these worries presuppose that mental states and events are non-physical. This supposition, Quine argues, is a scientifically superfluous idea which can simply be discarded.

Just as when ordered pairs get explicated in set-theoretical terms, Quine thinks the physiological explication of the mental involves the elimination of ontological commitments. It takes us from a dualist theory which postulates mental states and events in addition to physiological states and events, to a theory according to which mental states and events are nothing but physiological states and events. What underlies our perplexities was the dualist idea that mental states and events are somehow mysteriously non-physical. The perplexities disappear once we realize that making such an ontological distinction between the mental and the physical constitutes a scientifically needless way of talking.

3. Wittgenstein on the Connectives

Quine explicitly relates his discussion of explication to Wittgenstein's claim that philosophical problems should be dissolved rather than solved. "This doctrine," Quine says, "has its limitations, but it aptly fits explication. For when explication banishes a problem it does so by showing it to be in an important sense unreal; viz., in the sense of proceeding only from needless usages." (Quine 1960, p. 260) He then goes on to argue that most followers of Wittgenstein miss this point. Instead of appreciating the kinship between Wittgenstein's approach and explication, they complain that explications are unfaithful to ordinary language use. This complaint, Quine says, is completely misguided. In fact, he argues, it is the unfaithfulness to ordinary language which makes explication useful for the dissolution of philosophical problems, for "it is precisely by showing how to circumvent the problematic parts of ordinary usage that we show the problems to be purely verbal." (Quine 1960, p. 261)

I think Quine is wrong to suggest that Wittgenstein's approach to philosophical problems is similar to explication. This is not to say that Wittgenstein is an ordinary language philosopher in Quine's caricatured sense. Quine is right that Wittgenstein does not think that deviating from established language use is, in itself, a philosophical sin. The real difference between Wittgenstein and Quine is more intricate, but nonetheless deep and fundamental.

Quine's talk of "banishing" and "circumventing" philosophical problems already provides a clue to the nature of this difference. For banishing or circumventing a problem is different from dissolving it. While "dissolution" suggests disintegration under the pressure exerted by demands for clarification, "circumvention" and "banishment" suggest (not disintegration but) dismissal. And dismissing problems is precisely what Quine thinks explication allows us to do. The point of explication is *not* to make us realize that we have failed to give sense to the words we use to express the problems, but to make it clear that the ontological commitments against the background of which the problems appear important are superfluous. What explication is supposed to show is that, for the purposes of science, usages which involve these commitments can safely be left aside. According to Quine, being worried about, say, the possibility of causal interaction between a non-physical mind and a physical body is like being worried about how Santa Claus is able to deliver Christmas presents to all children in just one night, or about how witches can fly. His objection is not that such worries are nonsensical, but that they are scientifically futile.

To shed more light on how Wittgenstein's and Quine's conceptions differ, I will discuss an example from Wittgenstein's early writings. One reason why I use this example is that, on the surface, it may look rather similar to Quinean explication. But in fact the difference is significant, and I think pin-pointing this difference sheds much light on Wittgenstein's way of approaching philosophical worries. I am aware that using an example from the early writings in this sort of way means presupposing that there is a fundamental continuity between Wittgenstein's early and later writings as far as his conception of philosophy is concerned. This assumption needs to be argued for, but space does not allow me to do so here. Toward the end of the paper, I will briefly discuss an example from the later writings. I hope that discussion will clarify the continuity in question.

Already in 1913, Wittgenstein expresses dissatisfaction with existing logical notations, and, in particular, with the sentential connectives of *Principia Mathematica*. According to Wittgenstein, constructions such as '$\sim p$' and '$p \vee q$' wrongly suggest that negation and disjunction make some sort of genuine contribution to the content of sentences; the connectives look misleadingly similar to names or predicates. (Wittgenstein 1913, p. 98) In fact, Wittgenstein questions the very idea, clearly suggested by the Russell-Whitehead notation, that the role of the connectives is to form complex propositions by, as it were, gluing together elementary propositions. According to Wittgenstein, it might even be misleading to say that the simpler propositions on the basis of which complex propositions are generated "occur in" the complex propositions. He says, "a proposition cannot have to another *the* internal relation which a *name* has to the proposition of which it is a constituent, and which ought to be meant by saying it 'occurs' in it." (Wittgenstein 1914, p. 116) As Peter Sullivan remarks, the word 'name' in this passage should be construed broadly to mean any expression replaceable by a variable to yield a propositional function. Thus, Wittgenstein's point is that there are no propositional functions which take propositions as arguments. (Sullivan 2000, p.180) To be sure, the truth value of, say, the proposition expressed by '$\sim p$', is a function of the truth value of the proposition expressed by 'p', but the proposition expressed by '$\sim p$' is not itself a function of the proposition expressed by 'p'. A proper understanding of the connectives requires that we think of the relations between propositions not in functional but in *operational* terms.

This is intimately connected with Wittgenstein's introduction of the so-called TF-notation in the *Tractatus*. This notation is meant to clarify the nature of the logical interconnectedness of propositions. In this notation, a proposition is given by its truth-table, or, in a more condensed version of the notation, by a schema where the last column of the truth-table is given within parentheses before the elementary propositions are listed. In the condensed version of Wittgenstein's notation, '(TF)(*p*)' replaces '*p*', '(FT)(*p*)' replaces '~*p*', '(TFFF)(*p*,*q*)' replaces '*p*&*q*', '(TTTF)(*p*,*q*)' replaces '*p*∨*q*', and so on. It has to be kept firmly in mind that, for Wittgenstein, the role of truth-tables is not to calculate the logical properties of object language sentences. Rather, truth-tables simply *rearticulate* propositions in a way which makes their logical interrelations more perspicuous. A truth-table is itself a propositional sign, at the same level as the less clear sign which it is meant to replace (cf. Ricketts 1996, p. 81).

Sullivan uses the following example to illustrate Wittgenstein's point. In Wittgenstein's notation, '~(*p*⊃~*q*)' and '*p*&*q*' are both rendered as '(TFFF)(*p*,*q*)'. In order to negate the proposition expressed here, we do not attach a negation sign. Rather, we turn the proposition into a new one by replacing all 'T's with 'F's and all 'F's with 'T's. Starting from '(TFFF)(*p*,*q*)' we thus obtain '(FTTT)(*p*,*q*)'. This makes it clear that negating a proposition is not a matter of adding anything to it, but of using it as a base from which a new proposition is generated according to a determinate pattern of transformation. And the same is true of conjunction, disjunction, and so on. "An operation," says Wittgenstein, "is the expression of a relation between the structures of its result and of its bases. The operation is what has to be done to the one proposition in order to make the other out of it." (Wittgenstein 1974a, §5.22–5.23) This does away with any temptation to think that there is something essentially negative about propositions obtained by procedures of negation. "Negation", Sullivan concludes, "is characteristic only of the relation between two propositions, never of any proposition itself." (Sullivan 2000, p. 180) And similarly for the other operations of propositional logic.

Wittgenstein's TF-notation makes it less tempting to say that the elementary propositions on the basis of which a non-elementary proposition is generated "occur in" that non-elementary proposition. The signs '*p*' and '*q*' *are* parts of the sign '*p*&*q*', and this makes us inclined to think that the propositions they express are parts of the conjunction, in a similar sort of way as '*a*' and '*b*' are parts of '*a*R*b*'. By contrast, it is clear that the signs '(TF)(*p*)' and '(TF)(*q*)' are not, in this sense, parts of '(TFFF)(*p*,*q*)'. In the latter sentence, there is no occurrence of, say, '(TF)(*p*)' such that replacing it by a variable would yield a propositional function. This is one of the beauties of Wittgenstein's notation: on the one hand, it makes transparent how the truth-value of the proposition expressed by '(TFFF)(*p*,*q*)' depends functionally on the truth-values of the elementary propositions; on the other hand, the notation makes it clear that the way in which the proposition itself is constructed from the elementary propositions is not to be understood in functional terms.

4. The Dissolution of Philosophical Problems

I hope that what I said in the previous section makes clear the intimate connection between Wittgenstein's introduction of the TF-notation and what, in the *Tractatus*, he calls his "fundamental idea [...] that the 'logical constants' are not representatives; that there can be no representatives of the *logic* of facts" (§4.0312). Indeed, in §5.4, right after having discussed operations on propositions, Wittgenstein frankly concludes that "there are no 'logical objects' or 'logical constants' (in Frege's and Russell's sense)".

If I read them correctly, these two passages express the same point. What might be confusing is that in §4.0312, Wittgenstein uses the term 'logical constants' to refer to certain signs, whereas in §5.4 he uses the same term to talk about what Russell and Frege take such signs to represent. In fact, the difference in expression between these two paragraphs is directly relevant to the main topic of the present paper. Taken *ad notam*, §5.4 seems to say that the signs '~', '∨' etc., as standardly used in logic books, *purport* to represent but fail to do so because the entities which they are meant to stand for do not exist. This suggests a reading according to which Wittgenstein's introduction of the TF-notation is akin to Quinean explication. According to this sort of reading, Wittgenstein thinks that replacing the standard notation of the *Principia* with the TF-notation means abandoning a framework according to which there are such entities as negation, disjunction, and so forth, in favor of a framework which is not committed to the existence of such entities. Allegedly, the point of the TF-notation would be to show how logic can be done without this commitment, and thereby make it clear that the commitment is superfluous. If this is right, Wittgenstein's idea is that abandoning this commitment allows us to banish or circumvent many seemingly deep philosophical worries, such as "What is the subject matter of logic?" and "What justifies logical inference?" (Cf. §§6.124 and 5.132.)

The way Wittgenstein expresses himself in §4.0312, however, suggests a quite different sort of interpretation. According to this second kind of reading, Wittgenstein's aim is to make it clear that the signs '~', '∨' etc., as standardly used in logic books, do *not* purport to represent, even if their surface appearance may entice us to think so. Thus, the point of the TF-notation is not to dispose of ontological commitments inherent in the standard logical notation, but to make it clear that there are no such commitments to begin with. According to this reading, Wittgenstein's criticism of the *Principia* notation is not that it purports to talk about entities that need not be taken to exist – for it doesn't. Rather, Wittgenstein's diagnosis is that the notation's surface appearance makes it overwhelmingly tempting to think that the function of the signs '~', '∨' etc., is analogous to the function of signs which *do* play the role of representatives. According to Wittgenstein, there is nothing wrong as such with using the *Principia* notation; the use itself is perfectly all right. Rather, his point is that the way the notation looks on paper makes us prone to misconstrue its actual employment. The supposed advantage of the TF-notation is that it does not foster such misunderstanding.

This second reading seems much closer to Wittgenstein's actual intentions. What Wittgenstein offers is a diagnosis to the effect that philosophical confusion arises because of a conflict between the way in which we are accustomed to use the standard *Principia* notation, and the pictures or analogies prompted by the notation's surface appearance. Consider a very simple example. When we do classical propositional logic in the standard notation, we negate '*p*' by adding a negation sign: '~*p*'. Clearly, one wants to say, '*p*' and '~*p*' do not say the same thing – after all, they contradict each other. Since the only difference we are able to perceive between these two sentences is that one of them involves a negation sign, it seems natural to conclude that this negation sign makes a substantial contribution to the content of the sentence in which it occurs. We are tempted to think that '~*p*' says something else than '*p*' because the negation sign has a meaning of its own which is somehow added to the meaning of '*p*'. And then, we are not far from thinking that '~*p*' says something *about* the state of affairs described by '*p*', namely, that that state of affairs *is not the case*. In other words, we reason as if the proposition created by adding a negation sign to '*p*' describes, as it were, an essentially 'negative' fact. And once we have reached this point in our reasoning, it will be difficult to avoid the idea that the negation sign must represent some sort of real entity.

We run into difficulties, however, as soon as we consider the fact that negation cancels itself when doubled: '~~*p*' says the same thing as '*p*'. If the negation sign leaves a substantial contribution to the content of the sentences in which it occurs, then why doesn't it do so in its doubled form? How can the result which we get by adding another negation sign be equivalent to a sentence which contains no negation sign at all?

If Wittgenstein is read in the way I suggest that we read him, he would say that the root of this quandary is that we waver, unwittingly, between construing the use of the negation sign in two incompatible ways. On the one hand, our notation tempts us to think that the function of the negation sign must be analogous to the function of names or predicates. On the other hand, we remember, somewhat inchoately, that our customary use of the negation sign is such that '~~*p*' is equivalent to '*p*'. We think we have to give an account which does justice to both these construals, but at the same time we vaguely feel that this is impossible. Hence a sense of *paradox* or *contradiction*: the nature of negation seems elusive or mysterious, since an account which does justice to both construals seems required yet unachieveable.

Wittgenstein's TF-notation helps us dissolve this worry by showing that the idea that the negation sign leaves some sort of substantial contribution to the content of the sentences in which it occurs is an illusion created by orthographic features of the *Principia* notation. Our quandary is not circumvented or banished. Rather, it completely disintegrates once we realize that the "paradox" depended on our failure to see that we were hovering between two incompatible viewpoints.

The puzzle about double negation may seem like a too simple example of a philosophical worry, but I would argue that Wittgenstein offers a similar diagnosis of the more serious difficulties which his TF-notation is meant to help us dissolve. When he says that "[t]he propositions of logic [...] have no 'subject matter" (§6.124) or that "'Laws of inference' which are supposed to justify inferences [...] have no sense, and would be superfluous" (§5.132), he is not saying that commitments to a logical subject matter and to laws that justify inferences are scientifically unnecessary. Rather, his point is that these ideas are due to misleading pictures of what is actually and already involved in our daily activity of making logical inferences; pictures which are created by the surface structure of the notation we happen to be using, and which lead, not to meaningful but futile questions, but to seemingly irresolvable antinomies.

I cannot here discuss Wittgenstein's treatment of these more serious quandaries. Let me finish instead by briefly mentioning an example from Wittgenstein's later writings which may serve to further clarify the difference between his and Quine's conceptions of what it means to dissolve philosophical problems. What I have in mind is Wittgenstein's treatment of sensations in the *Philosophical Investigations*. He says, "if we construe the grammar of the expression of sensation on the model of 'object and designation' the object drops out of consideration as irrelevant." (§293) It would be a gross misunderstanding to read this passage as saying that since the grammar of the expression of sensation *is* to be construed on the model of object and designation, it is superfluous to postulate the existence of sensations. Rather, Wittgenstein questions the very idea that the role of an expression of sensation is to be construed according to a referential model. According to Wittgenstein, the temptation to think that expressions of sensations function in a similar sort of way as, say, names of ordinary physical objects, is due to the surface similarity between sentences such as "I have a toothache" and "I have a beetle". Taken seriously, this analogy leads to perplexities such as solipsism, and the way to dissolve those perplexities is to remind oneself of the difference in use between ordinary names and expressions of sensation.

The contrast with Quine is clear. As we have seen, Quine thinks the philosophical problems about sensations should be circumvented by giving a physicalistic reinterpretation of our sensation terms. If Wittgenstein is right, this leaves the fundamental illusion intact, namely, that the purported role of sensation terms is to refer. Unless this illusion is uprooted, Wittgenstein thinks the philosophical problems have not been adequately treated but swept under the rug.

5. Concluding Remarks

Wittgenstein says, "A philosophical problem has the form: 'I don't know my way about.'" (Wittgenstein 2001, §123) For him, it is a characteristic of philosophical worries that they have a strong taste of contradiction, paradox, irresolvable dilemma. According to Wittgenstein's diagnosis, the source of such a problem is a conflict of which the worried person is not clearly aware, between customary usage and the pictures and analogies which the surface appearance of words produces in the person's mind. Given this sort of diagnosis, it will be important to provide the worried person with a perspicuous overview of established usage, not because deviation from such established usage is wrong *per se* (it isn't), but because such an overview will help him become aware of the inconsistent hovering from which his perplexity arises. Giving such an overview will be a matter of reminding the person of something he, in a sense, already knows *qua* language user. Providing such reminders is what Wittgenstein is up to when he offers a new notation which does justice to how the relevant parts of language function in practice.

For Quine, philosophical worries are similar to scientific problems. Indeed, he thinks many philosophical problems *are* good scientific problems. But there are also philosophical problems which, Quine argues, are better left aside, and he thinks explication is the way in which they ought to be dismissed. What explication shows is not that such problems are incoherent or nonsensical, but that they involve usages and ontological commitments which are not needed for scientific theory-construction. Quine's conception of what it means to dissolve philosophical problems is thus congenial to his view of philosophy as continuous with science, for his whole approach is based on the idea that it is science which provides the yardstick determining whether or not a problem is worth caring about. By contrast, Wittgenstein thinks using science as a measure of relevance only makes it more difficult to identify the real source of our philosophical worries.

University of Stockholm

Acknowledgement

Work on this paper was financed by the Swedish Research Council.

References

Hertz, H. (1956), *The Principles of Mechanics*, New York, Dover.
Quine, W. V. (1960), *Word and Object*, Cambridge, MA, MIT Press.

Quine, W.V. (1969), *Ontological Relativity and Other Essays*, New York, Columbia University Press.

Ricketts, T. (1996), "Pictures, Logic, and the Limits of Sense in Wittgenstein's *Tractatus*", in H. Sluga and D. G. Stern (eds.), *The Cambridge Companion to Wittgenstein*, Cambridge, Cambridge University Press, pp. 59–99.

Sullivan, P. (2000), "The Totality of Facts", *Proceedings of the Aristotelian Society* 100, 175–192.

Wittgenstein, L. (1913), "Notes on Logic", in L. Wittgenstein (1979), pp. 93–107.

Wittgenstein, L. (1914), "Notes Dictated to G. E. Moore in Norway", in L. Wittgenstein (1979), pp. 108–119.

Wittgenstein, L. (1958), *The Blue and Brown Books*, Oxford, Blackwell.

Wittgenstein, L. (1974a), *Tractatus Logico-Philosophicus*, transl. B. McGuinness and D. Pears, 2nd ed., London, Routledge and Kegan Paul.

Wittgenstein, L. (1974b), *Zettel*, Oxford, Blackwell.

Wittgenstein, L. (1979), *Notebooks 1914-1916*, 2nd ed., Oxford, Basil Blackwell.

Wittgenstein, L. (2001), *Philosophical Investigations*, 3rd ed., Oxford, Blackwell.

Naturalism and the Normativity of Meaning[1]

Jussi Haukioja

This paper is concerned with some aspects of the following question: is there a conflict between the normativity of meaning and naturalism? That there is a conflict is generally taken to be one of the central claims in Kripke's interpretation of Wittgenstein (in Kripke, 1982). In particular, normativity considerations are supposed to refute *dispositional* accounts of meaning and content; that is, theories which would account for word meaning and mental content by relying, in one way or another, on our dispositions to apply words or concepts to objects in our surroundings. All the major candidates for naturalistic theories of meaning and content are, arguably, dispositional theories – hence, although the argumentation Kripkenstein (or "Kripke's Wittgenstein", *i.e.* Wittgenstein as read by Kripke) presents is seemingly directed on just dispositional theories, it is generally taken to put in question the entire possibility of naturalising meaning and content.

1. A Conflict Between Dispositionalism and the Normativity of Meaning?

The conflict between dispositional theories of meaning and the normativity of meaning is supposed to arise from the fact that dispositional facts and facts about meaning are facts of fundamentally different kinds. Dispositional facts are *descriptive* – they tell us what happens, or would happen, were certain conditions realised – while facts about meaning are *normative*, telling us what would be *correct* or what *ought* to be done. The charge against dispositionalism is that, because of this gap, a theory relying solely on dispositional facts will, as a matter of conceptual necessity, fail to account for the whole story about meaning.

Here is a clear and helpful formulation of the normativity argument, due to José Zalabardo (1997, pp. 469–470):

(A) The fact that determines the satisfaction conditions of a predicate licenses evaluative claims about ascriptions of the predicate.

(B) Dispositional facts are descriptive.

1 This work has been financially supported by the Academy of Finland (project 202513).

(C) Descriptive facts cannot license evaluative claims.

Therefore,

(D) Dispositional facts cannot determine the satisfaction conditions of predicates.

The following question becomes crucial: what are the "evaluative claims" mentioned in (A) and (C)? In other words, what, exactly, is meant by the "normativity of meaning"? There is no widespread agreement on this issue. Among the various formulations of the normativity thesis, I propose a distinction between claims of *minimal* and *substantial* normativity. I will return to Zalabardo's normativity argument below, after examining this distinction, and claim that on either reading of the claim that meaning is normative, one premiss will be false (or, at best, highly controversial).

2. Minimal vs. Substantial Normativity

Sometimes the normativity of meaning is understood as the claim that meaningful expressions and mental states with intentional content have *conditions of correct application* or *occurrence*. Nearly everyone agrees that meanings and intentional states are normative in this sense – in fact, this notion of normativity is often taken as a triviality. Indeed, it seems *constitutive* of the claim that a term is meaningful that it possess correctness conditions. Hence, we can call this the *minimal* normativity claim:

> *Minimal Normativity.* If S means something by predicate P, then there exists an unbounded range of actual and non-actual cases to which P correctly applies, and an unbounded range of actual and non-actual cases to which P does not correctly apply.

This claim can, in turn, be interpreted in (at least) two ways. Most people think that the correctness or incorrectness of particular applications is at least most of the time fully determined in advance of anyone's forming a judgement about the matter – that, say, the correctness of an application of "red" to a given, completely normal ripe tomato is already determined even if no one in our speech community has seen the tomato and formed an opinion about its colour. Surprisingly enough, some theorists disagree. Meaning *finitists* do not believe that meaning "reaches ahead of actual application" in this way (Barnes 1983; Bloor 1997; Kusch 2002). There is always, according to meaning finitists, an element of choice when we extend our usage of terms to new instances; the correctness or incorrectness of new applications is never fully determined in advance, but becomes determined when our community reaches an opinion about the new case.[2] One could say that these two views give different readings to the term "unbounded" in *Minimal Normativity*: the latter, finitist reading would be, roughly, "not limited to the cases familiar to S", whereas the former view would read it as "not limited to the cases familiar to someone in the S's community". This difference is irrelevant to my concerns in this paper, so I will ignore it in what follows.

A clear statement of the minimal normativity claim can be found in Paul Boghossian's important paper "The Rule-Following Considerations":

> Suppose the expression "green" means *green*. It follows immediately that the expression "green" applies *correctly* to *these* things (the green ones) and not to *those* (the non-greens). The fact that the expression means something implies, that is, a whole set of *normative* truths about my behaviour with that expression: namely, that my use of it is correct in application to certain objects and not in application to others. [...] The normativity of meaning turns out to be, in other words, simply a new name for the familiar fact that [...] meaningful expressions possess conditions of *correct use*. (Boghossian 1989, p. 513)

The claim that meaning is *minimally* normative seems, then, uncontroversial.

Claims of *substantial* normativity go beyond the minimal claim, and come in various forms. Typically, such claims assert that meanings and intentional states confer special semantic *obligations* on speakers and thinkers. Theories of meaning should then explain how it is that using a term with a certain meaning *commits* one to particular patterns of usage, or how it entails that one *ought* to use terms in certain situations and *ought not* do so in others. The charge against the naturalistic semantician is that the semantic is-ought gap cannot be bridged.

Unfortunately, many authors fail to distinguish between the two sorts of claim. Arguably, Kripke is a case in point. Robert Brandom, clearly a proponent of the substantial normativity claim, also runs the two claims together in *Making It Explicit*:

> [an adequate theory] must be able to fund the distinction between what is in fact done and what ought to be done [...] What is correct or appropriate, what is obligatory or permitted, what one is committed or entitled to do – these are normative matters. (Brandom 1994, p. 27)

---

2 For criticism of meaning finitism and of Bloor's argumentation for it, see Haukioja (2005a).

As mentioned above, we should accept the minimal normativity claim. At the same time, I think we should reject substantial normativity claims. There is no reason to believe in specifically semantic obligations – whatever obligations we have to use words and form beliefs in particular ways can be shown to follow from other, non-semantic obligations, and the minimal normativity claim. (See Wikforss 2001 and Hattiangadi (forthcoming) for detailed and persuasive arguments for this claim.)

Take "green", for example. To understand "green", one must somehow be sensitive to the fact that the word applies correctly to certain things and not to others; this is the minimal normativity claim. But no *obligation* to apply the word to anything follows from the fact that it has a *meaning*. True, one can have obligations to use words in particular ways, but I would claim that none of these stem solely from meaning. One might have a moral obligation to tell the truth about the colour of something; in this case one would be obliged to apply the word "green" to a green thing; but one might just as well, in some situation, have a moral obligation to *lie* about the colour of something. These obligations clearly do presuppose that the word "green" has correctness conditions – otherwise particular uses of the word would not count as telling the truth or as lying – but the *obligation* arises from morality, not from meaning. Similarly, I think, for rational, legal, and other obligations.

Minimal and substantial normativity claims are often conflated, and I can see at least two reasons for this. The first, rather uninteresting reason is the naturalness of using words such as "correct" and "right" – words with a normative ring – to speak about minimal normativity. The second, more interesting reason is the following. When discussing minimal normativity, many authors make claims about *hypothetical obligations* or *hypothetical imperatives*:

> ... if I mean something by an expression, then the potential infinity of truths that are generated as a result are *normative* truths: they are truths about how I *ought* to apply the expression, <u>if I am to apply it in accord with its meaning</u>. (Boghossian 1989, p. 509; underlined emphasis added)

> The dispositionalist gives a *descriptive* account of [the relation between an expression's meaning and the sum of one's dispositions to use it] ... But this is not the proper account of the relation, which is *normative*, not descriptive. The point is *not* that, if I meant addition by "+", I *will* answer '125,' but that, <u>if I intend to accord with my past meaning of "+"</u>, I *should* answer "125". (Kripke 1982, p. 37; underlined emphasis added)

It is tempting to drop the underlined antecedents and say things like "if I mean *green* by 'green', then I ought to apply 'green' to (all and) only green things", where what one really means is: "if I mean *green* by 'green', *and* I intend to accord with the expression's meaning, then I ought to apply 'green' to (all and) only green things". While the first one would say that claims about meaning imply semantic obligations, the second one only says that claims about meaning imply claims about *what one ought to do if one intends to use terms according to their correctness conditions*. But this is just to say that claims about the meaning of an expression imply claims about the correctness conditions for the use of that expression, and this is just an instance of the minimal normativity claim. No substantial normativity is implied.

It may be helpful to compare this to another situation. It feels quite natural to say "if it rains, one ought take an umbrella when going out". Yet, no one would think this to be a good reason to claim that meteorological facts are *normative* facts (or that meteorological facts alone imply normative facts). What one really means is: "if it rains, *and one intends to avoid getting wet*, one ought to take an umbrella when going out". The (rational) obligation is conditional on one's intention to avoid getting wet.[3]

To summarise the discussion so far, let us return to Zalabardo's formulation of the normativity argument. Depending on what we take it to be for facts to "license evaluative claims", we get two readings of the argument, both of them unsound. On the minimalist notion of the normativity of meaning, we take "licenses evaluative claims" to mean "entails facts about correctness conditions". On this reading, premiss *A* (the fact that determines the satisfaction conditions of a predicate licenses evaluative claims about ascriptions of the predicate) is trivially true but there is no obvious reason to think that premiss *C* (descriptive facts cannot license evaluative claims) is true. A substantive claim of the normativity of meaning, on the other hand, would take "licenses evaluative claims" to mean "entails facts about what the speaker is obliged to do". On this reading, premiss *C* is plausible but premiss *A* is not.

---

3 Antti Saaristo claimed in discussion that Robert Brandom would *accept* that facts about the weather imply, by themselves, normative facts, in his *Articulating Reasons* (2000). If Brandom really thought so, I guess my only reaction would be an incredulous stare. However, I could not find in Brandom a defence of such a claim. What Brandom does defend (pp. 84–89) is the view that to make sense of the goodness of inferences such as "It is raining, therefore I shall [note, *shall,* not ought to!] open my umbrella", one need not assume an implicit premiss about one's beliefs, desires, or intentions. While I have my doubts about this claim, too, it does not seem nearly as preposterous to me.

3. Why the Naturalist Should Still Take Kripke Seriously

I have claimed above that there are no strictly semantic obligations, and that the naturalist should not be worried by the argument (A)–(D). Nonetheless, I think that even the minimal normativity claim may cause some trouble for the naturalistic semantician, for reasons that may have also been pointed out by Kripke's Wittgenstein. In particular, I believe that a common quick-and-easy explanation of minimal normativity rests on an objectionable assumption. Fodor presents the point as follows.

> ... requiring normativity to be grounded suggests that there is more to demand of a naturalized semantics than that it provide a reduction of such notions as, say, extension. But what could that "more" amount to? To apply a term to a thing in its extension *is* to apply the term correctly; once you've said what it is that makes the tables the extension of "table"s, there is surely no *further* question about why it's *correct* to apply a "table" to a table. It thus seems that if you have a reductive theory of semantic relations, there is no job of grounding normativity left to do. (Fodor 1990, p. 135)

The final sentence is put forward as a reply to theorists such as Wright, Blackburn, and possibly Boghossian, who both attribute to Kripke and themselves endorse the claim that even if one had a naturalistic theory of predicate reference, it would not qualify as a full theory of meaning, because it would leave out normativity. Clearly, "normativity" is here taken in the substantial sense. Since I do not believe in the substantial normativity claim, I would agree with Fodor on this.

At the same time, I think that Fodor's response is problematic on other grounds. Fodor is here relying on the following implicit premiss: explanation of *extensions* is prior to the explanation of *correct use*. That is, he is assuming that the explanation of why a particular application of an expression is correct is to be based on a previous explanation of why the expression in question has a certain extension (plus the fact that the entity to which the expression applies is a member of that extension). Some theorists (Wilson 1994; 1998; Davies 1998) have read Kripke as arguing against precisely this picture. I cannot go into the issues of Kripkenstein exegesis here, but to illustrate the point, consider the following two claims:

(1) It is correct to apply "red" to a given ripe tomato *because* the tomato is in the extension of "red".

(2) A given ripe tomato is in the extension of "red" *because* it is correct to apply "red" to it.

Any theory of meaning is going to take only one of these to be plausible, when the "because" is read in the explanatory sense[4]. If a theory takes (1) as plausible, it will claim that the primary explananda of semantics are the relations between terms and *extensions*; the correctness of particular applications of terms follows from these relations. This is clearly Fodor's line in the quotation above. In contrast, if a theory takes (2) as plausible, it will claim that the primary explananda are the correctness conditions of particular applications; that is, that a theory of meaning should primarily explain the process by which a term gets to have correctness conditions in new, previously unencountered situations. We can then, if we wish, abstract away from particular situations to get extensions, but extensions play no role in explaining correct use.

I believe that Kripke's Wittgenstein provides a very powerful argument against theories which accept (1).[5] In particular, I think that the central problems Kripkenstein raises for dispositionalism – that the finiteness of our dispositional states will always leave us with indeterminacy, and that there is no non-circular way to filter out our dispositions to make mistakes – are insurmountable if we accept (1).[6] This is very important, because most current candidates for naturalistic theories of meaning subscribe to (1), implicitly or explicitly. Nevertheless, I am optimistic that a naturalistic theory of meaning can be formulated, but because of Kripkenstein's arguments, it has to subscribe to (2), not (1). This is not the place to present such a theory[7] – here I merely want to emphasise that the naturalistic semantician cannot simply ignore Kripke's argumentation, even if minimal normativity is the only sense in which meaning is normative.

---

4 I have studied the contrast between these two approaches to semantics in more detail elsewhere. In my (2004) and (2005b) I call these approaches *Platonism* and *non-Platonism* about meaning (these being committed to (1) and (2), respectively).

5 This is the case, I think, whether or not Kripke intended the arguments to be directed at such theories. Again, I want to stay clear of Kripkenstein exegesis; but I do think George Wilson (1994), in particular, has given good reasons to think this reading of Kripke is on the right track.

6 I obviously cannot defend this claim here; see Haukioja (2005b) and Haukioja (2004).

7 I think that a theory which sees our competence with basic concepts as response-dependent, roughly of the kind offered by Philip Pettit (see, *e.g.*, the papers collected in his 2002) is ideally suited to be a naturalistic theory of such kind. To make it very short, such a theory claims that to use a term correctly is to have one's disposition to use the term be manifested in conditions which would not give rise to intra- or interpersonal disagreement about the applicability of the term. See Haukioja (2005b) for reasons to think that such a theory can be non-Platonist (that is, committed to (2), not (1)); see Haukioja (forthcoming) for an extension of Pettit's theory, intended to avoid the problems pointed out by Frank Hindriks (2004).

4. Conclusion

It is sometimes claimed that, because meaning is normative, it cannot be fitted into a thoroughly naturalistic world view. In evaluating such claims, it is highly important to be clear about what, exactly, we mean when we claim that meaning is normative. On the minimal reading, this amounts just to the uncontroversial claim that meaningful expressions possess conditions of correct application. There is no reason to think that *minimal* normativity poses an insurmountable conceptual barrier to the project of naturalising meaning and mental content. Things might well be different if meaning were *substantially* normative – if using a word with a certain meaning alone entailed that one is under an *obligation* to apply it in a certain way; but there is no obvious reason to think that meaning *is* substantially normative.

The normativity of meaning cannot, then, be used as a knock-down argument against naturalism. Nevertheless, even the minimal normativity claim may have profound implications on what the order of explanation should be in naturalistic theories of meaning and mental content.

University of Turku

References

Barnes, B. (1982), "On the Extensions of Concepts and the Growth of Knowledge", *Sociological Review* 30, 23-44.
Bloor, D. (1997), *Wittgenstein, Rules and Institutions,* London, Routledge.
Boghossian, P. (1989), "The Rule-Following Considerations", *Mind* 98, 507-549.
Brandom, R. (1994), *Making It Explicit. Reasoning, Representing and Discursive Commitment*, Cambridge, Harvard University Press.
Brandom, R. (2000), *Articulating Reasons: An Introduction to Inferentialism*, Cambridge, Harvard University Press.
Davies, D. (1998), "How Sceptical is Kripke's 'Sceptical Solution'?", *Philosophia* 26, 119-140.
Fodor, J. (1990), *A Theory of Content and Other Essays*, Cambridge, MIT Press.
Forrai, G. (ed.) (2005), *Intentionality: Past and Future*, Amsterdam, Rodopi.
Hattiangadi, A. (forthcoming), "The Normativity of Meaning", *Mind & Language.*
Haukioja, J. (2004), "Kripke's Finiteness Objection to Dispositionalist Theories of Meaning", in M. Reicher and J. Marek (2004), pp. 137-139.
Haukioja, J. (2005a), "A Middle Position Between Meaning Finitism and Meaning Platonism", *International Journal of Philosophical Studies* 13, 35-51.
Haukioja, J. (2005b), "Normativity and Mental Content", in G. Forrai (2005), pp. 51-62.
Haukioja, J (forthcoming), "Hindriks on Rule-Following", *Philosophical Studies.*

Hindriks, F. (2004), "A Modest Solution to the Problem of Rule-Following", *Philosophical Studies* 121, 65–97.
Kripke, S. (1982), *Wittgenstein on Rules and Private Language*, Oxford, Basil Blackwell.
Kusch, M. (2002), *Knowledge by Agreement: The Programme of Communitarian Epistemology*, Oxford, Oxford University Press.
Pettit, P. (2002), *Rules, Rationality, and Civil Life: Selected Essays*, Oxford, Oxford University Press.
Reicher, M. and J. Marek (eds.) (2004), *Experience and Analysis. Papers of the 27th International Wittgenstein Symposium*, Kirchberg am Wechsel, Austrian Ludwig Wittgenstein Society.
Wikforss, Å. (2001), "Semantic Normativity", *Philosophical Studies* 102, 203–226.
Wilson, G. (1994), "Kripke on Wittgenstein and Normativity", *Midwest Studies in Philosophy* 19, 366–390.
Wilson, G. (1998), "Semantic Realism and Kripke's Wittgenstein", *Philosophy and Phenomenological Research* 58, 99–122.
Zalabardo, J. (1997), "Kripke's Normativity Argument", *Canadian Journal of Philosophy* 27, 467–488.

Induction, Naturalism and Laws

Lars-Göran Johansson

1. Introduction

Hume is recognised as the first to state the induction problem. He clearly saw that an inductive step from a number of particular cases to the general statement couldn't be justified; neither past experiences nor logic can provide any justification. Since these two are the only options available, inductive reasoning cannot be given any epistemological justification.

Popper claimed that science should be based on rationally justified methods, and since induction fails this criterion, science should do without such forms of reasoning. Instead we should adopt falsificationism as the only fully rational scientific method. We should do our best to try to falsify our hypotheses, and if the attempt failed we could say no more than that our hypothesis is corroborated. In Popper's terminology the word 'corroboration' means no more than 'so far not falsified'; hence, if we denounce inductive reasoning we cannot say that a corroborated hypothesis is probable, or could be believed to be true. However, Popper has been criticised for being inconsistent. He further argued that we have reasons to think that of two hypotheses, neither of which is falsified, the one which is better corroborated, i.e., it has survived more tests, is closer to truth, have more *truthlikeness*, than the other. In short, corroboration is a sign of truthlikeness. What is the argument for this thesis? No doubt, it is an inductive argument, as Newton-Smith once pointed out. Popper admitted that it is a 'whiff of inductivism' and Newton-Smith retorted 'This is a full-blown storm.' (1981, p. 68). Thus, Popper cannot coherently take the step from corroboration to truthlikeness. It is a telling fact that not even Popper succeeded in giving an account of science free from inductive reasoning. The induction problem is still with us and Quine once formulated the problem with his characteristic incisiveness: "The Humean predicament is the human predicament" (1969, p. 72).

But why do we need a general justification of induction? The usual answer is that it is the business of epistemology to provide a foundation for the sciences. Science should fulfil the highest standards of rationality; therefore we should be able to justify our scientific methods, in particular induction. Such a stance is part of a common view that epistemology is more basic and fundamental then the special sciences. Without a certain foundation of the general principles, science is built on the sands.

Epistemology tries to answer the question what we ought to believe and not merely what we in fact believe; thus it has a normative component. It results in recommendations how to proceed in scientific thinking. But quite often people don't care; it seems pretty clear that in cases where philosophically motivated epistemic principles conflict with actual scientific practice the latter usually wins. Scientists most often take little notice of what philosophers say. This I find entirely reasonable since scientific results, at least in the natural sciences, appear much less disputable than philosophical doctrines. Should we then stop doing epistemology as a pointless enterprise? No, a better option is to adopt a naturalistic stance, i.e., to view epistemology as part of our scientific knowledge of the world. Epistemology is the study of how the cognitive apparatus of human's works and under what conditions the resulting cognitive states represent real states of affairs.

Traditional epistemology results in epistemic norms. The critic might now claim that, as an empirical study, naturalised epistemology cannot entail any norms and so it cannot do its work. This is wrong. Epistemic principles have the form 'do so and so in order to get knowledge'. This can be reformulated as 'if you want to get knowledge, do so and so'. When the antecedent in this hypothetical is left unsaid as a tacit presupposition, we are left with a sentence of normative form. Thus, the normative form of epistemic principles can be explained as the result of a transformation from an explicit conditional to a categorical statement with a tacitly presupposed condition. Many norms, not just epistemic ones, have this character.

In the naturalistic view epistemology is fallible and revisable as all our knowledge. There is no vantage point of view from which to judge whether a particular method is bad or good; such a judgement can only be made from within the sciences. The naturalist move is to say that our formation of concepts is a natural process, which can be studied by science in the usual empirical way. We humans are part of the natural world and our cognitive capacities are our means of interacting with our environment. And our cognitive capacities, for example our ability to form useful concepts and to use induction, have no a priori status; they are features of the natural world. Hence, ultimate justification, in the sense of a priori arguments, cannot be had.

But, clearly, we cannot reject every demand of justification. On the contrary, the very idea of scientific knowledge is that we can give good reasons for it; any particular knowledge-claim must be justifiable. So what are the proper boundaries for justification in the naturalistic perspective? What is outside the domain of justification?

2. Justification in the Naturalistic Perspective

Justification is a relation between beliefs and when expressed in a language it becomes a relation between statements. One statement can contribute to the justification of another statement. No matter how we analyse the relation, it is obvious that a general demand for justification will result in an endless regress; if *B* justifies *A*, one will immediately ask for a justification of *B* etc. In practice we must stop somewhere and many have thought that the endpoint must be some kind of a priori and self-justifying statements. But is there a ground for empirical knowledge, a set of basic and self-evident statements, which do not require any justification? I don't think so; any statement can be doubted. Even the simplest observation or the most obvious logical principle can and has been doubted. Now the sceptic attacks. From the statement 'any particular statement can be doubted' it follows, he claims, 'All statements can be doubted'. But this is wrong. The premise can be paraphrased as 'For all *x*, it is possible that *x* is false' and the conclusion as 'It is possible that for all *x*, *x* is false'. This is an invalid inference, no matter which version of modal logic you adopt.[1]

Every chain of justification ends in what might be called animal beliefs; at every moment we unreflectively hold some beliefs while doubting others. This holds true even in logic; if we try to justify modus ponens we find ourselves using modus ponens, as is nicely shown by Lewis Carroll in the famous dialogue "What the Tortoise said to Achilles" (1895).

Our basic beliefs are mostly taken for granted as part of the background of our behaviour and our interaction with the environment, including other people. For example, we take it for granted that ordinary things in our vicinity, including other people, have roughly the same properties, both physical and mental, from one moment to another. We also take it for granted that other people in general mean something with their words and we do not doubt that our fellow humans think, have feelings and experiences more or less similar to ours. We follow a great number of rules in our thinking and behaviour without really justifying them. But we should carefully keep in mind that this use of the notion 'to follow a rule' has no normative force; it is just a description. Wittgenstein in his (1953) made this observation. In §§143-202 we find his famous discussion about the notion 'to follow a rule'. He arrives at the conclusion in §202: "And hence also 'obeying a rule' is a practice [...]" The point, as I interpret him, is that the request for general justification for a rule cannot be met.

I see a resemblance between Carroll's and Wittgenstein's stance on ultimate justification. The idea traces back to Hume's position in *Treatise*; discussing the

1 The same point has been made by Davidson on p. 194 in "Epistemology externalized". Reprinted in his *Subjective, Intersubjective, Objective*, Oxford, 2001.

sceptic's doubt about the veracity of our immediate experiences of external objects, Hume concluded that a convincing argument couldn't be given. But this does not lead to doubts about the existence of external objects:

> "Thus the sceptic still continues to reason and believe, even tho' he asserts, that he cannot defend his reason by reason; and by the same rule he must assent to the principle concerning the existence of body, tho' he cannot pretend by any arguments of philosophy to maintain its veracity. [...] We may well ask 'What causes induce us to believe in the existence of body?' but 'tis in vain to ask whether there be body or not? That is a point, which we must take for granted in all our reasonings." (1985, p. 238)

Thus Hume did not aspire to justify our experiences or our statements about cause-effect relations. Instead he stated that it is an empirical fact about us that we do believe that our perceptions are perceptions of external physical objects and we do believe that these objects may cause each other's motions. It belongs to our nature to assume that external objects exist and cause our impressions. What he did was to give a causal account how our beliefs were generated, not a justification.

The most explicit proponent of naturalism is Quine. The common trait in Hume's and Quine's position is the stance that justification of beliefs from a vantage point outside the realm of empirical knowledge is impossible. Epistemology is not the foundation for empirical knowledge, but an integrated part of it; as Quine puts it, "a chapter of psychology and hence of natural science." (1969, p. 82)

Some might claim that naturalism leads to a vicious circle in epistemology: we gain knowledge about our own knowledge process through precisely that very knowledge process we are describing. I agree that it is a circle, but it is not a vicious one. It is more like the hermeneutic circle: by an ever-increasing inquiry we constantly widen the circle in order to make it as vulnerable as possible to empirical constraints. And the constraints become visible as structural constraints: it is the structure of our theory, which must fit reality. We have arrived at a picture closely resembling Quine's view in *Two Dogmas.*

Summarising, justification of epistemic principles rests ultimately on an empirical theory about our cognitive processes, including our way of using language. It is grounded on a theory about how our thoughts represent the world. Like anything else in science it is fallible. No foundational argument can function as insurance against mistakes. In particular it is a mistake to require a general philosophical justification of induction.

3. Induction in the Naturalistic Perspective

In the naturalistic view the problem of induction is thus not that of justifying induction in general. But it is obvious that we do not consider all instances of inductive thinking equally good; we have strong intuitions that some conclusions are much more reliable than others. Hence, we should reformulate the induction problem as the task of describing more thoroughly our inductive practices and give an account of the methodological role of induction in our scientific work. We should try to explain why we think that certain inductions are trustworthier than others.

This is roughly Goodman's way of viewing the matter in his (1972). More precisely, he asked what kind of predicates is used in (normal) inductive reasoning. To illustrate the problem he construed the artificial predicate *grue*, defined as true of things examined before some time in the future, AD 3000 say, and found to be green, or examined after AD 3000 and found blue. All emeralds so far examined are thus both green and grue. Without further constraints simple induction tells us that we have equal reason to assume that the first emerald to be examined after the year 3000 will be green as well as grue, i.e. blue. One prediction, at least, will ultimately fail and we all believe that emeralds will continue to be green. But why? This is the induction problem in the new key. Goodman's formulation of the problem is that some predicates are *projectible* and some other not. Obviously, we need to know the conditions for a predicate being projectible. Goodman suggested that the notion of *entrenchment* could be used in order to distinguish between projectible and non-projectible predicates. But why do some predicates become entrenched? Goodman gave no answer.

Even though Goodman did not answer his own question, his analysis is a step forward, because he replaced the quest for general justification of induction with the more empirical question 'under what conditions can a particular instance of induction be expected to be successful?' However, stating the problem in terms of the distinction between projectible and non-projectible predicates, taken one at a time, is not entirely satisfying. Goodman overlooked a crucial component in describing the situation, viz. the identification of the referents of the singular terms used in our observation statements. When we for example ask which predicate to use in generalisations about emeralds, green or grue, we should also consider the rules we follow in identification of emeralds. The question is not, which *single predicate*, green or grue, to use in a particular case of inductive reasoning, but the correct *pairing of predicates*. In the sentence 'This emerald is green' we have two predicates, 'emerald' and 'green'. We use a predicate, 'emerald', in the identification of the referent of the noun phrase of this sentence and it is pretty obvious that that is generally the case in inductive reasoning. The induction problem can now be reformulated as: for which *pair of predicates A* and *B* is it reasonable to expect that if an object satisfies *A*, it also satisfies *B*?

In the case of grue or green emeralds it is rather simple. We know that emeralds consist of the mineral beryl contaminated with chromium. This metal makes the mineral green, according to fundamental physical laws. The necessary and sufficient condition for something to be an emerald is that it is a gem made up of beryl containing chromium. The same condition entails, via scientific laws, that it is green, independent of time. Hence if something satisfies the predicate 'emerald', it satisfies also the predicate 'green'.

This can be generalised. Suppose we have observed a regularity in nature: So far, all observed objects are such that if they satisfy a predicate *A*, they also satisfy another predicate *B*. Let us assume that both predicates are expressions taken from our vernacular without using scientific theory. We thus have two options: either to assume that the regularity so far observed is a mere coincidence, i.e., an accidental generalisation, or else to assume that it reflects a hidden structural feature. Taking the first option is to guess that sooner or later will we hit upon a counterexample. The second option is to guess that the generality 'for all *x*, if *Ax*, then *Bx*' is true. If this is correct, we have found a natural law.

Suppose we have found a natural law by using induction. Isn't the existence of such laws a bit astonishing: Why is it the case that an *indefinite number of objects* satisfies two logically unrelated predicates? Is not the most reasonable assumption that the probability for such a state of affairs is zero?

History of science suggests two ways of explaining such regularities. The first possibility is to derive the regularity, or some version close to it, from a set of more fundamental and independently acceptable principles. A telling example is the general law of gases. This law began life as Boyle's observation that the product of pressure and volume of a portion of gas is constant. Charles in 1787 and Gay-Lussac in 1808 found that this constant depends on temperature and still later the complete general law of gases was formulated when the concept of mole was available. For some time this law appeared to be an empirical regularity, a brute fact. However we now know that it can be derived from the principle of energy conservation, given the identification of absolute temperature with mean translational kinetic energy among the particles making up the gas. So it is not just an empirical fact that the two predicates '*x* is a gas under normal circumstances' and 'the pressure, volume and temperature of *x* satisfies the equation $pV=nRT$' are both satisfied by the same objects. It follows from a basic principle, given some auxiliary assumptions.

This brings us to the second way explaining the remarkable fact that an indefinite number of objects all satisfy two unconnected predicates. All scientific predicates start their lives as part of our vernacular. As science advances vague notions become sharpened and are changed into scientific predicates with explicitly defined criteria of application. And, of course, many new predicates are introduced by explicit definitions. The crucial point is that in this process of concept development a well-established regularity is normally not given up. Suppose we have such a well-established generality, 'for all x, if Ax, then Bx', and hit upon a putative counter example, an object which satisfies A but not B. Logically we have two options; either to drop the regularity as falsified, or to change the criteria of application of the predicate A so that the putative counter example can be rejected. A simple example of the latter is the history of the concept of fish. Although Aristotle knew that dolphins and whales are not fishes, his insights were forgotten and for a long time these mammals were thought to be fishes. Since it was accepted that all fishes have gills, our predecessors had a problem when learning that dolphins have no gills. But instead of rejecting 'all fishes have gills' they rejected 'dolphins are fishes' by sharpening the intuitive criteria for fishness, something like 'animal swimming in the seas with mouth, fins and eyes', with additional clauses.

Suppose we stick to the regularity by redefining the criteria of application of the predicate in the antecedent every time a new counter example pops up. The logical endpoint of this process is when the set of necessary conditions for satisfaction of the predicate in the consequent is a subset of those for the predicate in the antecedent; in such a case no further counter example is possible. We have arrived at a fundamental law, of which more later.

Inductive reasoning is intimately connected with theory development, but both inductivists and falsificationists have told a distorted story. The inductive process also involves concept development. Inductive reasoning is our way of finding out the structure of the world. The success of empirical science and in particular the usefulness of induction is explained fundamentally in the same way as other evolutionary processes; it is the result of adaptation and competition, in this case adaptation of concepts to the way world is and competition among theories.

4. Concept Formation, Theory and Reality

The story I have told clearly presupposes realism, i.e., the belief that the natural world is independent of our thinking, our theories and our concepts. When uttering an atomic declarative sentence we use a singular term and a general term; the singular term identifies the object talked about and the general term functions as predication. (I neglect mass terms in this context and consider only sortal predicates.) The content of the sentence could be described as follows: an object is referred to by the singular term and it is claimed that this object belongs to the class of objects satisfying the predicate. If our saying is true we succeed in identifying a particular object by the singular term and this particular object satisfies the general term.

General terms and their criteria of application are human inventions. Applying a general term is a sorting of all objects in two classes: those satisfying the term and those that don't. (In this context I neglect vague predicates and their associated 'fuzzy sets'. I also take for granted that restrictions on predicates are imposed in order to avoid paradoxes.) If we succeed in arriving at a true inductive conclusion (of the form 'all *A*:s are *B*:s') we have succeeded in finding a class structure: members of the *B*-class are members of the *A*-class.

Let us now direct our attention at the conceptual side of theory development. Successful theory construction consists in adapting the conceptual structure to empirical findings. Inductive generalisations directly supported by singular observations, i.e., sentences in which the general term express easily observable traits in the world, are integrated into a theory. It means that logical relations between such inductive generalisations and laws, postulates and other inductive generalisations are constructed. The aim is to make these inductive generalisations logical consequences of fundamental laws and postulates. This is not possible unless we use predicates with strictly defined criteria of application, both in the fundamental laws and in inductive generalisations. But predicates taken from everyday language never have strict criteria of application. Hence, the set of objects satisfying a predicate taken from everyday language is seldom well defined and possible to pick out by an explicit definition. This was, by the way, Wittgenstein's observation in (1953), §67 and later. He used the notion of *family resemblance* as a way of describing the situation. Any two objects satisfying a predicate share a number of characteristics, but there is no single characteristic shared by every object, and no other, satisfying the predicate. It is impossible to define a predicate occurring in natural language by giving necessary and sufficient criteria for its application. On this point Wittgenstein was right, I think. But that only shows us that real science cannot use such naturally occurring predicates.

It is clear that theory construction often forces us to change or sharpen the intuitive notions used as starting points. Suppose we start with two predicates A and B. The predicate *expressions* need not be changed, but only their criteria of application, in the light of a purported counter example. Or, perhaps, explicit criteria of application are introduced as a step towards a theory. In such a case one cannot literally say that the predicates, qua linguistic types, are changed, even though the set of objects satisfying respectively the predicates *A* and *B* might have been changed because of the change of criteria of application. It seems appropriate to describe the situation as a change of the *concepts* expressed by '*A*' and '*B*'. I here use the word 'concept' extensionally, denoting a predicate with its associated criteria of application.

This account of theory construction is not restricted to regularities expressed by observational predicates. Any regularity, independently of how observational or theoretical its general terms are, triggers the wonder why a possibly infinite number of objects satisfy two logically unrelated predicates. The answer given above is that the predicates are in fact logically dependent on each other, or can be so made; either by deriving the regularity from fundamental laws or else are the criteria of application for the consequent a subset of those for the antecedent. (Nothing of this applies, of course, to ordinary language, only to a well-structured scientific theory.)

It is obvious that this procedure, when repeated on more and more abstract and general levels, ends in a set of fundamental laws and principles, which are accepted as fundamental without being derived from other principles. Now the same question recurs, with still stronger force: how is it possible that an indefinite number of objects satisfy two logically unrelated predicates? Wittgenstein suggested an answer to this question in (1953), §79. He writes that what today is seen as an empirical regularity can tomorrow be conceived as a definition of the theoretical concepts involved. I think this suggestion is partially true and I have developed the idea in my (2005).

5. Induction as a Heuristic Device

The picture emerging from all this is that induction should not be seen as particular form of reasoning for which one need independent and non-empirical justification, but as an heuristic device in theory construction, in particular construction of theoretical predicates. We observe, in a number of cases, a regularity using two more or less well-defined predicates. Sometimes we believe that the observed regularity reflects a structural feature in nature. This naturally induces the scientist to try to invent a theory, which reflects this structural feature and the goal is reached when the theory entails the empirical regularity, or some formulation reasonably close to it.

What I have just said resembles to some extent Aristotle's stance in (1960). According to Hankinson (1998, p. 168), Aristotle's word 'επαγογε' (epagoge) which usually is translated 'induction', should not be interpreted as an inference principle, but rather as a causal term. Hankinson writes:

> "The method in which we arrive at first principles is called by Aristotle 'epagoge'. ... Starting from individual perceptions of things the perceiver gradually, by way of memory, builds up an experience (empeiria), which is 'the universal in the soul, the one corresponding to the many' (Posterior Analytics 2.19.100^{a}6-8); and it is this which provides the *arche*, or first principle:
>
> > 'these dispositions are not determinate and innate, or do they arise from other more knowledgeable dispositions, but rather from perception, just as when a retreat take place in battle, if one person makes a stand, another will too, and so on until the arche has been attained.' (2.19.100^{a}9-13)

> This process gives us universals (such as 'man') without which we cannot utter assertoric sentences, which in turn lead to higher-order universals, such as 'animal' from particular species. (2.19.100ᵃ15–ᵇ3). It is described in causal, not inferential terms (which is why 'induction' is misleading): the world simply impresses us in such a way that we come to internalize ever wider and more inclusive concepts. We are by nature equipped to take on form in this way; if we are diligent and unimpaired, our natural faculties will see to it that we do so. Thus in a relatively literal sense we just come to see that Callias is a man and, ultimately by the same process, what it is to be a man."

In conclusion, the demand for a general justification of induction should be rejected as incompatible with naturalism. Inductive reasoning is a natural process. Sometimes it goes wrong, but who had thought otherwise? And who is to tell in advance when we will succeed?

Uppsala University

References

Aristotle (1960), *Posterior Analytics*, Cambridge MA, Loeb Library, Harvard University Press.

Carroll, L. (1895), "What the Tortoise Said to Achilles", *Mind* 4, No. 14, 278–280.

Goodman, N. (1972), "The New Riddle of Induction", in *Problems and Projects*, Indianapolis & New York, Bobbs-Merrill Company.

Hume, D. (1739), *A Treatise of Human Nature*, London, Penguin Classics, 1985.

Hankinson, R. J. (1998), *Cause and Explanation in Ancient Greek Thought*, Oxford, Oxford University Press.

Johansson, L.-G. (2005), "The Nature of Natural Laws", in Faye, Needham, Scheffler and Urchs (eds.), *Nature's Principles*, Logic, Epistemology and the Unity of Science, vol 4., Dordrecht & Boston, Springer, pp. 171–187.

Newton-Smith, W. (1981), *Rationality of Science*, London, Routledge.

Popper, K. (1959), *Logic of Scientific Discovery*, London, Unwin.

Quine, W. V. (1961), "Two Dogmas of Empiricism", in his *From a Logical Point of View*, Cambridge MA, Harvard University Press.

Quine, W. V. (1969), "Epistemology Naturalised", in his *Ontological Relativity and other Essays*, New York, Columbia University Press, pp. 69–90.

Wittgenstein, L. (1953), *Philosophical Investigations*, Oxford, Blackwell.

The Problem of Referential Meaning in Quine's Philosophy of Language

Antti Keskinen

My paper will focus on Quine's thesis of the inscrutability of reference. I will sketch Quine's argument from his basic epistemological views to the inscrutability thesis. I also attempt to show why this thesis is problematic – even for Quine himself. I try to show how referential semantics constitutes a problem within Quine's philosophy of language. In the end, I will briefly consider a couple of possible solutions to the problem. My purpose is not to give a detailed criticism or refutation of Quine's views, but only to point out some problems and possibilities for further research.

1. Introduction: From Naturalized Epistemology to Semantic Indeterminacy

Quine's thesis of the inscrutability of reference derives from his project of naturalized epistemology. In a rough outline, we get from the basic assumptions of naturalized epistemology to Quine's inscrutability thesis as follows:

Naturalized epistemology studies *the acquisition of language/theory* (natural language, for Quine, is also a kind of theory, "rudimentary science" about physical and abstract objects). Naturalized epistemology consists in the study of the relation between evidence and theory, that is, the relation between a stimulus event and a verbal response. The relation between empirical evidence and language/theory is the basis of linguistic meaning – whether *holophrastic* (sentence meaning) or *referential* meaning (meaning of terms). Thus, epistemology and philosophy of language are closely connected – in fact, almost identical – in Quine's philosophy. According to Quine, naturalized epistemology is *scientific* in spirit. It studies the foundations of science by means of science itself. Quine says that the only way to approach questions about linguistic meaning in the spirit of empirical science is *behaviorism*:

> In psychology one may or may not be a behaviorist, but in linguistics one has no choice. Each of us learns his language by observing other people's verbal behavior and having his own faltering verbal behavior observed and reinforced or corrected

> by others. [...] As long as our command of our language fits all external checkpoints, where our utterance or our reaction to someone's utterance can be appraised in the light of some shared situation, so long all is well. Our mental life between checkpoints is indifferent to our rating as a master of the language. (Quine 1992, pp. 37–38)

One of the "two cardinal tenets" of Quine's external empiricism says that all inculcation of meanings of words must rest ultimately on sensory evidence (Quine 1969b, p. 75). In the above quotation, which lays down the basic position of Quine's behavioral theory of meaning, we can see the repudiation of mentalistic semantics. Quine (1975, p. 87) distinguishes three levels of semantic theorizing: The mental, the behavioral and the physiological. Of course, the level of microstructure – physiological description – is according to Quine the most fundamental level also in the study of linguistic meaning. The macrophysical (behavioral) level and the microphysical level are compatible. Theory of meaning can well be stated in behavioristic terms – *even if* we had full knowledge of the physiology underlying the behavioral level. After all, "each of us learns his language by observing other people's verbal behavior and having his own faltering verbal behavior observed and reinforced and corrected by others". Only the uncritical mentalistic talk about meaning is to be avoided. Meanings as mental or intensional entities do not have clear criteria of identity and mentalistic meaning-talk involves us into *opaque* constructions or contexts that must be avoided within *extensional* scientific discourse. Quine's extensionalism thus supports his linguistic behaviorism – and vice versa. These theses support, and, are supported by, exclusive emphasis on the theory of reference as distinguished from theory of (intensional) meaning. And, as I indicated above, this whole web of philosophical beliefs derives from Quine's basic epistemological views, from naturalized epistemology.

This brief account of Quine's epistemology provides some background for the more specific problem that is the focus of my paper. Already in the classic "Two Dogmas" -article Quine expresses his misgivings about the common-sense notion of linguistic meaning, discussing synonymy as the basis of analyticity. Later in *Word and Object* he developed the thesis of indeterminacy of translation and showed that this indeterminacy applies also at home, within the home language. Later on Quine came to appreciate that there are two different kinds of indeterminacy of meaning. On one hand, we have the indeterminacy of sentence meaning, on the other hand, the indeterminacy of referential meaning, also known as the inscrutability of reference or ontological relativity. I'm not certain which one of these indeterminacy theses is stronger. Quine himself seems to think that the former, indeterminacy of sentence meaning, is "more serious" and stronger, and perhaps also more controversial philosophically. Anyway it seems to me that at least some sort of indeterminacy of sentence meaning follows from the inscrutability of reference, but I cannot argue for this point in detail here. Thus, I will leave open the question about the relation between the two forms of semantic indeterminacy; we can well follow Quine in keeping these two forms apart and independent of each other. In what follows I will concentrate on the indeterminacy of reference.[1]

2. Quine's Argument for the Inscrutability of Reference

Quine himself joins some of his critics in admitting that the forms of semantic indeterminacy are a consequence of his linguistic behaviorism. What he does not admit is that the indeterminacy theses were a *reductio ad absurdum* of his behaviorism. Since the inscrutability of reference is our main issue here, let's have a closer look at the arguments Quine offers for it.

Quine's behavioral theory of meaning is based on the concept of the *stimulus meaning* of a sentence. Stimulus meaning can be roughly characterized as truth conditions of a sentence stated in terms of perceptual physiology – in terms of firing-patterns of receptor cells. Stimulus meaning states the conditions that must obtain in the case of an affirmative or a negative response to a sentence. At this *holophrastic* level one cannot determine the ontology of a language – stimulus meaning does not determine the references or extensions of terms. A sentence has a stimulus meaning only when it is considered as an undivided whole. If we approach *referential* meaning in the spirit of *Word and Object*, that is, in terms of translation, we come to the task of dividing the foreign language into units which we then correlate with the terms and grammatical particles of our own language. At this stage the translator comes up with *analytical hypotheses*, to use the terminology of *Word and Object*. These hypotheses, however, are not real hypotheses, because they do not concern the realm of facts in physical world. There is no fact of the matter in choosing between empirically equivalent analytical hypotheses. Later on Quine came to call this point Ontological relativity, or the Inscrutability of reference.[2]

There are two different kinds or levels of inscrutability of reference, and Quine advances two different arguments for these two levels. First, there is the *proxy function argument* which is directed at object theories presented in the form of

---

1 Orenstein (2002, pp. 142–146) presents a good reconstruction of Quine's arguments concerning the two forms of semantic indeterminacy.

2 The indeterminacy-argument which refers to the thought-experiment of *radical translation* is sometimes called "the argument from below", to distinguish it from "the argument from above" which relies on the holistic thesis of underdetermination of theory by evidence. See e.g. Wright (1997).

canonical notation, that is, formalized in first-order predicate logic. This argument does not establish inscrutability of reference across the board; it only concerns object theories with a well-defined primitive vocabulary and logical or grammatical constructions. The proxy function argument is stated from the point of view of a background theory whose referential apparatus and truth-concept are taken as a primitive and unquestioned starting point. Briefly put, the idea is that by means of our background theory we can define a function that assigns a proxy to each object of the domain of the object theory and reinterpret its predicates so that the stimulus meanings of the sentences of our object theory remain unaltered. Thus, we come up with two different theories of interpretation for our object theory. Quine claims that the proxy function argument offers a *constructive proof* of the inscrutability – to establish this point he refers to the reduction of arithmetic to set theory: there we have an example of a successfully defined proxy function. The possibility to translate talk about numbers into talk about sets shows that it is in a sense meaningless to ask what objects arithmetic really deals with. Here we see the beginnings of Quine's radical structuralist thesis that objects are nothing but neutral nodes in the structure of a theory. However, this argument from the case of number theory, I think, is a dubious one: the case of mathematics is extended to a whole host of "more empirical" theories, where the actual stimulus meanings carry more weight. But I can't argue for this point here. It suffices here to say that a proxy function offers a translation manual that brings out the inscrutability of reference in the case of logically regimented object theories.

The other, more drastic case of inscrutability is that which obtains within the home language, or, if you like, the ultimate background theory. For this kind of all-around inscrutability Quine does not, and cannot, offer a conclusive argument. What I call "The Grand Permutation Argument" is intended by Quine to have only suggestive power. It makes us *appreciate* that the inscrutability of reference applies across the board. The Grand Permutation Argument is in its essentials given already in *Word and Object*, where Quine takes some time to consider "the impractical joke" of devising a translation manual for a wild reinterpretation of the speech of our compatriot. This grand permutation within our background theory, our natural language, cannot be defined as a proxy function, so there is no *proof* for the all-around inscrutability, even if one is willing to grant the full force of the proxy function argument. To make a long story short, once the rabbits and the undetached rabbit parts of the radical translation argument get mixed up in the case of our next door neighbor, we realize that we have had no evidence all along of the fixity of reference in his verbal behavior. There might exist such an exotic intralingual translation manual. Quine says:

> Our advantage with the compatriot is that with little deviation the automatic or homophonic hypothesis of translation fills the bill. If we were perverse and

> ingenious we could scorn that hypothesis and devise other analytical hypotheses that would attribute unimagined views to our compatriot, while conforming to all his dispositions to verbal response to all possible stimulations. [...] [T]he main lesson to be derived concerns the empirical slack in our own beliefs. For our own views could be revised into those attributed to the compatriot in the *impractical joke* imagined; no conflicts with experience could ever supervene[.] (Quine 2001, p. 78, italics added)

What was an impractical joke in *Word and Object* was later taken very seriously by Quine:

> We can reconcile all this with our neighbor's verbal behavior, by cunningly readjusting our translations of his various connecting predicates so as to compensate for the switch of ontology. In short, we can reproduce the inscrutability of reference at home. [...] We begin to appreciate that a grand and ingenious permutation of [...] denotations, along with compensatory adjustments in the interpretations of auxiliary particles, might still accommodate all existing speech dispositions. (Quine 1969a, pp. 47-48)

These reflections led Quine to his thesis of global structuralism and motivated his attempts to rescue determinate reference within the home language or, in other words, the ultimate background theory.

3. Problems of Global Inscrutability

There seems to be a tension between Quine's inscrutability thesis and certain other semantic views he adheres to. Besides linguistic behaviorism, the inscrutability argument presupposes extensionalism: there is nothing more to the meaning of terms than their reference, the objects or collections of objects they refer to. However, the arguments for the inscrutability presuppose a level of linguistic meaning that lies beyond the clarity and austerity of extensionalism. The truth predicate of the background theory plays a central role in the permutation arguments. Not just any translation manual for an object theory constitutes a proxy function. Sentences of the object theory have to remain *true* in spite of the change of extensions, that is, the reinterpretation of terms. The concept of holophrastic stimulus meaning rests on the fact that we have a grasp on, or, in other words, we *understand* the truth predicate of our background theory. But, as Quine himself makes very clear in many of his writings, due to well-known paradoxes the extension of this predicate must remain indeterminate. Still, he claims, the Tarskian T-schema suffices as an explication of this concept – we understand what "true" means, but we cannot give its extension without falling into paradox. The truth predicate of the ultimate background theory has an *intensional ring*, to use a phrase from Lieven Decock's (2002, pp. 158–159) book *Trading Ontology for Ideology*.

In his later writings Quine seems to be bothered by his global inscrutability thesis. One might say that in order to make sense the argument for the inscrutability demands that reference is not indeterminate within the background theory. After all, the domain of the background theory has to include the objects the object theory purports to talk about, *plus* the proxies of those objects – a meaningless distinction, according to the inscrutability thesis itself. Maybe this is the reason why Quine says for example the following:

> We seem to be maneuvering ourselves into the absurd position that there is no difference on any terms, interlinguistic or intralinguistic, objective or subjective, between referring to [objects and their proxies]. [...] Reference would seem to be nonsense not just in radical translation but at home. Toward resolving this quandary, begin by picturing us at home in our language, with all its predicates and auxiliary devices. [...] This network of terms and predicates and auxiliary devices is, in relativity jargon, our frame of reference, or coordinate system. Relative to *it* we can and do talk *meaningfully* and *distinctively* of rabbits and parts, numbers and formulas.
>
> [I]n practice we end the regress of background languages, in discussions of reference, by acquiescing in our mother tongue and taking its words at face value. (Quine 1969a, pp. 47–49, italics added)

Here we see how Quine proposes that by *acquiescing* in our home language we are able to avoid the troubles of inscrutability. This idea of acquiescence has been criticized for example by Barry Stroud (1995). Since Quine's epistemological-linguistic behaviorism entails the inscrutability of reference, and since behaviorism is the only scientifically respectable way of approaching meaning also in the case of the ultimate background theory, one is tempted to think that Quine postulates here some non-scientific (and non-extensional) aspect to the meaning of terms. But since Quine has always been an enthusiastic promoter of extensionalism, we are left in the dark about the dimension of linguistic meaning that is supposed to fix the extensions of terms and make the truth predicate understandable in the ultimate background theory. I would say that this is the real absurdity of Quine's position: on one hand the inscrutability of reference affects also the ultimate background language, on the other hand Quine tries to escape this all-around inscrutability, perhaps because it is intuitively and philosophically unacceptable.

In addition to the obscure idea of acquiescence, Quine also tries, in *Pursuit of Truth*, to give a more "technical" argument for the fixity of reference in the home language. This solution is based on the idea that we *choose* the identity function as our only translation manual within our home language. I have argued elsewhere (Keskinen 2004) that this attempt at a solution fails. The so-called "identity transformation" namely "'x' refers to x" ('rabbit' refers to rabbits) is a totally trivial maneuver. Anyone could do it on any given language, as long as he can distinguish the singular and general terms of the language. The identity transformation does not provide us with a "cosmic exile", a vantage point from where we could make words stick to objects.[3]

4. Solution to the Problems?

Quine's idea of scientific semantics stresses the *public character* of language. For him, this means the repudiation of mentalistic semantics. The mental level, first-person conscious experience, is totally *insignificant* when it comes to linguistic meaning (Quine 1970, p. 5). All meaning there is is out there in observable behavior. And, as we have seen, this view leads to the inscrutability thesis. John Searle points out the inconsistencies in Quine's semantic views and proposes a solution to the problem in terms of his philosophy of mind. Searle claims that once we give up all vestiges of behaviorism and recognize conscious experience as scientifically respectable, the semantic problems of behaviorism also disappear. According to Searle the intentional, first-person experience is of crucial importance to linguistic meaning. The intensional aspect of meaning of terms, dismissed by Quine, is taken care of by the concept of *aspectual shape* of objects, and objects and their proxies are clearly two different things in our conscious experience. Searle's approach shows that one possible solution to the Quinean problem of referential meaning might be sought in the direction of philosophy of mind, by taking conscious experience seriously. He claims that fixed linguistic reference is a *fact* that needs to be *explained*, not *eliminated* by appealing to empiricism or scientific semantics. According to Searle, when one studies consciousness and intentional phenomena such as linguistic reference, one should be sensitive to the nature of the phenomena studied. And he says that his brand of biological naturalism accomplishes this. Needless to say, Searle's basic position is unacceptable to Quine, and vice versa – even though both are avowed naturalists. It is Searle who thinks that the inscrutability thesis amounts to a *reductio ad absurdum* of linguistic behaviorism and hence of Quine's naturalized epistemology. We need not take a stand on this question here, but nevertheless I think Searle is quite right in his criticism of the inconsistencies in Quine's semantic views:

> We cannot, on the one hand, insist on a rigorous behaviorism that implies that there is no fact of the matter, and then, when we get in trouble, appeal to a naïve notion of

---

3 Glock (2003, pp. 203–205, 217–218) has criticized Quine's conception of the inscrutability of reference in a similar vein as I do.

> a mother language with words having a face value in excess of their empirical behavioral content. (Searle 1987, p. 133)

Searle's program is meant to free us from the often unnoticed spell of behaviorism. However, when we restrict our attention to questions of linguistic meaning, Quine's behavioral approach has its undeniable virtues. If we give up behaviorism, as Searle is willing to do, we have to find a way of answering those who claim, in the spirit of Quine, that we have given up the empirical spirit of natural science in semantics. The thesis of the public character of meaning that is essential to Quine's behaviorism is appealing, for a simple reason: that is how we learn language. From the moment of first driving a wedge to the side of our mother tongue we rely on observing the verbal behavior of others. Quine stresses that behaviorism is the only option to those who wish to avoid uncritical mentalism and the illusion of explaining something when in fact one's theory has no explanatory value. So, if we are not willing to practice Searle's science of first-person conscious experience, can we still find a way to avoid the problems of Quine's position? Of course, one could adopt Davidson's view that there is no way to escape the indeterminacy of reference.[4] According to him, there is no sense in talking about ontological *relativity*. Ontology is not relative to any interpretation given in a background language; rather, ontology is simply inscrutable. Now Davidson is not a behaviorist in quite the same sense as Quine is, but he accepts the validity of Quine's permutation argument. So, Davidson's way out of the Quinean problem of referential meaning is to deny that there was a problem in the first place. We can only accept the fact that reference is left indeterminate by all possible semantic evidence. But is it possible to hold on to linguistic behaviorism while taking the Quinean problem seriously? I would like to suggest one such possibility, which I have not been able spot in the literature on Quine.

I have already pointed out that the proxy function argument and the grand permutation argument should be seen as separate arguments. I will not focus here on the question whether the case of reduction of number theory to set theory should be considered as constructive proof of the inscrutability for all kinds of logically regimented object theories. For the sake of argument, let us grant that. Ultimately the Quinean problem of referential meaning comes down to the inscrutability of reference within the home language, the ultimate background theory. And as an argument for *this* kind of inscrutability Quine offers only the suggestion about the possibility of a "grand and ingenious permutation". This kind of permutation cannot be accomplished within the ultimate background language and so the all-around inscrutability cannot be as conclusively argued for as the more restricted case of object theories. So, we may accept the behavioristic starting point of Quinean

---

4 See Davidson (1984a).

semantics and also accept the proxy function argument, but at the same time deny the force of the permutation argument in the case of our home language. We might say that it is part of the nature of natural language that it includes means of blocking each attempt at giving an ingenious permutation a la Quine. But, our argument will remain as inconclusive and as suggestive as Quine's idea of a grand permutation. However, I think it is worth emphasizing that we do not have to accept the global inscrutability on the basis of Quine's argument. This approach might offer a way out of the problem Quine was struggling with, and it does not, at least *prima facie*, force us to give up linguistic behaviorism or to resort to some mysterious form of acquiescence in our home language. Of course, if we wish, for example, to hold on to the concept of truth in the way Quine does, we are still left with the problems I referred to a while ago.

Another way of keeping a third person behavioral perspective on linguistic meaning while denying the force of Quine's inscrutability arguments is to insist that Quine does not get the nature and breadth of the relevant behavioral evidence right. Perhaps we should not confine ourselves merely to sentence meaning and assent-dissent conditions of declarative sentences in studying the empirical checkpoints of linguistic meaning. Perhaps term meanings, that is, linguistic reference, can be determined independently of sentence meanings and analytical hypotheses. It has lately been argued that Quine's thought experiment of radical translation misses some relevant anthropological facts about human interaction. When we look at the actual practice of linguistic and other kinds of interaction between agents and reconsider the situation of radical translation from a more plausible anthropological point of view, we may notice that Quine's arguments for the primacy of sentence meaning and the possibility of incompatible translation schemes are not very well established.[5]

To summarize my points, we have distinguished four different solutions to the Quinean problem of referential meaning:

1. There is the Davidsonian view that Quine's arguments show (or should we say: prove) that fixed reference turns out to be a myth when semantic questions are approached in a scientific spirit. Quine's own views of relativizing ontology to a primitively adopted background theory can then be dismissed as a failure of nerve.
2. We can deny Quine's argument for the global inscrutability. The grand permutation argument is not suggestive enough for us to appreciate that such a permutation is possible. This does not prevent us from accepting the argument from

5 Glock (2003, ch. 6, 7) has presented this kind of critique towards Quine's semantic ideas.

proxy functions in the case of object theories. However, this solution, as the idea of a grand permutation itself, is much more a matter of faith than argument.
3. We can argue that Quine does not get the nature of the relevant behavioral evidence right in his thought-experiment of radical translation. Quine's ideas about stimulus meaning and the primacy of sentence meaning might be based on implausible anthropological assumptions. Perhaps there are ways of finding out the references of certain terms wholly independently of the Quinean stimulus meanings of sentences. Perhaps the domain of semantically relevant behavioral evidence goes beyond the assent-dissent conditions of declarative sentences, for example in that we are allowed to put more weight on ostensive teaching of words than Quine does. On this view Quine's thought experiment of radical translation misses an anthropological perspective about cross-cultural human interests and the nature of human interaction in such a situation.
4. We can take a wholly different stand on the basic tenets of Quinean philosophy of language. One example of this is Searle's view. Perhaps behaviorism and extensionalism should be abandoned in favor of a different conceptual framework – one that takes seriously the role of conscious experience (Quine's "mental level") in the formation of linguistic meaning. This approach may allow us to take into account the non-extensional aspects of meaning. But this means giving up the empiricism and the "scientific spirit" of the Quinean project.

University of Tampere

References

Decock, L. (2002), *Trading Ontology for Ideology: The Interplay of Logic, Set Theory and Semantics in Quine's Philosophy*, Dordrecht, Kluwer.

Davidson, D. (1984a), "The Inscrutability of Reference", in Davidson (1984b), pp. 227-245.

Davidson, D. (1984b), *Inquiries into Truth and Interpretation*, Oxford, Oxford University Press.

Glock, H.-J. (2003) *Quine and Davidson on Language, Thought and Reality*, Cambridge, Cambridge University Press.

Guttenplan, S. (ed.) (1975), *Mind and Language: Wolfson College Lectures 1974*, Oxford, Oxford University Press.

Hale, B. and C. Wright (eds.) (1997), *A Companion to the Philosophy of Language*, Oxford, Blackwell.

Keskinen, A. (2004), *W. V. Quinen kielifilosofia* ("The Philosophy of Language of W. V. Quine"), Master's thesis, Department of Mathematics, Statistics and Philosophy, University of Tampere.

Kiefer, H. and M. Munitz (eds.) (1970), *Contemporary philosophic thought part I: Language, belief, and metaphysics*, Albany, State University of New York Press.
Leonardi, P. and M. Santambrogio (eds.) (1995), *On Quine: New Essays*, Cambridge, Cambridge University Press.
Orenstein, A. (2002), *W. V. Quine*, Bucks, Acumen.
Quine, W. V. O. (1969a), "Ontological Relativity", in Quine (1969c), pp. 26-69.
Quine, W. V. O. (1969b), "Epistemology Naturalized", in Quine (1969c), pp. 69-90.
Quine, W. V. O. (1969c), *Ontological Relativity and Other Essays*, New York, Columbia University Press.
Quine, W. V. O. (1970), "Philosophical Progress in Language Theory", in Kiefer & Munitz (1970), pp. 3-18.
Quine, W. V. O. (1975), "Mind and Verbal Dispositions", in Guttenplan (1975), pp. 83-97.
Quine, W. V. O. (1992), *Pursuit of Truth. Revised edition*, Cambridge, Harvard University Press.
Quine, W. V. O. (2001), *Word and Object*, Cambridge, The MIT Press.
Searle, J. (1987), "Indeterminacy, Empiricism, and the First Person", *The Journal of Philosophy* 84, 123-147.
Stroud, B. (1995), "Quine on Exile and Acquiescene", in Leonardi and Santambrogio (1995), pp. 37-53.
Wright, C. (1997), "The indeterminacy of translation", in Hale and Wright (1997), pp. 397-427.

Quinean Naturalism and the Metaphysical Turn[1]

Heikki J. Koskinen

Willard van Orman Quine's (1908–2000) thought is closely intertwined with the twentieth century analytic tradition of philosophy that has often preferred to call itself scientific. Quine published his first book in 1934, and his last one came out as recently as in 1995. Through seven decades of publishing books and articles, of teaching and lecturing, Quine has left lasting marks on the pages of the history of philosophy. The influence of Quine is widely acknowledged, but the evaluation of this influence is another matter. From a metaphilosophical point of view, Quine's legacy is an issue of heated debate. According to the worst imaginable scenario, Quine's naturalism leaves no room for the autonomy of the conceptual, nor for any kind of typically philosophical perspective, and thus results in the destruction of philosophy as a recognizable and worthwhile discipline. The harbingers of doom are likely to think that in the hands of Quinean naturalism, *logic* is spoiled by some sort of psychologism, or trivialized by making it pseudo-empirical. *Epistemology* in turn is naturalized into a chapter of psychology or some other kind of empirical cognitive science, leaving no meaningful role for philosophy or normativity.[2] To complete the gloom, what is left of *metaphysics* is turned over to the special sciences. One by one, it seems to the defeatist, the central fields of theoretical philosophy are blown away by the dreadful winds of naturalism. In the following, I shall try to offer some evidence against this kind of interpretation. My general approach is metaphilosophical, for I am mainly interested in the nature and role of philosophy within Quinean naturalism (cf. Koskinen 2004a).[3] Of the main subdivisions of philosophy, my emphasis shall be on metaphysics.

---

1 This paper is based on an earlier presentation at *the 12th International Congress of Logic, Methodology, and Philosophy of Science*, Oviedo, Spain, 7–13 August 2003. I wish to express my gratitude to Alfonso and Heidi Rottensteiner.

2 Cf. Markus Lammenranta's "Is Naturalism Question-Begging?" and Pentti Määttänen's "Naturalism: Hard and Soft", both included in this volume.

3 Cf. also Martin Gustafsson's paper "Quine and Wittgenstein on the Dissolution of Philosophical Problems" in this volume.

1. Quinean Naturalism and its Metaphilosophical Interpretation

Roger F. Gibson – one of the leading contemporary Quine scholars – has (Gibson 1992) insisted that a key to unlocking a correct interpretation of Quine is his commitment to naturalism, which is taken to consist of the pair of claims that (i) there is no first philosophy and that (ii) it is up to science to determine what there is. In a later elaboration on these two claims, Gibson (1998, p. 668) has also written:

> Naturalism consists of a pair of theses, one negative, one positive. The negative thesis is that *there is no successful first philosophy*, that is, there is neither an a priori nor an experiential foundation outside of science upon which science can be grounded (i.e., justified or rationally reconstructed). The positive thesis is that *science is the measure both of what there is (ontology) and how we know what there is (epistemology)*.

Supposing now that Gibson manages to capture the essential features of Quinean naturalism, from a metaphilosophical point of view there are two directly opposing ways of interpreting Quine's contribution to the twentieth century analytic tradition. Let us call these the *gloomy* and the *glorious* one.

Although Gibson speaks of the negative and positive theses of Quinean naturalism, the gloomy metaphilosophical interpretation takes the implications of both to be negative: the denial of first philosophy eradicates the possibility of any specifically philosophical perspective (cf. Haaparanta 1999) and the acceptance of science as the final arbiter simply hands all the remaining problems over to the special sciences.

There are many variations of the gloomy interpretation in the literature from which to choose. Ilkka Niiniluoto (1999, p. viii) takes the influence of the towering figure of Quine to be strong but pernicious. He (ibid., p. 17) also attributes to Quine the "positivist" view urging that science may be a child of philosophy, but has since grown completely independent of her mother, i.e. mature science has happily got rid of metaphysics and epistemology. Sami Pihlström (1998, p. xii) in turn considers Quine an example of *extreme scientistic post-philosophy*, suggesting that Quine's "extreme scientism" gives disproportionate importance to science and thus leads to a "post-philosophical" situation, where philosophy is in effect over. Various critical assessments of Quine's metaphilosophical influence have also been offered e.g. by Michael Friedman (1997), P. M. S. Hacker (1998), and Peter Hylton (1998) (cf. also Bealer 1996).

There is, however, another more positive and constructive way of viewing the metaphilosophical implications of Quine's thought. This approach emphasizes the glorious metaphysical turn effected by Quinean naturalism (cf. Benardete 1989, p. 5; Glock 2003, p. 40; Kim & Sosa 1999, p. x; Koskinen 2004b; Loux & Zimmerman 2003, p. 2; Orenstein 2002, p. 67).

2. The Linguistic Turn

A metaphysical tradition traceable to Aristotle views philosophy as the most general of all the sciences because it studies the fundamental features of reality. According to one standard historical interpretation, rationalists took metaphysics to uncontrollable excesses, empiricists criticized this, and Kant then tried to settle the dispute with his notion of synthetic a priori. With the help of the analytic/synthetic distinction, Carnap in turn thought he could rid philosophy of metaphysics altogether by effecting a clear demarcation between various empirical sciences studying the world on the one hand, and the analytic disciplines that were purely linguistic or conceptual on the other. Since metaphysics was taken to be neither an empirical science nor an analytic discipline, it would have to be completely overthrown (cf. Carnap 1932), or at the very least demoted to a non-cognitive status (cf. Carnap 1956, pp. 205–221).

The gloomy metaphilosophical interpretation of Quinean naturalism is related to a conception of philosophy as completely distinct from science, and this conception in turn is related to the so-called *linguistic turn* (cf. Hochberg 2003; Rorty 1992). In sharp contrast to the Kantian solution, Carnap's answer to the foundational metaphilosophical question was based on a strict rejection of the synthetic a priori together with an exclusive division of the field of knowledge into the analytic a priori and the synthetic a posteriori (cf. Carnap 1935). For the Vienna Circle and the adherents of logical empiricism, there was no possibility of a third way politics within the domain of meaningful discourse. The field of synthetic a priori was considered empty, and the Kantian role of metaphysics was rejected with great determination (cf. Neurath *et al.* 1929, p. 308). At the same time, philosophy was clearly demarcated from the empirical special sciences, which alone had any authority on the truth value distributions of the synthetic a posteriori claims. Philosophy itself was to be a purely conceptual enterprise dealing with linguistic frameworks. Philosophers could not, and should not even try to say anything about the empirical world. In Carnap's (1935, p. 84) own words:

> …there can be no such thing as a philosophical investigation of nature, because whatever can be said about nature, that is about any events in time and space and about their connections, has to be said by the scientist on the basis of empirical investigation. There remains nothing for the philosopher to say in this field. Metaphysicians do, indeed, venture to make a lot of statements about nature, but such metaphysics is, as we have seen, not theory, but rather poetry.

The notion of analyticity was seen to eradicate any reference to vague "intuitions" as a justification for a priori knowledge, and to offer a better epistemological solution (cf. Creath 1987; 1990). Analyticity was also what characterized logical as opposed to empirical investigation, made the linguistic turn possible, and expressed the distinctive character of philosophy itself (cf. Friedman 1998). It is against this background of pure philosophy clearly distinguished from science that Quine's critique of analyticity, his acceptance of science as the final arbiter, and his naturalization of epistemology appear to lead to the gloomy conclusion of the destruction of philosophy. To see the glorious metaphilosophical path opened up by Quinean naturalism, we have to realize that the point is not so much that all these things central to the conception of pure philosophy get rejected, but that the relations between science and philosophy are simultaneously radically reconceived (cf. Koskinen 2004b). This reconception is what makes it possible to rediscover metaphysics (cf. Linsky 1997, p. 108ff.) as well as to give philosophy a positive and constructive role.

3. The Metaphysical Turn

Quinean naturalism rejects the analytic/synthetic distinction, and it is committed to a holistic form of empiricism that undermines the theoretical foundations of a pure philosophy in the logical empiricist vein. It also blurs the supposed boundary between speculative metaphysics and natural science (cf. Quine 1980, p. 20). The result of all this is a *metaphysical turn*: metaphysics returns to the philosophical scene as a central subdivision. Despite certain important differences between Aristotle and Quine, the latter's naturalism could be seen to imply a return to metaphysics in the grand tradition of the former (cf. Koskinen 2004a). Quinean naturalized metaphysics is abstract through being very general, and it belongs on the same gradualistic continuum with the rest of science.

Repatriating the once expelled metaphysical questions is made possible because within the Quinean position, analyticity no longer functions as an epistemic or a metaphilosophical demarcator. Ontological issues are not seen as illusory pseudo-problems or non-cognitive external questions, when all truth is on a par, and the holistic web of belief covers everything from the most casual observation sentences to the truths of logic and the deepest metaphysical assumptions. Once holism and gradualism are accepted, abstract ontological issues slip back into the sphere of science and philosophy.

In "On Carnap's Views on Ontology" Quine (1976, pp. 203–211) notes that Carnap disapproves of his giving meaning to the word 'ontology' which belongs to traditional metaphysics and should therefore be meaningless. However, Quine also points out that according to his own understanding, Carnap does accept the

Quinean values-of-bound-variables standard for judging whether a given theory accepts given alleged entities. Their disagreement on the status of metaphysics centers on the possibility of distinguishing between certain types of questions in connection with linguistic frameworks. The wanted contrast is said to be clinched by the analytic/synthetic distinction, and it is *this* distinction that Quine takes to be the basic point of disagreement (cf., however, Alspector-Kelly 2001).

After posing the ontologically and epistemologically central question of what counts as evidence for existential quantifications, and also stating that there is no simple general answer to be found, Quine (1969, p. 97) considers the examples of quantifiable open sentences like '*x* is a rabbit', and '*x* is a unicorn'. In these cases the evidence – if such there be – is largely *the testimony of the senses*. On the other hand, if our open sentence is something like '*x* is a prime number between 10 and 20', then the evidence will obviously have to lie in *computation*. But what about the typically abstract philosophical or metaphysical types of existence statements? On these Quine writes:

> If the open sentence is merely 'x is a number', or 'x is a class', or the like, the evidence is much harder to pinpoint. But I think the positivists were mistaken when they despaired of evidence in such cases and accordingly tried to draw up boundaries that would exclude such sentences as meaningless. Existence statements in this philosophical vein do admit of evidence, in the sense that we can have reasons, and essentially scientific reasons, for including numbers or classes or the like in the range of values of our variables. And other existence statements in this metaphysical vein can be subject to counter-evidence; we can have essentially scientific reasons for excluding propositions, perhaps, or attributes, or unactualized bodies, from the range of our variables. Numbers and classes are favoured by the power and facility which they contribute to theoretical physics and other systematic discourse about nature. Propositions and attributes are disfavoured by some irregular behaviour in connection with identity and substitution. Considerations for and against existence are more broadly systematic, in these philosophical examples, than in the case of rabbits or unicorns or prime numbers between 10 and 20; but I am persuaded that the difference is a matter of degree. Our theory of nature grades off from the most concrete fact to speculations about the curvature of space-time, or the continuous creation of hydrogen atoms in an expanding universe; and our evidence grades off correspondingly, from specific observation to broadly systematic considerations. Existential quantifications of the philosophical sort belong to the same inclusive theory and are situated way out at the end, farthest from observable fact. (Quine 1969, pp. 97–98)

Talk of "the same inclusive theory" here is consistently in line with the idea of "a single unpartitioned universe of values of bound variables" (Quine 1960, p. 230), and the associated egalitarian notion of existence. In this sense, Quine is a metaphysical liberal who does not want to impose any ontological apartheid policies based on distinctions between types, or between existence and subsistence. The philosopher and the scientist are thus in the same boat, and philosophy is not to be distinguished in essential points of purpose and method from good and bad science (cf. ibid., pp. 3–4).

4. From a Metaphilosophical Point of View

The two cardinal tenets of empiricism that remain unassailable for Quine (1969, p. 75) are (1) whatever evidence there is for science is sensory evidence, and (2) all inculcation of meanings of words must rest ultimately on sensory evidence. Even at the level of observation sentences, however, there is a certain bipolarity (cf. Quine 1993, p. 109), for observation sentences enjoy a peculiar kind of double life: *holophrastically* construed, they are conditioned to patterns of proximal stimuli, while *analytically* construed, they are linked to other statements, including theoretical ones, by virtue of a shared vocabulary (cf. Gibson 1996, p. 104). Towards the *outside* world, therefore, our system of beliefs faces the tribunal of experience via monolithic sentences, while towards the *inside* of the system the sentences are analyzed into conceptual components interlocking with various other sentences. Quine himself also clearly recognizes the contrasting forces felt within the web of belief when he (Quine 1981, p. 31) writes:

> A good scientific theory is under tension from two opposing forces: the drive for evidence and the drive for system. ... If either of these drives were unchecked by the other, it would issue in something unworthy of the name of scientific theory: in the one case a mere record of observations, and in the other a myth without foundation.

The drive for evidence and the drive for system are related to the data-driven bottom-up perspective of holistic empiricism and the conceptually driven top-down approach of systematic philosophy (cf. Koskinen 2004a, p. 249ff.).

In addition to naturalism, another main tenet of Quine's philosophy is his *extensionalism*. Lieven Decock (2002, p. 220) writes that Gibson offers an adequate presentation of the naturalistic aspect of Quine's thought, but that this is a one-eyed reading of Quine in which some vital elements in the structure of Quine's philosophy are neglected. Quine's philosophy is said (*ibid.*) to be presented in an impoverished way because his sense of precision is of little importance for naturalists. It could also be said that *naturalism* is generally dominated by the bottom-up drive for *evidence*, while *extensionalism* and metaphysics emphasize the top-down drive for *system*. The reader of *Word and Object* will easily see the emphasis move from the former towards the latter as she proceeds through the chapters of Quine's philosophical *magnum opus*. It is the one-eyed reading of

Quine that also threatens to lead to the gloomy metaphilosophical conclusion. By becoming aware of the systematic top-down aspect of Quinean naturalism, we can better appreciate the nature and role of philosophy in relation to the special sciences.

Quine has had a strong interest in formulating a criterion of ontological commitment according to which *to be is to be the value of a variable*. He has also repeatedly emphasized his standard of ontological commitment, according to which there is *no entity without identity*. Quine's extensionalism (cf. Koskinen 2004a, p. 126ff.; p. 254ff.) and his resistance to extensions of first-order predicate logic based on adding devices like modal operators (leading to problems concerning possible individuals, essentialism, trans-world identity, etc.) are also metaphysical by their very nature. Even Quine's influential critique of analyticity and meaning can be seen as an extended study into *the metaphysics of meaning*. Moreover, Quine shows his strong commitment to the project of ontology or *metaphysica generalis* by offering his own system of the world, whose official ontology consists of just physical objects and classes. With good reason, then, Quine has been called a "key contemporary American metaphysician" (Canfield 1997b, p. 9).[4]

The metaphysical turn effected by Quinean naturalism has important consequences for the fate of philosophy. Because metaphysics has a central role in seeking the broad outlines of the whole system of the world (cf. Magee 1978, pp. 14–15), it connects philosophy with the rest of science (cf. Koskinen 2004b). Secondly, the possibility of metaphysics as a field of inquiry that studies the highest kinds or categories of being (cf. Loux 2002; Lowe 1999) shows that naturalism does not destroy philosophy, as some critics of Quine have claimed. Third, metaphysics gives systematic structure to important philosophical issues and helps to characterize the relations between philosophy's central subfields like epistemology, logic, and the philosophy of language (cf. Koskinen 2004a, p. 160). And finally, metaphysics gives a clear identity to philosophy, connecting it with the discipline's historical tradition and functioning as a firm ground from which to answer the challenges that the special sciences might be seen to issue to philosophy. For all these reasons, the metaphysical turn is indeed a glorious one.

Academy of Finland
University of Helsinki

---

4 On possible complications of the scenario presented here, see Antti Keskinen's paper "The Problem of Referential Meaning in Quine's Philosophy of Language" included in this volume (cf. however Koskinen 2004a, pp. 233–248).

References

Alspector-Kelly, M. (2001), "On Quine On Carnap On Ontology", *Philosophical Studies* 102, 93-122.

Bealer, G. (1996), "On the Possibility of Philosophical Knowledge", *Philosophical Perspectives* 10, 1-34.

Benardete, J. A. (1989), *Metaphysics: The Logical Approach*, Oxford, Oxford University Press.

Biletzki, A. & A. Matar (eds.) (1998), *The Story of Analytic Philosophy: Plot and Heroes*, London, Routledge.

Canfield, J. V. (ed.) (1997a), *Routledge History of Philosophy, Volume X: Philosophy of Meaning, Knowledge and Value in the Twentieth Century*, London, Routledge.

Canfield, J. V. (1997b), "Introduction", in Canfield (1997a), pp. 1-10.

Carnap, R. (1932), "Überwindung der Metaphysik durch logische Analyse der Sprache", *Erkenntnis* 2, 219-241.

Carnap, R. (1935), *Philosophy and Logical Syntax*, 1996 Reprint, Bristol, Thoemmes Press.

Carnap, R. (1956), *Meaning and Necessity: A Study in Semantics and Modal Logic*, Second Edition, Chicago, The University of Chicago Press.

Creath, R. (1987), "The Initial Reception of Carnap's Doctrine of Analyticity", in Føllesdal (2001), pp. 321-343.

Creath, R. (1990), "Introduction", in Richard Creath (ed.), *Dear Carnap, Dear Van: The Quine-Carnap Correspondence and Related Work*, Berkely, University of California Press, pp. 1-43.

DeAngelis, W. J. (1997), "Metaphysics I (1900-45)", in Canfield (1997a), pp. 76-107.

Decock, L. (2002), *Trading Ontology for Ideology: The Interplay of Logic, Set Theory and Semantics in Quine's Philosophy*, Dordrecht, Kluwer.

Friedman, M. (1997), "Philosophical Naturalism", *Proceedings and Addresses of the American Philosophical Association*, Vol. 71, No 2, 7-21.

Friedman, M. (1998), "Logical Positivism", in Edward Craig (ed.), *Routledge Encyclopedia of Philosophy*, Version 1.0, London, Routledge.

Føllesdal, D. (ed.) (2001), *Philosophy of Quine, Volume 1: General, Reviews, and Analytic/ Synthetic*, New York, Garland.

Gibson, R. F. (1992), "The Key to Interpreting Quine", in Føllesdal (2001), pp. 25-38.

Gibson, R. F. (1996), "Quine, Wittgenstein, and Holism", in Føllesdal (2001), pp. 102-118.

Gibson, R. F. (1998), "Quine's philosophy: A Brief Sketch", in Lewis Edwin Hahn and Paul Arthur Schilpp (eds.) (1998), *The Philosophy of W.V. Quine*, Expanded Edition, La Salle, Open Court, pp. 667-683.

Glock, H.-J. (2003), *Quine and Davidson on Language, Thought and Reality*, Cambridge, Cambridge University Press.

Haaparanta, L. (1999), "On the Possibility of Naturalistic and of Pure Epistemology", *Synthese* 118, 31-47.

Hacker, P. M. S. (1998), "Analytic Philosophy: What, Whence, and Whither?", in Biletzki and Matar (1998), pp. 3-34.

Hochberg, H. (2003), *Introducing Analytic Philosophy: Its Sense and its Nonsense 1879-2002*, Frankfurt am Main, Ontos Verlag.
Hylton, P. (1998), "Analysis in Analytic Philosophy", in Biletzki and Matar (1998), pp. 37-55.
Kim, J. and E. Sosa (eds.) (1999), *Metaphysics: An Anthology*, Malden MA, Blackwell.
Koskinen, H. J. (2004a), *From a Metaphilosophical Point of View: A Study of W. V. Quine's Naturalism*, Acta Philosophica Fennica 74, Helsinki, The Philosophical Society of Finland.
Koskinen, H. J. (2004b), "On Some Metaphilosophical Implications of Quinean Naturalism", in J. C. Marek and M. E. Reicher (eds.), *Experience and Analysis: Papers of the 27th International Wittgenstein Symposium*, Kirchberg am Wechsel, Austrian Ludwig Wittgenstein Society, pp. 178-180.
Linsky, B. (1997), "Metaphysics II (1945 to the present)", in Canfield (1997a), pp. 108-133.
Loux, M. J. (2002), *Metaphysics: A Contemporary Introduction*, Second Edition, London, Routledge.
Loux, M. J. and D. W. Zimmerman (eds.) (2003), *The Oxford Handbook of Metaphysics*, Oxford, Oxford University Press.
Lowe, E. J. (1999), *The Possibility of Metaphysics: Substance, Identity, and Time*, Oxford, Oxford University Press.
Magee, B. (1978), "The Ideas of Quine: Dialogue with W. V. Quine", in Føllesdal (2001), pp. 13-23.
Neurath, O., H. Hahn, and R. Carnap (1929), "Wissenschaftliche Weltauffassung: Der Wiener Kreis", English translation in M. Neurath and R. S. Cohen (eds.) (1973), *Otto Neurath: Empiricism and Sociology*, Dordrecht, Reidel, pp. 299-318.
Niiniluoto, I. (1999), *Critical Scientific Realism*, Oxford, Oxford University Press.
Orenstein, A. (2002), *W.V. Quine*, Bucks, Acumen.
Pihlström, S. (1998), *Pragmatism and Philosophical Anthropology: Understanding Our Human Life in a Human World*, New York, Peter Lang.
Quine, W. V. (1960), *Word and Object*, Cambridge MA, MIT Press.
Quine, W. V. (1969), *Ontological Relativity and Other Essays*, New York, Columbia University Press.
Quine, W. V. (1976), *The Ways of Paradox and Other Essays*, Revised and Enlarged Edition, Cambridge MA, Harvard University Press.
Quine, W. V. (1980), *From a Logical Point of View*, Second Edition, Revised, Cambridge MA, Harvard University Press.
Quine, W. V. (1981), *Theories and Things*, Cambridge MA, Harvard University Press.
Quine, W. V. (1993), "In Praise of Observation Sentences", *The Journal of Philosophy* 90, 107-116.
Rorty, R. M. (ed.) (1992), *The Linguistic Turn: Essays in Philosophical Method*, with two Retrospective Essays, Chicago, The University of Chicago Press.

Non-Scientific Naturalism?

Jonathan Knowles

Many, perhaps most contemporary philosophers like to see themselves as naturalists – as opposing Cartesian egos, divine purpose and other supernatural entities and phenomena that philosophical reflection at times might tempt one into positing. But does naturalism mean being a scientific naturalist, with a wholly scientific view of the world? Some think this is not a necessary or indeed tenable implication. In this paper, I want to present the ideas of some of these, as I call them, *non-scientific naturalists*, and then sketch a dialectical path through a set of issues leading to the conclusion that this is a very difficult position to sustain. In the short space I have available, I will not be able to present all the arguments necessary to establish this conclusion (some of these, I, and others, have offered in other places). My main aim is to sketch a dialectical backdrop to the debate in a way that brings out what is a stake and also, I hope, goes some way to suggesting where the burden of proof lies. The final part also offers substantive arguments aimed at rebutting one particular line of defence that has been offered by some supporters of non-scientific naturalism (*NSN* henceforth).

There are several sources of NSN within the standard disciplinary matrix of analytical philosophy, but I am most familiar with two in particular, one deriving from philosophy of mind (and to an extent, metaphysics more broadly), one from epistemology; it is hence these (for the most part) I will concentrate on. According to the first source, a naturalistic view of human beings and, in particular, their distinctive mental capacities, permits, indeed, demands an understanding of these capacities that is other than scientific (cf. e.g. McDowell 1994; Hornsby 1997). According to the second source, one need not hold there is any first philosophy – any autonomous, *a priori* set of epistemic norms or principles, in the way Quine (1953; 1969) has famously attacked – to think that epistemological theorising has life as an autonomous, distinctively philosophical discipline (cf. especially Haack 1993). Though these two different forms of NSN have been discussed largely independently of one another, they are close in spirit and there is a degree of overlap between the ways they are defended.

In the following, I will first present what I call 'the basic argument' for NSN, which sums up what I see as a fundamental line of thought of many its supporters, both from epistemology and philosophy of mind; I will then offer some first pass criticism of this argument (section 1). I will then consider two possible ripostes to this critique (based on extant ideas in the literature), and offer my counter-ripostes to these. The first riposte and counter-riposte constitute section 2; the second riposte and counter-riposte are the subject of the third and final section of the paper.

1. The Basic Argument for NSN

The following argument is inspired most directly by my reading of Jennifer Hornsby (op. cit., forthcoming) and the epistemologist Richard Feldman (2001, Pt. A.). Though its exact formulation here is my own, its basic dialectical thrust is one which these, and many other supporters of NSN, would undoubtedly subscribe to:

> [Basic argument for NSN]
> According to a scientific view of the world, everything that is true is ultimately a physical truth – that is, it is entailed by or implicit in some conjunction of physical, descriptive truths, concerning the basic physical make-up of the universe (cf. e.g. Jackson, 1998). Thus, if intentionality, morality, epistemic normativity etc. are real, the facts they enunciate are going to have to reduce to physical truths in this way. Call this view *physicalism*. However, if as part of our common sense we are antecedently convinced that some range of phenomena are real and indispensable, e.g. that there are truths about beliefs, justification, morality etc., and, moreover, it seems highly unlikely that these will reduce to physical truths (on some suitable conception thereof), then physicalism seems committed to saying, contrary to appearances, that there are no facts of the relevant kind. But this line is radically and implausibly revisionist. The only reasonable alternative is thus to reject physicalism. But this rejection is not motivated by appeal to facts that are not naturalistic. So scientific naturalism, in the sense of physicalism, can be rejected without reneging on naturalism *per se*.

The most egregious problem with this argument is that though physicalism, as understood above, probably is false, it is not plausibly part of science. Therefore, it is no part of a position worthy of the title 'scientific naturalism', nor something which a position wishing to distinguish itself from such a view can do so by opposing. Current physics offers no final list of physical properties, nor an overall conception of how different tracts of reality hang together. To assume that physics is complete, in the way physicalism presupposes, is thus either false or trivial – trivial in case what we mean by it is just whatever account we will eventually need of the world to explain everything (cf. Crane 1991). Nor is it at all obvious that biology or even chemistry as they are practised today in any sense aim – even ideally – at reduction to physics (cf. Dupré 1993).

Further, it is not clear, as some suggest in response to these considerations (cf. Papineau 1993, ch. 1; Jackson 1998, ch. 1; Fodor 1987), that at least *the mental* is not part of 'basic science', and hence can serve to define physicalism – that is, negatively, as a view which restricts the realm of basic facts to, at least in the first instance, non-mental and non-intentional ones. Though taken as an intuitively reasonable starting point, little by way of argument is offered to back up the restriction. This is strange given there is no coherent conception of scientific law that rules out the psychological *as such* as law-like (as Davidson, 1971, tried to argue): Laws in much non-physical science, and perhaps even physical science (cf. Cartwright 1983), are non-strict, but such laws exist in psychology (cf. Fodor 1974). Given this, there is no reason to think there can't be autonomously causally efficacious psychological properties (as Fodor, 1990, admits); and also, especially in view of the autonomy of biology, no good argument for thinking that, even though we might admit there are such properties, they need somehow to be grounded in physical mechanisms (cf. Knowles 2001 – the latter is something Fodor, fallaciously I argue, does think is the case). In short, there are no good metaphysical or scientific reasons for thinking that psychological properties should not count amongst the basic ones as much as any other kinds of properties.[1] This doesn't mean we need to deny, what would be equally implausible, that *local* reductions of particular psychological models of cognition, say, do occur and are legitimately sought, or that a general goal of *coherence* between psychological and computational and/or neurophysiological theories is defensible. However, no version of physicalism based on the idea of the basic facts excluding the mental ones *a priori* can be motivated. (Cf. Strawson 2003 for a different argument aimed at much the same conclusion.)

2. The First Riposte from NSN

I take is as read, then, that physicalism, understood as above, is a non-starter within science, and hence not in a position to give content to the idea of a scientific naturalism which NSN can define itself as opposed to. But this is not all supporters of NSN have to say for themselves. The first riposte I glean from their writings goes as follows:

---

1 Burge (1993) and Hornsby (1997, ch. 8) offer somewhat similar lines of defence of the autonomy of intentional properties, though see this as favouring a kind of non-scientific form of naturalism. This is because they presuppose that physicalism is part of science, which is what I am questioning here.

> [R1]
> *There is something special that distinguishes at least everyday intentional psychology ('common sense psychology', or 'folk psychology', as I will call it) from all other special sciences which makes it* a priori *that folk psychology could never be reduced to any more basic science, whereas it is not* a priori *that these other sciences could not be so reduced (cf. Davidson 1987; Child 1994, ch. 2.1). Folk psychology explains behaviour and thought in relation to* constitutive *principles of rationality that are also* normative *(cf. Davidson 1971; 1987; McDowell 1985). Connectedly, folk psychological explanation is* holistic *in nature, and involves* taking a perspective *on what is explained, both of which features explanations of physical events standardly lack. Finally, folk psychology is concerned with a* 'common sense world', *i.e. the world of macroscopic objects and their properties, including human agents and their actions, whereas science is fundamentally concerned with the underlying microstructure of such objects and properties (cf. Hornsby 1997).*

The dialectical move of this riposte is similar to that discussed in section 1 in relation to the attempt to define physicalism negatively: In the absence of a positive demarcation of the physical *vis à vis* everything else, one looks instead for a demarcation of the psychological *vis à vis* everything else. However, in contrast to those who seek to use this strategy to formulate a defensible form of physicalism – viewing the psychological as the paradigmatically *non-physical* – defenders of NSN seek to offer a *positive* conception of the psychological which, moreover, is meant to make the prospect of reduction to anything *non-psychological* unthinkable – something we can exclude *a priori*. Folk psychology as a scheme of explanation has properties which distinguish it from all special sciences – but cannot, without prejudice, to be written off as 'unnatural' for that reason.

A big problem with R1 is that there is, arguably, no clear demarcation between folk psychology and scientific psychology: there is no clear, *de facto* divide between a realm of explanation that focusses on rationality (folk psychology), and one which focusses on purely causal explanations of behaviour (scientific psychology) (cf. Rey 2001; Morris and Richardson 1995 for arguments to this effect; I say 'purely' in order to avoid any implication that rational explanations cannot also be causal). A more fundamental objection is that the features mentioned above are not even in principle capable of demarcating any kind of psychological explanation from scientific ones (in a broad sense of 'scientific'). The various arguments for this claim I have spelled out fully elsewhere (cf. Knowles 2002); the following is a brief overview of the main ideas of this paper (plus one more) in relation to the different points mentioned above:

i) *Constitutive principles*: Though it might be true that folk psychological explanation involves principles the holding of which are an *a priori* condition for the application of the predicates the principles contain ('belief', 'desire' etc.), the same is true of much explanation in science. Thus, for example, the predication of some or other mass to an ordinary object presupposes its satisfaction of the laws of physics concerning mass. (More needs to be said about this view of scientific concepts and, in particular, the impact of the possibilities of conceptual change, but none of this seems to enunciate any principled difference between physics and folk psychology – cf. ibid., pp. 207–210.)[2]
ii) *Normative principles*: The idea that we explain human action in relation to rational norms has a venerable tradition in both human science and philosophy (cf. Hempel 1965 (for discussion of such views), McDowell 1985). However, it has never been made clear what role *normative* statements – statements about what *ought* to be the case – can coherently play in the explanation of any *actual* behaviour (cf. also Henderson 1993, esp. ch. 5, 6 for extended discussion of this point).[3]
iii) *Holism*: It is often said that everyday explanation of mental events is an 'open-ended' affair, that it presupposes a background of beliefs and attitudes that is difficult or perhaps even impossible to specify completely and explicitly (cf. e.g. Davidson 1971, pp. 216–217; McGinn 1979, p. 39). The problem for a supporter of NSN is that a parallel point can be made about the explanation of everyday physical events, such as the sinking of a ship: We might first cite its collision with an ice-berg as the cause, precisifying with factors of its material constitution, the weather, and so on and so forth in a potentially endless spiral (insisting we might come to an end here, but not in the psychological case would just be question-begging).

---

2 Note that these points do not compromise Davidson's argument in 'Mental events' (1971) for anti-reductionism, which basically builds on the fact that the mental and the physical involve different constitutive constraints or 'ideologies'. However, reductionism is not is what is at issue at present, but rather the prospect of giving some reason to think reductionism is ruled out *a priori* in the case of the mental, but not that of other special sciences (a point which Davidson 1987 sees).

3 This point is not meant to impact on the idea that norms might justiably be applied in giving an interpretation of an agent and her behaviour. However, those who invoke norms in support of NSN have always done so also in relation to the *explanation* of behaviour, and it is this that seems incoherent (what implications this has for the interpretationist project I leave for others to ponder).

iv) *Perspective-taking*: Many say that explanations of human behaviour essentially involve use of first-personal capacities – to see the behaviour as reasonable, by one's own lights. However, to the extent that this point does not merely reiterate one or more of the ideas above, it plausibly enunciates only a contingent feature of *how we give* explanations, and does not impact on the nature of the explanation given (a verdict echoed by that of many commentators on the significance of the so-called *simulation theory* of folk psychology – e.g. Davies 1994).

v) *Common sense world*: (This suggestion is discussed in Knowles 2002a). The idea that a 'common sense world' is somehow at odds with science's conception of the world immediately pales when one considers biological science (biological organisms generally and their behaviour are presumably just as much part of any 'common sense world' as we are). It is, also, ultimately unclear that there is any sharp or absolute divide between the common sense and scientific worlds.

In short: Though, as we saw in section 1, psychology does not have to reduce to physics or some other more basic science to be scientifically respectable (hence physicalism is no part of anything worthy of calling scientific naturalism), it is not the case that psychology (or some aspect of it) *a priori* has *not* so to reduce, either. There is simply no principled distinction between folk psychology and other special sciences for supporters of NSN to base this claim on.

3. The Second Riposte from NSN

I call this *R2*, and it is inspired by Feldman (ibid., Pt. B), and Haack (ibid., ch. 6):

> [R2]
> *Once one has given up on the idea of unified metaphysical conception of what there is, and given also that one, as a good naturalist, relinquishes the idea of an* a priori *scientific method, the idea that everything should be scientific loses content. The idea of there being no absolutely* a priori *givens doesn't enunciate a peculiarly* scientific *way of understanding and conceiving of things. Insisting on scientific understanding thus becomes either prejudicially scientistic or empty. The viability of what we intuitively see as non-empirical, philosophical questions about e.g. justification,*

evidence and knowledge are consequently completely unthreatened by a naturalistic view of the world.[4]

I note first that R2 would seem to be in tension with R1: If certain special features demarcate the mental as necessarily amenable only to a special, non-scientific understanding, it seems there must be a usefully determinate sense of what kinds of understanding science can offer such that these features are not uncovered by them. But let us now consider R2 in its own right.

Is it true that, if we empty science of a metaphysical fundament and an *a priori* method, that we empty it of its role to provide rational constraints on any other theories (or 'vocabularies')? It is unclear why this should be. If there are extant scientific programmes aimed at understanding the phenomena of mind, knowledge etc. as an integral part of a world that can also be understood biologically and physically, it is unclear how some other kind of understanding of these phenomena – which acknowledges the value of such scientific understanding – can remain autonomous from it.

This in itself might seem simply to lead to a stand-off: On the one hand, we have the friends of those who see extant paradigms in cognitive science as in principle apt to explain everything about the mind – or at least up to (and including) the fact that there will forever remain certain mysteries about it. On the other hand, we have the supporters of autonomous philosophy of mind and epistemology (of NSN) who just 'carry on as before'. A conciliatory response might be that these two lines may illuminate each other, but that there will be no question of replacement either way. Now, since there's lots of good work we still need philosophers to do, I am in some sense very sympathetic to this latter line. However, saying this is compatible with saying that what philosophers are *really* involved in is an inquiry which the scientific conception defines; and also with saying that, to the extent this work does not make sense within the framework of this enquiry, it is *prima facie* a suitable object for scepticism. I think there is reason to uphold the idea of such an asymmetrical dependence of philosophy on science, insofar as philosophers have no good response to scientifically motivated scepticism – something that is underscored by the historical record of science repeatedly 'trumping' philosophy

4 Some might wonder if Rorty's (and, to an extent, Davidson's) 'multi-descriptionist', pragmatist position would be an instance of such a line (cf. e.g. Rorty 1982). I do not think it is. It is not clear to me whether Rorty is really a hard-line naturalist, who thinks all there is really is physical stuff, as understood by physics, and the rest is essentially more or less felicitous word-play; or whether he is a kind of idealist/sceptic who thinks talk of the physical, along with everything else, is also word-play. Either way, however, he is not a non-scientific naturalist, that is, someone who wants to motivate a *robustly realistic* stance on various forms of distinctively *non-scientific* discourse.

– whilst scientists can offer some explanation, at least in principle, of what philosophers are traditionally engaged in.

What do philosophers essentially do, in terms of positive work? In the Anglo-American tradition I am familiar with, the bread-and-butter activity is that of giving definitions or, better, *analyses* of concepts like belief, knowledge, truth, right and so on. (I consider a view that would dissent from such a line below.) I contend that a thoroughly scientific naturalism can account for this kind of exercise and explain its value. Firstly, a supporter of scientific naturalism (henceforth *SN*) can allow that our concepts, whether or not they have real representational significance ('latch on to the world'), are apt to be studied systematically, and *as such* (by which I mean, not merely in terms of their causal history and/or any biological function they might subserve).[5] Such concepts might for example instantiate a complex mental structure whose interconnections we can map out and perhaps explain by reference to deeper underlying principles. This mapping-out would, however, ultimately be conceived as a scientific study into some kind of human cognitive structure *à la* Chomsky's conception of grammar (cf. e.g. Chomsky 1986, chs. 1–2), and not as a kind of insight into any peculiarly philosophical realm.[6] On the other hand, such a study would be able to acknowledge the significance of philosophical work already carried out – in something analogous to the way Christian thinkers could acknowledge the importance of earlier non-Christian philosophy (I don't mean to be understood as subscribing to the Christian reinterpretations). The philosophical work is seen as important, but the new conception of it reveals a need to adjust our understanding of its import.

SN can also allow, and the explain the fact that some of our common sense concepts and our analyses of them can capture significant tracts of *non-mental* reality, without being, at least in the first instance, part of science. Here I draw on ideas of Hilary Kornblith and others (cf. especially Kornblith 2003), who hold that many of our concepts can plausibly be viewed as veiled *natural kind* concepts. According to Kornblith, this is at least the case for *knowledge*: We have made progress at a common sense – if you will, philosophical – level, but in fact the progress we have made is due to our partial grasp of a natural kind, which our intuitive characterisations allow us significantly to latch onto, though without giving a deep understanding of. Kornblith sees contemporary epistemological debates, after having made progress for some decades, as now by and large

5 This does not mean such a study could not illuminate a concept's *correctness* conditions, if it had them (see below in relation to ethical concepts).

6 I talk of a 'philosophical' rather than 'Platonic' realm to avoid broaching issues about abstract entities, the viability of which concept strikes me as thoroughly orthogonal to the current issue.

stagnating. To go further, we need to look at how knowledge *the phenomenon* really is.

(It is noteworthy that this strategy has also been suggested as applicable to moral concepts – cf. e.g. Boyd 1988. This does not seem unreasonable: If morality is both real and a peculiarly human phenomenon, tied to our needs and purposes, then presumably a deeper understanding of these needs and purposes will tell us more about what is truly good. This may be partly achieved through scientific psychology, but also empirical studies of actual political and ethical systems will illuminate what is ultimately 'of value'. It is also worth noting that it would seem to be precisely those who have studied ethics at a philosophical level who would best placed to enrich their understanding through empirical studies; it needn't, in other words, be a case of natural scientists taking over ethics departments. In fact, I do not feel able to endorse the idea that ethical concepts are natural kind ones, a suggestion which, though tantalizing, I still find somewhat unclear. What does seem correct is that it is hard to see a principled reason for denying that empirical studies should substantially impact on our conception of ourselves as ethical beings, and thus also on what we regard as right or wrong to the extent these predicates have cognitive significance.)

In the case of what is often seen as our intuitive grasp of belief-desire-action psychology (which forms at least a central part of our folk psychology – see section 2, above), we have, arguably, a case which illustrates a combination of the two ways so far discussed of understanding the significance of traditional philosophical work. Here there is evidence that humans possess some domain-specific, probably innate understanding of *real* complex interrelations between belief, desire and action. Given this scientific background, one can do justice, not merely to philosophical analysis construed as an empirical study of a conceptual structure, and not merely to the idea of significant, albeit desultory, insights into a natural kind, but also to the idea of a study aimed at eliciting the underlying structure of a systematic common sense understanding of some part of world.

The dialectical situation is then this. From the perspective of SN, much or even most of what supporters of NSN see as distinctive of and valuable in common sense-based philosophy can be accounted for from within a perspective that is true to SN. By contrast, it is hard to see how SNs scepticism to the significance of common sense conceptual practices that are not in some further manner accounted for can be convincingly demurred at or adequately explained away by supporters of NSN – given, after all, that they accept the validity of the scientific practices which tend to engender this scepticism, as well the historical record of science's prevalence over philosophy. On balance, SN would seem to be the more internally consistent position.

In closing, I would like to consider the view that what really distinguishes philosophy from science is the fact that it seeks to given an understanding of our experience of life which is somehow consonant with the mode of that experience itself. Science, and indeed much of what goes under the name of philosophy, especially in the analytic tradition, employs a mode of understanding experience whose adequacy is at best only contingently linked to what it seeks to understand; but real philosophy, it might be said, must not merely account for experience, but seek to articulate it in terms that are themselves recognizably experiential – it must, to put it in a word, be *phenomenologically* valid.

This is too big a topic for me to seek discuss in depth here, but a few closing remarks are worth registering. I would admit that, if music, visual art, poetry, fiction and so on can literally *be* philosophy, then at least some philosophy is thoroughly unscientific, without being in any way non-naturalistic. However, it strikes me that philosophy itself is not identifiable with such arts, even though philosophical views may be communicated by the latter, and even though many works of philosophy are themselves works of art. In seeking understanding, philosophy must objectify in roughly the same way science does.

My view, then, is that naturalism *just is* scientific naturalism (and, moreover, that we should be naturalists). However, this does not sound the death-knell for philosophy, for everything I have said above is quite consistent with philosophers and philosophical activity continuing to play a central role in development of empirical theories; in a word, much of science is 'armchair science'. And in any case, the significance of philosophy is not restricted to such theorising. I have said almost nothing about the ethical concerns that are central to our rational mode of existence, and here it strikes me that exercising philosophical competence will, as ever, continue to be of the essence (I allude here first and foremost to first- rather than second-order ethical problems). Again, however, non-scientific naturalism is not a position we need to hold to justify such a line.[7]

Norwegian University of Science and Technology

---

7 This paper was conceived and written whilst a fellow at the Centre for Advanced Studies (CAS) at the Norwegian Academy of Science and Letters, whose stimulating environment I gratefully acknowledge. Thanks to the audience at CAS who heard and responded to a short version of this talk presented as a lunch seminar. Thanks especially to Jennifer Hornsby, Carsten Hansen, Olav Gjelsvik and Magne Dybvig for helpful comments at a seminar based on the paper. Thanks finally to audiences at Lund University (October 2003) and at the Internordic symposium in Helsinki for their responses.

References

Boyd, R. (1988), "How to be a moral realist", in G. Sayre-McCord (ed.), *Essays on Moral Realism*, Ithaca NY, Cornell University Press.

Burge, T. (1993), "Mind-body causation and explanatory practice", in J. Heil and A. Mele (eds.), *Mental Causation*, Oxford, Oxford University Press.

Cartwright, N. (1983), *How the Laws of Physics Lie*, Oxford, Oxford University Press.

Child, W. (1994), *Causality, Interpretation, and the Mind*, Oxford, Oxford University Press.

Chomsky, N. (1986), *Knowledge of Language*, New York, Praeger.

Crane, T. (1991), "Why indeed? Papineau on supervenience and the completeness of physics", *Analysis*, 51.

Davidson, D. (1971), "Mental events", reprinted in his *Essays on Actions and Events*, Oxford, Oxford University Press, 1980.

Davidson, D. (1987), "Problems in the explanation of action", in P. Pettit, R. Sylvan and J. Norman (eds.), *Metaphysics and Morality: Essays in Honour of J. J. C Smart*, Oxford, Blackwell.

Davies, M. (1994), "The mental simulation debate", *Proceedings of the British Academy*, 83, 99-127.35-49.

Dupré, J. (1993), *The Disorder of Things: Metaphysical Foundations of the Disunity of Science*, Cambridge MA, Harvard University Press.

Feldman, R. (2001), "We are all naturalists now", ms., delivered at APA meeting 2001 (available at http://www.ling.rochester.edu/~feldman/papers/naturalism.htm).

Fodor, J. (1974), "Special sciences", reprinted in his *Representations* (1981), Cambridge MA, MIT Press.

Fodor, J. (1987), *Psychosemantics*, Cambridge MA, MIT Press.

Fodor, J. (1990), "Making mind matter more", in his *A Theory of Content and Other Essays*, Cambridge MA, MIT Press.

Hempel, C. (1965), "Reasons and covering laws in historical explanation", in S. Hook (ed.), *Philosophy and History*, New York, New York University Press.

Henderson, D. (1993), *Interpretation and Explanation in the Human Sciences*, Albany, State University of New York Press.

Hornsby, J. (1997), *Simple Mindedness*, Cambridge MA, Harvard University Press.

Hornsby, J. (forthcoming), "Physicalism, conceptual analysis and acts of faith" (in collection on Jackson 1998).

Haack, S. (1993), *Evidence and Inquiry*, Oxford, Blackwell.

Jackson, F. (1998), *From Metaphysics to Ethics: A Defence of Conceptual Analysis*, Oxford, Oxford University Press.

Knowles, J. (2001), "Does intentional psychology need vindicating by cognitive science?" *Minds and Machines* 11, 347-377.

Knowles, J. (2002), "Is folk psychology different?", *Erkenntnis* 57, 199-230.

Knowles, J. (2002a), "Actions and events in common sense and scientific psychology: some critical reflections on Hornsby", ms., presented at the Birkbeck College Philosophical Society, University of London, May.

Kornblith, H. (2003), *Knowledge and its Place in Nature*, Oxford, Oxford University Press.

McDowell, J. (1985), "Functionalism and anomalous monism", in B. McLaughlin and E. Lepore (eds.), *Actions and Events*, Oxford, Blackwell.

McDowell, J. (1994), *Mind and World*, Cambridge MA, Harvard University Press.

McGinn, C. (1979) "Problems in the explanation of action", in N. Bolton (ed.), *Philosophical Problems in Psychology*, London, Methuen.

Morris, W. & R. Richardson (1995), "How not to demarcate cognitive science and folk psychology: A response to Pickering and Chater", *Minds and Machines* 5, 339–355.

Quine, W. V. (1953), "Two dogmas of empiricism", in his *From a Logical Point of View*, Cambridge MA, Harvard University Press.

Quine, W. V. (1969), "Epistemology Naturalised", in *Ontological Relativity and Other Essays*, Cambridge MA, Harvard University Press.

Rey, G. (2001), "Physicalism and psychology: A plea for a substantive philosophy of mind", in C. Gillet and B. Loewer (eds.), *Physicalism and its Discontents*, Cambridge, Cambridge University Press.

Rorty, R. (1982), *Consequences of Pragmatism*, Minneapolis, University of Minnesota Press.

Is Naturalism Question-Begging?

Markus Lammenranta

Naturalistic epistemology relies on the results of psychology and other empirical sciences in trying to solve epistemological problems. This appeal to science would be condemned as circular if the task of epistemology were to justify our beliefs about the external world, or to show how we can have any knowledge about the world at all. Of course, we cannot rely on science in accomplishing these tasks. Science relies essentially on perceptual evidence concerning matters in the external world, and the point in question is whether this sort of evidence is to be trusted.

Naturalists think that these tasks are connected to a certain conception of knowledge and justification that can be called classical Cartesian foundationalism. According to this conception, we have direct, non-inferential knowledge only about the contents of our own mind and obvious necessary truths. All other knowledge is inferential and must be inferred from this sort of basic knowledge. This conception rationalizes the attempt to justify our beliefs about the external world allowing only premises based on introspection and rational intuition. If it were successful, it would also explain how we can have any knowledge about the world at all.

There are two responses to Cartesian foundationalism corresponding to two ways of naturalizing epistemology. One of them is given by W. V. Quine (1969) who argues, in his classical paper "Epistemology Naturalized", that the Cartesian demands for knowledge cannot be satisfied. The quest for finding a certain foundation for knowledge is a lost cause. Yet, he says, epistemology goes on. It "simply falls into place as a chapter of psychology and hence of natural science" (p. 82). Instead of trying to explain how experience justifies our beliefs about nature, it describes how sensory experience – or rather the stimulations of sense organs – causes our beliefs. The question of justification is replaced by the question of the causal origin of beliefs.

Quine's (1969, p. 74) reaction can be read as a concession to Cartesian skepticism. Because we cannot justify our beliefs, the Humean predicament remains unaltered. The other naturalistic reaction denies the skeptical moral. The Cartesian skeptic is wrong. We do have knowledge and justified beliefs about the external world. Skepticism fails because it relies on false epistemology. Rather than accept skepticism, we should reject classical foundationalism. Indeed, since the sixties, the main project in analytical epistemology has been to give an account of knowledge and justification that eases our skeptical worries.

The first naturalistic accounts replaced the traditional justification condition of knowledge with a naturalistic one. What makes true belief knowledge is – rather than justification – some natural relation between belief and truth. An accidentally true belief cannot be knowledge. In order to be knowledge, a belief must be connected to reality by a reliable causal or modal relation.[1]

As a second step, under the lead of Alvin Goldman, the justification condition was retained but understood in reliabilist terms. Goldman (1986, pp. 58–121) gave a reliabilist account of justification that can be expressed roughly as follows:

> A belief is justified if and only if it is produced or sustained by sufficiently reliable psychological processes.

This account does not require that we have arguments for the truth of our beliefs, or that we can justify them in any way. It just requires that they be appropriately caused, whether we know it or not.

We thus have two different ways of naturalizing epistemology. Following Goldman (1999, pp. 2–4), we may call them scientific naturalism and moderate naturalism:

> Scientific Naturalism: Epistemology is a branch of science, and uses the empirical method of science.

> Moderate Naturalism: (A) All knowledge and justification is a function of psychological processes that produce and sustain belief. (B) Epistemology needs help from science, especially cognitive science.

The latter position, accepted by Goldman, is more moderate, because it preserves a role for philosophical epistemology that is independent of empirical research. The task of this part of epistemology is to specify the content and criteria of epistemic evaluation. Empirical science helps then in the attempt to decide whether the criteria are satisfied.

The standard critique of Quine and scientific naturalism, by Goldman and others,[2] is that this sort naturalization of epistemology rules out the normativity of epistemology. Psychology and other empirical sciences are descriptive in content. No branch of science takes on the normative task of specifying the criteria of epistemic evaluation.

---

1 See, for example, Goldman (1967) and Nozick (1981, pp. 172–196).

2 See Goldman (1986, pp. 2–3; 1999, p. 2), Kim (1988, pp. 388–391), and Kornblith (1985, pp. 3–7).

Quine (1998, pp. 664–665; 1990, pp. 19–21) has since rejected this interpretation of his project. He insists in his more recent work that naturalistic epistemology does not rule out the normative dimension of epistemology. Insofar as theoretical epistemology gets naturalized into a chapter of theoretical science, normative epistemology gets naturalized into a chapter of engineering:

> It is the technology of truth-seeking... There is no question here of ultimate value, as in morals; it is a matter of efficacy for an ulterior end, truth or prediction. The normative here as elsewhere in engineering, becomes descriptive when the terminal parameter is expressed. (1998, p. 665.)

Let us assume that the terminal parameter is truth. Then the descriptive ground of normativity is truth-conducivity or reliability. If we want truth, we should use reliable processes and methods. The most important norm of Quine's naturalized epistemology that we should rely on the senses in testing our theories derives from the finding of science that our information about the world comes only through the senses. Science tells us that our senses are reliable.

There is thus a common core in the two types of naturalistic epistemology: the question of the reliability of our sources of belief. The difference is that while scientific naturalism raises the question for guiding inquiry, moderate naturalism raises it for determining the scope of knowledge and justified belief.

They also have a common problem: in order to determine the reliability of our sources of belief, they rely on science and thus on these very same sources. It is often claimed that this makes the project viciously circular. To evaluate the charge, I will distinguish between circularity and begging the question. The former is a flaw in reasoning, the latter is a fault in a dispute. When we make this distinction, we can see that naturalists are right when they say there need not be any vicious circularity in their reasoning, but that they are wrong if they think that they can avoid the dialectical fault of begging the question without serious costs. Finally, I will evaluate what consequences this has to the naturalistic project.

When we evaluate the reliability of our sources of belief, we must rely on reasoning that at some point or another depends on premises that are themselves based on the sources in question. William Alston (1989) calls this sort of reasoning epistemically circular. For example, in order to show that sense perception is reliable, we may use the following inductive argument:

At t_1, S_1 formed the perceptual belief that p_1, and p_1.
At t_2, S_2 formed the perceptual belief that p_2, and p_2.
⋮
At t_n, S_n formed the perceptual belief that p_n, and p_n.
Therefore, sense perception is reliable.

There is no formal circularity, because the conclusion does not appear as one of the premises. But because we must use perception to determine the truth of the premises, we presuppose that sense perception is reliable when we use this argument to justify the conclusion. The argument does not justify the conclusion unless we are justified in accepting the premises, and we are not justified in accepting the premises unless sense perception is reliable. So the truth of the conclusion is a necessary presupposition for our being justified in accepting premises. This is what, according to Alston, makes the argument epistemically circular.

Yet, Alston argues, epistemic circularity need not make our reasoning defective. If sense perception really is reliable, we are justified in accepting the premises; and if inductive reasoning is also reliable, we are justified in drawing the conclusion from the premises. Goldman's position in *Epistemology and Cognition* (1986, pp. 118–121) is similar. He points out that it is logically possible to apply the criterion of reliability justifiably, and thus to determine which psychological processes are reliable, because it is logically possible that we have cognitive processes relevant to the task and that they are sufficiently reliable.

For Quine, who does not accept reliabilism, the situation is slightly different. Science tells us how sensory stimulation causes our scientific beliefs about nature, and because these beliefs compose our very science, it tells also that these beliefs are true, and that sensory perception is thus reliable.

There is no problem of vicious circularity, because Quine looks at things from within science. Science tells both how things are in the world and how science itself is so successful in describing the world. If there is any internal conflict between these two stories, coherence can be attained by revising either of them. We are like sailors who can only rebuild their intellectual ship at sea while floating on it, as Quine's favorite simile from Otto Neurath puts it. Sometimes, also Goldman (1986, p. 3) appeals to this Neurathian picture. The message is that if we want to attain a broadly coherent view of the world including ourselves in it, circularity is no problem. It helps to achieve the aim.

Both responses show that there need be nothing viciously circular in the reasoning process as such. The problem is rather that somebody else can use similar reasoning in defending the reliability of some alternative sources of belief. That person can also point out that it is logically possible that these sources are reliable, or construct a system of beliefs that is equally comprehensive and coherent. How can naturalists then defend their list of reliable processes against that person without begging the question? So the problem is dialectical rather than some flaw in reasoning.

Let us take a supernaturalist system as an example. William Alston (1991) argues that a theist can defend the reliability of religious experience in the same way as a naturalist defends the reliability of sense perception. Alvin Plantinga (1993, ch. 12; 2002) argues that a theist can boost the coherence of her system of belief even further by being able to explain how her sources of belief are so reliable. This is because she has been created in the image God. This divine origin guarantees the reliability of her faculties. Indeed, Plantinga thinks that a naturalist cannot attain a comparable degree of coherence.

How does a naturalist then defend herself against this sort of a supernaturalist? The naturalist may say that the sources of belief substantiated by scientific research really are reliable, while those substantiated by theistic religion are not. The naturalist epistemology is true, she may say, while the supernaturalist epistemology is false. There simply is no God. So beliefs about God based on religious experience are false, and religious experience is not a reliable source of information.

It is not difficult to see that this response begs the question against the supernaturalist. It just assumes that naturalistic ontology is true without arguing for it. Of course, the naturalist could defend her ontology in terms of her epistemology, but then in order to defend her epistemology she has to appeal to her ontology once again. Circularity – whether formal or epistemic – is surely vicious in this kind of dialectical context. Circular arguments cannot convince an opponent who does not already accept the conclusion.

How damaging is this for the naturalist? No more damaging than for the supernaturalist. Neither can defend her position against the other without begging the question. How should we then react to this sort of irresolvable dispute?

Some may think that all genuine disagreements are resolvable in principle. So perhaps we can say that both parties in this sort of dialectical position are right and thus not in genuine disagreement. Both are right relative to their own standards or conceptual schemes. But this would mean that both the naturalist and the supernaturalist should give up their central metaphysical commitments and become relativists or antirealists. That would be damaging for both.

Equally damaging would be the option of suspending judgment about matters under irresolvable dispute. This would also mean that both give up their positions. To avoid these destructive options, the naturalist could, perhaps, simply point out that once there is a fundamental disagreement, we have no choice but to go ahead and beg the question.[3] What else could we do but rely on our own deepest convictions? So she simply claims that she is right and even knows this, and that anybody who disagrees is wrong. To be sure, this is a dogmatic position. But, perhaps, there is nothing wrong with dogmatism.

There is a less dogmatic strategy that appeals to Neurath's metaphor. A naturalist thinks that, once the Cartesian dream of certainty is discarded, there is no choice but to work within our current science and current beliefs and to correct them when conflicts arise. This is to settle for appearances, for how things appear within our science and our beliefs. And it does not beg the question against the

3 This seems to be Roderick Chisholm's (1982, p. 75) suggestion.

supernaturalist who is allowed to keep her own appearances. This option resembles the decision of the mature Pyrrhonian skeptic to live by appearances.

There are thus two ways of responding to the dialectical problem that are open for the naturalist: the dogmatic one and the skeptical one. Neither may be really attractive, but they are the only ways of preserving the naturalist position. The naturalist had better learn to live with them – or stop being a naturalist.

Academy of Finland

References

Alston, W. P. (1989), "Epistemic Circularity", in *Epistemic Justification*, Ithaca, Cornell University Press.

Alston, W. P. (1991), *Perceiving God*, Ithaca, Cornell University Press.

Chisholm, R. (1982), *The Foundations of Knowing*, Brighton, The Harvester Press.

Goldman, A. (1969), "A Causal Theory of Knowing", *The Journal of Philosophy* 64, 357-372.

Goldman, A. (1986), *Epistemology and Cognition*, Cambridge, Harvard University Press.

Goldman, A. (1999), "A Priori Warrant and Naturalistic Epistemology", in J. E. Tomberlin (ed.), *Philosophical Perspectives* 13, *Epistemology*, Oxford, Blackwell, pp. 1-28.

Kim, J. (1988), "What Is 'Naturalized Epistemology'?", in J. E. Tomberlin (ed.), *Philosophical Perspectives* 2, *Epistemology*, Atascadero, Ridgeview Publishing Company, pp. 381-406.

Kornblith, H. (1985), "Introduction: What Is Naturalistic Epistemology?", in H. Kornblith (ed.), *Naturalizing Epistemology*, Cambridge, The MIT Press.

Plantinga, A. (1993), *Warrant and Proper Function*, Oxford, Oxford University Press.

Plantinga, A. (2002), "Introduction: The Evolutionary Argument against Naturalism", in J. Beilby (ed.), *Naturalism Defeated? – Essays on Plantinga's Evolutionary Argument against Naturalism*, Ithaca, Cornell University Press.

Nozick, R. (1981), *Philosophical Explanations*, Cambridge, Harvard University Press.

Quine, W. V. (1969), *Ontological Relativity and Other Essays*, New York, Columbia University Press.

Quine, W. V. (1998), "Reply to Morton White", in E. Hahn and P.A. Schilpp (eds.), *The Philosophy of W.V. Quine*, Chicago, Open Court.

Quine, W. V. (1990), *Pursuit of Truth*, Cambridge, Harvard University Press.

Naturalism: Hard and Soft

Pentti Määttänen

W. V. O. Quine's philosophical naturalism is based on natural science. Epistemology is, on this view, only a chapter of psychology understood as a natural science, hard science, so to say. This naturalism might be called hard naturalism. Of course, Quine was not the first one to talk about naturalism. John Dewey's earlier version of naturalism was based only on the assumption that we are products of natural and cultural evolution and, therefore, we don't have any supernatural capacities for cognition or acquiring knowledge. Therefore, no *a priori* methods are available, but this does not include commitments to any particular method of science or to any type of science. This is why it might be called soft naturalism. In what follows I shall try to point out some inconsistencies in Quine's naturalism and discuss some problems from the viewpoint of soft naturalism.[1]

1. Hard Naturalism: W. V. O. Quine

Quine's definition of naturalism is problematic because of its emphasis on natural science. On one hand, the move that epistemology is a part of psychology entails that epistemological problems are empirical problems – which is what one would expect of naturalism. But on the other hand, what empirical evidence there is for the claim that epistemology is or should be a part of *natural* science? It seems to be an *a priori* decision that all empirical problems should be dealt with the methods of natural science, although these methods have not been so successful in social sciences and humanities. And *a priori* decisions are *not* what one would expect of naturalism. It would be more consistent to say that each science has its problems and is allowed to use any methods that prove to be useful in solving these problems. The basic idea of naturalism, the view that there is no need to appeal to supernatural elements in explaining human knowledge, does not require this kind of commitments to natural science.

Quine is often characterized to be a pragmatist. His philosophy is, however, quite different from the classical pragmatism. He calls Peirce "an old empiricist" and analyses experience in terms of "stimulus-meaning" quite in the spirit of classical empiricism (Quine 1969, p. 79). The practical bearings mentioned in Peirce's

---

1 This paper is written in the project SA 201402.

pragmatic maxim are changed into evidence (ibid., p. 80) which is specified to consist of stimuli and observation sentences just as an old empiricist (but not Peirce) would have it. Quine ignores completely the basic idea of Peirce's pragmatism, which is: We must find out how action changes the concept of experience; "the concept of experience is broader than that of perception, and includes much that is not, strictly speaking, an object of perception" (CP 1.336). Experience consists of perception and action, and the appeal to action is the point from which pragmatism got its name in the first place.

The core idea of Quine's naturalism seems to be empiricism combined with the (aprioristic) assumption that the methods of natural science are the only adequate way to do any scientific research. This commitment to a method is something that Dewey called the supremacy of method (Dewey 1984, ch. 9). The problem of this approach is that if one method is taken to be *the* method, then this method determines beforehand what kind of problems are judged to be scientific, namely those problems that are thought to be solvable by this method. Issues that seem unfit to such a "rigorous science" are doomed to be unscientific. This is why Quinean naturalists want to get rid of the so-called folk-psychology and stick to neuroscience. The mental should be reduced to the physical, to the brain, and so it is "explained away" (Quine 1995, p. 86). All sciences referring to mental contents are changed into natural science referring to neural mechanisms. This kind of reduction of the mind to the brain is quite common in cognitive and neuroscience (see Bennett and Hacker 2003).

The attempt to "explain away" by reduction is problematic. The model of theoretical reduction is taken from natural science, and the common example is the reduction of the caloric theory of heat into a theory about molecular movement. However, the reduction of folk-psychology into neuroscience cannot be compared to this example. One reason is that folk-psychology is not a theory (ibid., pp. 367–370).

Theories typically refer to entities that are outside the scope of observation and common experience (like caloric fluid or phlogiston). To assume that folk-psychology is a theory that refers to non-observables entails that psychological predicates are attributed to something else than human persons behaving in a certain way. Quinean naturalists typically attribute them to the brain. They say that the brain thinks, has beliefs and desires, makes decisions, gives orders to the body and so on. This is what M. R. Bennett and P. M. S. Hacker call the mereological fallacy (ibid., pp. 68–107). However, it is not the brain that thinks, but a human being thinks with the brain, just like it is not the legs that run, but a human being runs with the legs. The brain will think in a vat just as well as a pair of legs cut off from a body will run. Actually, as Bennett and Hacker argue (ibid., pp. 111–114), the identification of the mind and the brain is a form of Cartesianism: what René Descartes attributed to the soul is attributed to the brain. The framework of the formulation of the problem of mind is Cartesian (as explained also in Määttänen 1995, ch. 5). So, naturalists trying to explain folk-psychology away in fact adopt a Cartesian view about the mind – which is not what one would expect of a naturalist.

What actually takes place in folk-psychology is that psychological predicates are attributed to human beings behaving in a certain way. The behavior of people is understood by attributing to them predicates like beliefs and desires. The important point is that behavior is the criterion for attributing these predicates, nothing else. Folk-psychological predicates are predicates of behaving people, and this behavior is something that can be observed. It is part of our everyday experience.

If psychological predicates like beliefs and desires are really dropped away (like the predicates 'caloric fluid' and 'phlogiston') we have no behavior left. Folk-psychology has no concepts to attribute to anything, and it is with the folk-psychological predicates that we understand people's activities as behavior. If these predicates are replaced by neurological predicates we simply don't understand people's activities as behavior. We might have descriptions like: "Some C-fibers were firing which fact caused the body to move its organs to the effect that some lumps of molecules were transferred inside the body through a hole." But we don't understand from this that a person was eating because she was hungry.

Actually, even neuroscience does not get on without folk-psychological predicates. These predicates belong as a built-in feature to the method of neuroscience. The only way to find out that some neural process can be connected to hunger is to correlate this process with the use of the folk-psychological predicate 'hungry', and this is in fact what neuroscience does, or tries to do. For example, someone's brain is stimulated and the person reports that she is hungry. It is always the case that some human being, the scientist or the test subject, interprets the external part of the correlation, and if it is about a mental predicate, that human being uses her own folk-psychological predicates to interpret the situation, which interpretation is then correlated with that neural process. Neuroscience makes use of folk-psychological predicates in research.

Even if folk-psychology could somehow be reduced away, this reduction would not be the kind of reductionism that takes place in natural science. Take the example of the caloric theory of heat. It is a fact the science of physics does not need anymore the caloric theory, but heat is not explained away. Heat is just understood in a new way. It is not the case that we get rid of the term 'heat' when we talk about the movement of molecules instead of caloric fluid. The predicate 'hot' is attributed to an object that is experienced to be hot, and the predicate 'hungry' is attributed to a person who is observed to be eagerly eating her meal. Theoretical reduction in natural science does not even try to get rid of the predicate 'hot', and neither should anybody try to get rid of predicates like 'hungry'. We understand heat better when we talk about molecular movement rather than about caloric fluid, but if we drop altogether predicates like 'hungry' our understanding of human behavior is gone.

And is the behavior of a hungry person better understood if we know that certain neural process takes place in her head? Not necessarily. Actually we know beforehand that some nerves are firing in her head. What use could other than neuroscientists have with the additional information of the exact location and nature of her brain processes in understanding her behavior? Probably nothing, except in the case that those neural mechanisms are malfunctioning. Then we might understand that some odd behavior is caused by these defects. What could a neuroscientist do with the information? Probably a lot, but then she is interested of something different. The scientist is not interested of the question why a person went to a restaurant. The scientist takes it as it is, as a simple fact. The scientist is interested of the role of neural processes in performing this behavior, and this is a completely different issue to understand.

2. Soft Naturalism: John Dewey

A looser version, a soft version of naturalism is based only on the view that we are products of natural and cultural evolution, and everything we have and are capable of, including language, logic and mathematics, should be explained on empirical grounds.

In John Dewey's soft naturalism culture is a product of nature; it is a product of biological organisms being in continuous interaction with their environment. From this point of view culture is not something distinct from and opposed to nature. The natural and cultural features and capacities belong to the same creature and have both evolved down here, on planet Earth. As cultural beings we cannot have access to any non-natural sphere of being. Our ability to think and use natural language should be explained without any reference to such modes of being. The method of classical philosophy, that of aiming at eternal conceptual truths completely independent of any empirical knowledge, is useless in naturalism. Naturalism is not the end of philosophy, however, because traditional philosophical problems do not disappear. The solutions of these problems just are not intended to be eternal conceptual truths that are completely independent of empirical experience. The solutions are always hypothetical and tentative. Future experience can, in principle, force us to change our present conceptions. This is what Charles Peirce called fallibilism.

From a methodological point of view this entails that everyday experience, science and philosophy are not completely different ways of acquiring knowledge. The task of philosophical naturalism is to explicate a general methodological framework that can be applied to all these fields. One candidate for this framework is pragmatism, which was defined as the science of abduction by Charles Peirce. Shortly speaking, abduction is search for explanations. These explanations are typically conceptions about the real conditions of experience, of perception and action. We encounter a certain kind of experience, and we have reason to believe that if the world were such and such, then such an experience would take place. Therefore we can assume that the world is such and such. And as we know from the logic of hypothetico-deductive inference, there are always several possible explanations. This is why a naturalist cannot claim to have reached any one and only eternal truth about the world. It is always possible, in principle, that an alternative explanation is found. The evaluation of these explanations is based on the success of future action that is performed on the ground of them. Habits of action form an essential part of the abductive method, and this is why Peirce called them *logical* interpretants.

One specific consequence of this approach is that there is no particular method that would take the dominant position as in Quine's hard naturalism. A consistent naturalist cannot claim to have too much to say without some acquaintance with actual scientific research and its specific problems and methods. Conceptual analysis of philosophical problems should be sensitive to empirical findings. As an example of the dangers of insensitive discussion one might take the claim that the brain is an organ for manipulation of symbols. The so-called Language of Thought hypothesis maintains that the brain computes symbols according to certain rules. Empirical brain research has not, however, found anything of the kind. According to Merlin Donald (2001, p. 102) it can perhaps be said that the brain is a computing device, but only in the sense of analogue computation. And analogue devices do not employ symbols.

Naturalist philosophers, however, have still much to do in clarifying their use of philosophical terms. Metaphysics is a good example. There are at least two senses of this word that we must be aware of. First, metaphysics can mean a general conception about the basic features of reality. Second, empiricists have often used the word in a negative sense, holding that metaphysical views are not based on experience, that they are unscientific. This negative attitude is quite justified in the face of the history of philosophy. Metaphysical views are often presented as an outcome of *a priori* conceptual reasoning based on some kind of non-empirical access to objective ideas, immaterial and immutable meanings or other abstract and independent entities. If some form of aprioristic epistemology is taken as an essential feature of metaphysics, then the expression 'naturalist metaphysics' is a contradiction in terms. A naturalist is committed to the view all genuine knowledge is, in the last instance, based on experience. There is no *a priori* epistemology. Indeed, how could natural beings have achieved supernatural powers for *a priori* reasoning?

It is, of course, possible to use the term '*a priori*' in a relative sense. Konrad Lorenz used the expression 'biological *a priori*' referring, for example, to the fact that our biological structure to some extent determines what kind of colors we can perceive. It is a kind of *a priori* condition of our experience here and now. But in the final instance the theory of evolution has the task of giving an empirical explanation of why and how our ability to perceive colors emerged.

It is also useful to carry on some conceptual and theoretical analysis before going to look for empirical information, and concepts and theories are needed for the interpretation of experience. In this relative sense conceptual conditions are sort of *a priori* conditions of experience. But in the final instance the explanation of how we got our ability to use natural language, logic and mathematics must be empirical. Natural language is natural, not supernatural.

A naturalist may, of course, use the term '*a priori*' in this relative sense, but there is always the risk of misunderstanding. The same holds for the term 'metaphysics'. It is perfectly possible, for a naturalist, to discuss the basic features of reality in general without committing oneself to aprioristic epistemology in the classical sense of the expression. But the possibility of misunderstandings should always be taken into account.

There are similar difficulties with the word 'transcendental'. It is possible to follow Immanuel Kant and say that experience is transcendental if it is directed not to the objects but to the ways of experiencing the objects (*KdrV*, B 25). Transcendental philosophy would then simply be "an inquiry into the human cognitive capacity" (Pihlström 2003, p. 95). But what kind of inquiry? According to Quine epistemology should be a chapter in empirical psychology, which is, indeed, an inquiry into the human cognitive capacity. Kant, however, continues the definition by a qualification: a transcendental doctrine deals with the question how experience is *a priori* possible. And Kant's *a priori* is not relative in any sense. It is the *a priori* laws that make the empirical nature possible, in the first place (*KdrV*, A 216, B 263). The pure synthesis of understanding is the *a priori* basis of the empirical (*KdrV*, B 140). In other words, nature is for Kant constituted by the pure forms of sensibility, pure concepts and categories of understanding, by pure synthesis that comes from over and above nature.

A naturalist is committed to the view that human understanding with its cultural capacities is a product of nature. The very possibility of nature is not given by *a priori* features of human understanding, even if it is a fact that features of human understanding affect the way we experience nature. Nature, the real world, is what it is quite independently what we say and think about it or how we experience it. On the other hand, the way we experience it depends not only on the real features of nature. It depends also on us, but we, as products of nature, possess nothing that comes from over and above nature. An empirical realist need not be a transcendental idealist in Kant's sense.

A misleading dichotomy between culture and nature – which dichotomy may partly be a consequence of the domination of Quinean hard naturalism – may lead to a conception according to which culture as symbolic activity indeed is over and above of nature and thus transcendental in a certain sense. Combined with a view that human cognition is basically determined by culture we get the conclusion that cultural practices form a transcendental basis of our experience.

Richard Rorty seems to be of this opinion. He writes papers like "The World Well Lost". The reason for the loss is the fact that nature is always experienced as being ordered in a certain way. For Kant the structure and the order of things in nature are given by pure forms, concepts and categories. Rorty apparently thinks that language does the job. "But nature undescribed in any human language is simply Kant's unknowable thing-in-itself – an utterly useless notion, the plaything of philosophical skeptics, a toy rather than a tool" (Rorty 1997, p. 17). Then he sums up: "the more one thinks about language, the less need there is to think about nature" (ibid.). The question that he does not pose (let alone answer) is how we got the ability to use language in the first place.

Rorty is right in maintaining that nature "under a description will always exhibit an order" (ibid.), but it does not follow that order is given *only* by some description in a human language. We as human beings experience us as living bodies that are distinguished from tables, trees and so on. Is this order of experience a product of language, a product of a description in a human language? Then what is language? We can distinguish between letters, words and sentences. Where did this ability come from if not from the previous ability to perceive order in nature? Concepts? Then what are concepts? The celebrated phrase that the reality is conceptually categorized, that concepts carve up the world and give it its structure is in deep trouble as soon as we ask what the concepts are independently of our being biological organisms in nature, that is in a world of middle-sized three-dimensional objects. What are concepts independently of our ability to distinguish between letters, words and sentences? Note that the usual expression "concepts are ways of categorizing the world" does not help here because the result "the world is categorized by ways of categorizing the world" is not particularly informative.

And from where does the order of the conceptual realm come from? There seem to be two alternatives. The nature just is ordered in some way and our knowledge must accommodate to that order, or the concepts just are ordered in a certain way and the nature is carved up accordingly by some unknown mechanism that nobody has succeeded to describe in any detail. Those who think that the nature is carved up according to our concepts should give some explanation of how this takes place and what concepts are independently of our already being products of nature.

While waiting for the answer it is good to notice that this line of thought is not the only alternative. Soft naturalism maintains, following John Dewey, that culture is a product of nature. Cultural, linguistic, discursive practices are practices of living organisms in constant interaction with their environment. We are embodied beings capable of using natural language, but this does not entail that we are embodied beings by virtue of our being capable of using natural language. What would that even mean? Soft naturalism does not need help from any kind of idealism, transcendental or whatever.

Cultural practices are not over and above nature, meanings are not ideally existing abstract entities outside the scope of empirical experience. "Meaning is use" is quite a simple principle, and from it follows that meanings exist as ongoing cultural practices in which embodied beings use different kinds of sign-vehicles in different contexts. Meanings are cultural practices of a certain kind of biological organisms and as such they are within the reach of empirical investigation (for some details see Määttänen 1993). These practices are there open for anyone to find out by empirical methods even though, for example, the average use of a word in a linguistic community cannot be immediately observed. No supernatural powers are needed for that. In other words, meanings are not transcendental conditions of experience in Kant's sense, that is, in the sense that meanings were absolute preconditions of the constitution of nature. Meanings emerge inside nature. Nature is the factual precondition of our development into cultural beings that can understand meanings.

The ultimate background of Kantian transcendental conditions is René Descartes who split the world in two substances, material and mental. From this viewpoint it is possible to think that we as thinking subjects are somehow inside the immaterial substance, and that the nature is somehow out there. A Rortyan conception of language as the only tool that gives structure to experienced nature, the conception that we as thinking beings are somehow inside language, continues this tradition. And so does Hilary Putnam's internal realism, which maintains that concepts, which are (somehow) inside something, carve up (somehow) the (somehow) external world into experienced objects.

This Cartesian dichotomy of internal and external looses its strength as soon as we realize that cognition is always embodied. The cognizing body is already in here, within this material world. External to the Cartesian mind is not the same as external to the body. And the only way to conceive the body (and the brain, of course) is to conceive it as a product of nature. Cognition has emerged and takes place within nature.

Rorty rejects explicitly the significance of embodiment by underrating those who slam a table and claim that they are expressing something significant with that. (By the way, with this statement Rorty takes distance to the pragmatism of Peirce who writes repeatedly about muscular effort and resistance.) According to Rorty to slam a table is not a way to get outside language. Those who claim so are essentialists who, erroneously, think that they can have access to an object's intrinsic features. An antiessentialist can pick up an object only as articulated by a certain set of sentences (Rorty 1999, p. 56).

It is, however, a mistake to call a proponent of embodied experience an essentialist. By tapping some table one gets knowledge not about the table's intrinsic features but about certain objective and real conditions of action, that is, one comes to know that muscular effort meets resistance. And that's all there is to it. One way to express the idea is to say that by tapping a table one gets knowledge about *interactional* features instead of intrinsic features. It is hard to take this aspect of experience into account if one ignores our embodied existence and thinks that language is the only vehicle of thought and the only tool that gives structure and order to the experienced world.

Peirce did not share this view. According to him habits of action are beliefs, and this holds also for bodily behavior. Most of us have the habit of using the door and not the window when exiting a room. This habit is a structured series of acts, and this structure accommodates to the real and objective structure of the world quite forcefully. A necessary but not transcendental condition of our survival is the accommodation of our habits to the real structure of nature, to the real conditions of action as opposed to the real but unperceivable causes of perceptions, as "an old empiricist" (but not Peirce) would have it.

Also discursive habits are habits of living organisms. Not a word is emitted without some bodily behavior. I don't see any serious alternative to the view that we are embodied beings in continuous interaction with our natural and cultural environment. The limits of our experience are not limits of language; they are limits of this interaction.

University of Helsinki

References

Bennett, M. R. and P. M. S. Hacker (2003), *Philosophical Foundations of Neuroscience*, Oxford, Blackwell.

Dewey, J. (1984), *The Quest for Certainty: The Later Works 4*, ed. by J. A. Boydston, Carbondale and Edwardsville, Southern Illinois University Press.

Donald, M. (2002), *A Mind so Rare: The Evolution of Human Consciousness*, New York, Norton.

Kant, I. (KdrV), *Kritik der reinen Vernunft*, Frankfurt, Suhrkamp, 1968.

Määttänen, P. (1993), *Action and Experience: A Naturalistic Approach to Cognition*, Annales Academiae Scientiarum Fennicae B 64, Helsinki.

Määttänen, P. (1995), *Filosofia. Johdatus peruskysymyksiin* [Introduction to Philosophy], Helsinki, Gaudeamus.

Peirce, C. (CP), *Collected Papers*, ed. by Charles Hartshorne and Paul Weiss, Cambridge, Harvard University Press, 1932–1958.
Pihlström, S. (2003), *Naturalizing the Transcendental: a Pragmatic View*, Amherst, Humanity Books.
Quine, W. V. O. (1969), *Ontological Relativity and Other Essays*, New York, Columbia University Press.
Quine, W. V. O. (1995), *From Stimulus to Science*, Cambridge, Harvard University Press.
Rorty, R. (1997), *Truth, Politics and 'Post-Modernism'*, Assen, Van Gorcum.
Rorty, R. (1999), *Philosophy and Social Hope*, London, Penguin Books.

Early Cognitive Science – A Challenge to Analytic Philosophy?

Ahti-Veikko Pietarinen

1. Introduction

In her masterful historical exposition, Liliana Albertazzi (2001) recently revealed a rich conceptual repertoire and thematic complexity in early Central-European theories concerning the accounts of cognitive science developed in 1870–1930. These theories, she argues, anticipated much of the content of the contemporary cognitive sciences as well as many of the modes of research characteristic of current mainstream research in cognitively-oriented sciences and philosophy.

I wish to substantiate and supplement these findings by pointing out that a revival of these early endeavours in developing conceptual categories of the mind and psyche is currently underway at the intersections of artificial intelligence (AI), logic and cognitive science. The cases in point are the following. (i) Instead of predicates, many logical theories, especially computational ones dealing with practical reasoning in rational agenthood, have preferred iconic – pictorial, visual, graphical and spatial – modes of representation in order to capture different qualities associated with assertions and objects. (ii) A related tendency to (i) is the dispensing with the traditional notion of logical constants and the inclination to substitute an assortment of iconic, topological and similarly Gestalt-oriented notions of dynamic theories of action and experimentation concerning the relations involved in the representations. (iii) Instead of the accustomed vernacular that speaks of individuals, scientists have resorted to the more phenomenological terminology of objects of cognition and consciousness across a wide range of inquiry.

Largely as a result of investment in the philosophical foundations of the computational sciences, early European theories are beginning to be discernible in embodied and enactive minds, open-systems approaches to verification, the refinement and composition of computational processes and programs, as well as in the interactive, emergent and synthetic notions of meaning entertained in cognitive semantics and in evolutionary approaches to diachronic linguistics.

This revival is, I shall maintain, due to two principal factors: the decline of the mainstream logico-formal and analytic paradigm that dominated the better part of twentieth-century philosophy, and the revitalisation of the pragmatistic attitude, the origins of which coincide with those of the early European contributions to cognitive science.

2. The Intrusion of Hardcore Ideology

To understand the interplay between cognitive science and philosophy, one needs to be aware of the development of their common intellectual ideas. For instance, characterising cognitive science as 'naturalised philosophy' or 'naturalised epistemology' suggests that it has long subsisted alongside major philosophical theories, including Cartesian interactionism, empiricism, or their Kantian synthesis.

Even more markedly, it seems that cognitive science tried to identify itself as a self-standing scientific discipline long before the much-highlighted events that took place during the 1950s. Such events were driven by 'hardcore ideology'. Examples include the inauguration of AI – boasting the new science of symbolic thought in the brain, the emergence of transformational grammar which provided grounds for the intuitive judgement of grammaticality and processing by Turing machines, and applications of Shannon's information theory to human cognitive processes and experimental psychology.

But there are also some recent examples in the cognitive and computing sciences that owe nothing or very little to the tradition associated with hardcore ideology. These examples bear a closer intellectual allegiance to developments that preceded the era than with those influenced by Minsky, Chomsky or Shannon.

Accordingly, the landmarks of the 1950s may be re-read as signs of a relapsing state of affairs that the alliance between science and philosophy, in the first instance the alliance of symbolic logic and analytic philosophy, had reached after the mid-20th century. I need not venture into criticism of what is dubious in such an alliance with regard to the aims of cognitive science, since my purpose is to substantiate Albertazzi's argument on the existence of a rich network of theories, mutual influences, conceptual repertoire and thematic complexity that by and large avoided the fate of first-generation AI and its spin-offs.

This evasion succeeded because of the reluctance evinced in these theories to join forces with some of the characteristic features of what may be conceived as 'analytic' philosophy. At the same time, the early cognitive theories did not turn a blind eye to many of the virtues of the analytic movement, including the rigour provided by mathematics and the use of well-identified scientific methodology.

Many of the tools that these early-modern – pre-1950 – theories resorted to were inadequate and underdeveloped to such an extent that they soon had to give way to the hardcore ideology welding cognitive science and philosophy. But that welding also suffered from oversight, including the illicit use of computer analogy and functional explanations. It is these, to some extent patchy, streams of thoughts antedating hardcore ideology that might be identified as falling under the umbrella of 'the early cognitive sciences', some key components of which Albertazzi has identified, linking them with the Central-European tradition in philosophy.[1]

It is therefore the ambivalent relationship between the early scientific and conceptual issues in the study of cognition and the analytic movement, together with the resumption of interest in those issues under new guises, which define the senses in which the early cognitive sciences may be taken to constitute a continuing challenge to analytic philosophy and the hardcore ideology.

Critics might hold that the reason for this challenge is simply that early European thought should be seen as continental philosophy (Chrudzimski & Huemer 2004). While this is a self-appointed, ideological claim, it holds a gist of truth. Early cognitive sciences are strongly grounded in the same roots as phenomenology. It is hardly the case, however, that phenomenology is exclusively of continental proprietary. It used to be rooted in the logical principles common to both continental and analytic thought. A historical case in point is the conceptual repertoire of pragmatist philosophers. Charles S. Peirce's phenomenological concept of the "Phaneron" was "the collective total of all that is in any way or in any sense present to the mind, quite regardless of whether it corresponds to any real thing or not" (CP 1.284, *Adirondack Lectures*, 1905).[2] To analyse and understand Phaneron, however, one needs "logical analysis" (MS 283, *The Basis of Pragmaticism*, 1905).

Peirce's programme of logical analysis was to uncover the essence of the mind in thought by improving upon the logic used by Kant. That essence was expressed in both declarative and non-declarative assertions. Peirce studied a whole range of different kinds of assertions, first in terms of his algebraic logic of relatives via valental analysis, and later on by using diagrammatic logical systems (Pietarinen 2003; 2005b; 2005d). The role of the Phaneron was pivotal in providing universes of discourse that were not confined to any objective part or situation about the world. A striking illustration of the merger between phenomenological and semantic topics is Peirce's proposal of how models are constructed by two "collaborating" parties that undertake to build a Phaneron against which assertions are then evaluated (CP 4.538; 4.552, *Prolegomena to an Apology for Pragmaticism*,

1 'Soft computing' is another pet term for cognitively- or neuroscientifically-oriented computing sciences that derive inspiration from the behaviour of the human mind and which might be seen as a counterbuff to hardcore ideology. Such coinage is, however, much more recent.

2 The references CP are to Peirce (1931–58) by volume and paragraph number; the references MS are to Peirce (1967) by manuscript number; the references SS are to Peirce (1977) by page number.

1905). On balance, in Peirce's approach both the seeds of phenomenology and those of analytic philosophy grew in peaceful coexistence.

A caveat: the intention of the present article is not to characterise what analytic philosophy actually is, or what it is not. The Hon. Rt. Lord Quinton thinks it began when Wittgenstein arrived in Cambridge in 1912 to study with Bertrand Russell.[3] Sir Michael Dummett prefers Frege and the Context Principle. But we also know that Wittgenstein's early encounters with Frege were, at best, inconclusive, and that before Frege, many thinkers including Wilhelm Wundt expressed principles similar to those expounded by Frege. Accordingly, Dummett's characterisation amounts to the anomalous position that both psychologism and anti-psychologism would have to be admitted by analytic philosophers.

There are further historical curiosities, many of them vital both in biographical and systematic terms. Wittgenstein soon associated with C. K. Ogden, who had been Victoria Welby's assistant during the 'signific' era in Cambridge. The purpose of significs was to analyse meaning and signification in language in order to achieve better communication and understanding by using logical methods and functional analysis of the 'language strata' (Pietarinen 2004c; Ogden 1911/1994). Apparently, no requirement to confine this task to symbolic communication such as natural language was felt at that time. Indeed, Otto Neurath's later *International System of Typographic Picture Education* (ISOTYPE) was an intrepid attempt towards genuinely iconic forms of communication, and to provide a standard for such communication (Nemeth and Stadler 1996). Proposals of this type did not bear fruit in the interwar atmosphere dominated by discrete-symbolic logic.[4]

As such, Peirce and his followers represented no drastic departure from the main ideals of early analytic philosophy. The ideas of Welby and her correspondent Peirce converged. Excerpts from that correspondence were reprinted in the first, 1923, edition of Ogden and Richard's *The Meaning of Meaning*, a work that was ready for print in the early 1910s. Wittgenstein repudiated the body of this text, but

---

3 In *Oxford Companion to Philosophy* (Oxford University Press, 1995), p. 28. The correct date for Wittgenstein's arrival is October 1911, not 1912.

4 Neurath (1936/1980) writes: "Pictures, whose details are clear to everybody, are free from the limits of language" (p. 18). "The ISOTYPE system makes use of the connection of parts not only in one direction only, but in two, and the effect is a *language picture*" (p. 62). Almost three decades earlier, in relation to his graphical logic, Peirce noted that, "Three dimensions are necessary and sufficient for the expression of all assertions; so that, if man's reason was originally limited to the line of speech (which I do not affirm), it has now outgrown the limitation." (MS 654: 6–7, 19 August 1910, *Preface to Essays on Meaning*, 1st draft; see Pietarinen 2005b).

not, apparently, its appendices.[5] He might have approved the signific ideas about meaning. (Pietarinen 2004b; 2004c). There were also some other, predominantly Dutch, philosophers and cultural figures who were more or less following Peirce's lead and came to influence Wittgenstein.[6]

One upshot of this historical interlude is simply that the origins of analytic philosophy are so convoluted as to be unworthy of pursuing in isolation. One might characterise the situation by stating that, rather than being identified with the so-called 'linguistic turn', analytic philosophy sympathised with something akin to a 'linguistic trap', in other words replacing both functional analyses of language and descriptive approaches to the categories of the mind and psyche with symbolic and discrete studies of language and communication.

3. The Resurgence of the Early Cognitive Sciences

In a nutshell, my purpose in what follows is to argue that the revival of the early endeavour to develop conceptual categories of the mind is discernible in several intersections of AI, logic and cognitive science. Summaries of three cases will, I hope, serve to illustrate the point.

Firstly, instead of 'predicates', many logical theories, especially the computational ones that deal with practical reasoning in rational agenthood have switched to non-symbolic – pictorial, visual, graphical, spatial and indexical – modes of representation in order to capture different notions of the 'qualities' associated with assertions and objects. In particular, these methods articulate continuity in a manner that is by and large ignored by the tradition of symbolic logic.

The second, related tendency has been the dispensing with the traditional notion of 'logical constants', substituting for it an assortment of iconic and topological notions of dynamic theories of action and experimentation concerning the relations involved in the representations.

Thirdly, instead of 'objects', many scientists were accustomed to nominalistic talk about 'individuals', but have subsequently switched to phenomenological terminology, including the involvement of modes of content in cognition and consciousness.

I will now elaborate on these three interconnected points using two examples. In Peirce's logic, predicates were understood in spatial terms, as bounded regions on a surface that differ in quality from other bounded regions on that surface (CP

---

5 In addition to the Peirce/Welby correspondence, the appendices included Bronislaw Malinowski's *The Problem of Meaning in Primitive Languages*, which resulted in a lasting occupation with Wittgenstein (Pietarinen 2005a).

6 Pietarinen (2004b, 2005b) documents some of these connections in greater detail.

4.416, *Existential Graphs*, 1903). Moreover, one of the grand mysteries in Peirce's later logical writings has been what he may have meant when speaking about "continuous predicates" in relation to his iconic logic of diagrams:

> A predicate which can ... be analyzed into parts all homogeneous with the whole I call a *continuous predicate*. It is very important in logical analysis, because a continuous predicate obviously cannot be a *compound* except of continuous predicates, and thus when we have carried analysis so far as to leave only a continuous predicate, we have carried it to its ultimate elements. I won't lengthen this letter by easily furnished examples of the great utility of this rule. (SS, 71-72).[7]

As far as I know, no explanation of what Peirce could have meant by continuous predicates is forthcoming elsewhere in his writings. However, an enlightening possibility here is to think of predicates as subsets A of X in the mappings from a space X to the set $\{0, 1\}$, where X is an arbitrary topological space. Continuity of the predicate would now depend on the topology defined for $\{0, 1\}$. Since suitable continuous functions from X to a two-element lattice would now correspond to open sets of X, we have established a relationship between predicates and open sets.

While this is a way to derive continuous predicates, is it something that Peirce could have envisioned? The answer is yes, because as Peirce stated in the above quotation, a continuous predicate is also "a predicate which can ... be analyzed into parts all homogeneous with the whole". But this is simply another way of saying that continuous predicates can be characterised as the set of all predicates consistent with it.

Moreover, the iconic, or diagrammatic, method of representing assertions and reasoning about them is, as Peirce noted, best conducted by establishing logical principles upon topological, continuous principles. At that time, the state of these principles was fairly elementary, but also rich in not being preoccupied with the set-theoretical continuum, unlike the events that occurred during the early phases of the symbolic era in logic (Pietarinen 2005b).

This explication may appear somewhat cursory or overly specific for the overall purpose of charting the propagation of wider intellectual ideas. But I believe it also illustrates that the early contributors to cognitive science raised more general and serious concerns than the mere technicalities would have us believe. In a nutshell, the point is that semantics could be based on principles which have a continuous core. This may happen either in the semantic attributes (as in modal logics using Bergson-Hamblin interval semantics) or in describing the parts of representations

7 14 December 1908, *A Letter to Welby*. See also CP 4.438, *On Existential Graphs, Euler's Diagrams, and Logical Algebra*, c. 1903.

(as in diagrammatic or metaphoric mental images) as being continuously connected ('path-connected').

To substantiate such associations, we may allude to claims such as that in Wolfgang Köhler's *Field Semantics* for the theory of perception dating back to the 1920s, namely that "neural functions and processes with which the perceptual facts are associated ... are located in a continuous medium" (Köhler 1940, p. 55). A notable connexion here is with Peirce's remark in his statement quoted above that predicates, propositions and inferences are to be continuously connected with one another. Topologically, this expresses connectivity between different parts of the surface, thus forming propositions, and in terms of propositions being connected (in the equally topological sense) under continuous transformations from one graph to another, thus forming inferential arguments.

The conceptualisation of a continuous predicate and, more generally, the continuous connections between separate parts of the representations already carried the traces of later Gestalt or perceptive schemes rather than anticipating anything such as the Tarski-type logical semantics that had an inherently compositional and recursive nucleus. Peirce, however, did not make the connection with the later truth-conditional and semantic issues an explicit one, since he was first and foremost striving to account for what would count as necessary reasoning in terms of a dynamic understanding of 'motion' through continuous morphing from one diagram to another according to given rules of transformation. These wider concerns were therefore easily sidestepped by the advocates of discrete-symbolic logic who began to gain ground shortly after these early contributions.

Related to the idea of continuity in topological semantics is the viewing of logical constants as icons of some concrete perceptual processes. Instead of obeying some rules of inference, logical constants obey what Peirce called 'illative' transformations, in other words transformations that, besides being sound and complete for given diagrammatic theories, denote the locations, movement and the spatial orientation that the inference is going to take (CP 4.505, *On Existential Graphs, Euler's Diagrams, and Logical Algebra*, c. 1903). For example, conjunction is juxtaposition of graphs on a topological (open-compact) manifold. Disjunction is derived by making cuts around graphs to iconise negation. Introduction rules follow from the fact that in a negative area (an area enclosed within an odd number of cuts) any assertion may be added (symbolically, e.g. $A \rightarrow A \vee B$), while elimination follows from the fact that from a positive area (an area enclosed within an even number of cuts) any assertion may be erased (e.g., $A \,\&\, B \rightarrow A$). The standard quantifiers *Any* and *Exists* are identified with continuous connections between predicates and outermost negative (*Any*) or positive (*Exists*) areas. In this manner, concrete cognitive content is given to the meaning of logical constants.

Illative transformations also answer the question of what are the privileged logical constants. For example, it is impossible to iconise the infamous connective *tonk*, defined by the two rules $A \rightarrow A$ *tonk* B and A *tonk* $B \rightarrow B$, on topological manifolds because of the different polarities associated with the two rules. Since no conceivable illative transformation rule characterises such a connective, it is not among the privileged logical constants at all.[8]

Congeniality between the early cognitive sciences and pragmatism is clearly evident. The characteristically Central-European discipline, much of which evolved from Franz Brentano's 'experimental metaphysics', was reflected in Alfred Whitehead's process ontology that surpassed many of the Gestalt theories on the mind and perception. In keeping events and processes distinct from things that participate in those processes, even making them the primary ontological entities, Whitehead was following Peirce.

Looked at from the contemporary point of view, one needs to recognise the extent to which Whitehead's dynamic ontology was successfully infused into the projects of designing knowledge-based computer systems and database and web technology – perhaps more comprehensively than any other ontological system that has been proposed by a philosopher. One should also be mindful of those physical theories that appeal to the idea of dynamical processes rather than fixed space-time structures, concepts that highlight the evolutionary nature of the universe with its changeable physical laws (Smolin 2001).

Many more contemporary links between early ideas concerning cognitive science and recent developments exist than the space available to describe them permits. In summary, largely as a result of investment in the philosophical foundations of computational sciences, early European theories are beginning to flourish in the concepts of embodied and enactive minds, game-theoretic and open-systems approaches to computational tasks such as the verification, refinement and composition of computational processes and computer programs, as well as in the interactive, emergent and synthetic notions of meaning entertained in cognitive semantics and the allied dynamic, evolutionary approaches to semantic change and diachronic linguistics.[9]

The early paradigm in the Gestalt theory of "thinking in images" therefore strikes back in the wake of the storm triggered by the scornful attitude of the hardcore advocates of AI who viewed thinking primarily as symbol-crunching. The key notion for representing linguistic meaning in the revised agenda is an image

---

8 The *tonk*-connective was fabricated by Prior (1960) to argue against the inferential thesis according to which the meanings of logical constants are completely determined by the rules of inference associated with them.

9 Pietarinen (2005c) addresses the evolutionary emergence of language in view of the philosophical aspects of evolutionary game theory. See Pietarinen (2004d) on the history of the notion of the common ground, the particular intellectual idea that is re-emerging in cognitive linguistics.

schema which consists of abstract pictures constructed from elementary topological and geometrical structures common to perception, memory and semantic meaning, replacing uninterpreted symbols with iconic signs, namely images, diagrams and metaphors. It is my belief that, for logically-oriented developments of such image schemas or cognitive spaces, the diagrammatic representational system suggested by Peirce for the analysis of assertions will prove to be ground-breaking.[10]

4. Conclusion

The revival of interest in the common themes in early cognitive science seems to follow the decline of the mainstream logico-formal and analytic paradigm that dominated twentieth-century philosophy. It is also a result of revitalisation of the pragmatistic attitude in logical philosophy. Significantly, the origins of the latter coincide, both in terms of their timing and the common network of influences, with the origins of the early European contributors to cognitive science.

Under this spotlight, analytic philosophy may be seen as a wasted attempt at placing some *a priori* constraints on the study of mind such as disembodiedness – a move that would not have been approved by the foregoing pragmaticists. As far as some of the specific examples are concerned, the emerging representational approaches to logical and mathematical problems in cognition and computation are among the first in the long overdue resurgence of the idea of how the mind, as one might say, leaks back into the world.

University of Helsinki

References

Albertazzi, L. (2001), "Back to the origins", in L. Albertazzi (ed.), *The Dawn of Cognitive Science: Early European Contributors*, Dordrecht, Kluwer, pp. 1-27.

Chrudzimski, A. and W. Huemer (2004), *Phenomenology and Analysis: Essays on Central European Philosophy*, Frankfurt, Ontos Verlag.

Köhler, W. (1940), *Dynamics in Psychology*, New York, Liveright.

Nemeth, E. and F. Stadler (eds.) (1996), *Encyclopedia and Utopia: The Life and Work of Otto Neurath (1882-1945)*, Dordrecht, Kluwer.

---

10 Peirce's logical system is also illuminating in regard to neurophysiological research, for instance to the Binding Problem (Pietarinen 2004a).

Neurath, O. (1936/1980), *International Picture Language: The First Rules of Isotype*, London, Kegan Paul. (Reprint by University of Reading, Department of Typography & Graphic Communication, 1980.)
Ogden, C. K. (1911/1994), *From Significs to Orthology*, ed. by W. T. Gordon, London, Routledge/Thoemmes Press.
Peirce, C. S. (1931–58), *Collected Papers of Charles Sanders Peirce*, 8 vols, ed. by C. Hartshorne, P. Weiss, and A. W. Burks, Cambridge, MA, Harvard University Press.
Peirce, C. S. (1967), "Manuscripts", The Houghton Library of Harvard University, identified by R. Robin (1967), *Annotated Catalogue of the Papers of Charles S. Peirce*, Amherst, University of Massachusetts Press.
Peirce, C. S. (1977), *Semiotic and Significs: The Correspondence between Charles S. Peirce and Victoria Lady Welby*, ed. by C. S. Hardwick and J. Cook, Bloomington, Indiana University Press.
Pietarinen, A.-V. (2003), "Peirce's Game-Theoretic Ideas in Logic", *Semiotica* 144, 33–47.
Pietarinen, A.-V. (2004a), "Logic, Phenomenology and Neuroscience in Cahoots?", in G. Büchel, B. Klein and T. Roth-Berghofer (eds), *Proceedings of the 1st International Workshop on Philosophy and Informatics, CEUR Workshop Proceedings* 112, Technical University of Aachen (RWTH).
Pietarinen, A.-V. (2004b), "Logic, Language-games and Ludics", *Acta Analytica* 18, 89–123.
Pietarinen, A.-V. (2004c), "Wienin piiri, kieli ja signifistiikka" (in Finnish, "The Vienna Circle, language, and significs"), *Ajatus* 60, 327–353.
Pietarinen, A.-V. (2004d), "Agenda Cognitive Linguistics: C. S. Peirce and the Emergence of the Common Ground", *Text Processing and Cognitive Technologies* 9, 208–214.
Pietarinen, A.-V. (2005a), "Ludwig Wittgenstein", in C. Routledge and S. Chapman (eds.), *Key Thinkers in Linguistics and the Philosophy of Language*, Edinburgh, Edinburgh University Press, pp. 271–278.
Pietarinen, A.-V. (2005b), *Signs of Logic: Peircean Themes on the Philosophy of Language, Games, and Communication*, Dordrecht, Springer.
Pietarinen, A.-V. (2005c), "Evolutionary Game-Theoretic Semantics and Its Foundational Status", in N. Gontier, J. P. van Bendegem, and D. Aerts (eds), *Evolutionary Epistemology, Language and Culture*, Dordrecht, Springer.
Pietarinen, A.-V. (2005d), "Peirce's Magic Lantern: Moving Pictures of Thought", forthcoming in *Transactions of the Charles S. Peirce Society*.
Prior, A. (1960), "The Runabout Inference-Ticket", *Analysis* 21, 38–39.
Smolin, L. (2001), *Three Roads to Quantum Gravity*, New York, Basic Books.

Intentional Action and the Limits of Scientific Naturalism: Davidson's Alleged Refutation of the Logical Connection Argument[1]

Antti Saaristo

1. Introduction

We all know the standard story. In the good old times philosophers followed common sense and thought that our reasons are causes of our actions. Then in the 1950s something went horribly wrong. Perfectly sane philosophers got overexcited by reading too much Wittgenstein and started to revolt against the received wisdom by defending the so-called Logical Connection Argument, which states that reasons cannot be causes, for reasons and actions are logically or conceptually connected, and already Hume taught us that a cause must be logically distinct from its effect. However, the gospel of contemporary action theory continues, just when the Wittgensteinian folly was getting out of control, Donald Davidson restored peace and harmony with his 1963 article "Actions, Reasons, and Causes", which demonstrated that the Logical Connection Argument is invalid and reasons are causes – and hence intentional explanations are causal explanations – after all.

This standard story is today almost universally accepted. As Alfred Mele puts it, "Davidson's greatest contribution to the philosophy of action is his resurrection of causal theories of action and action explanation. As long as Davidson's challenge to noncausalists remains unmet [...], causalism will be the biggest game in town, if not the only one." (Mele 2003, p. 82.) Similarly, social scientists and even philosophers of social science routinely state that they follow Davidson in taking reasons to be causes and intentional explanations of action to be causal explana-

1 I am grateful to the participants of the XV Internordic Philosophical Symposium *Science – A Challenge to Philosophy?* and the 2005 Columbia/NYU Graduate Conference in Philosophy for comments and criticisms. The paper has also benefited from comments by J. McKenzie Alexander, Caroline Baumann, Richard Bradley, Damien Fennell, Mauro Rossi and Stuart Yasgur. My research for this paper has been supported by the Helsingin Sanomat Centenary Foundation, the Jenny and Antti Wihuri Foundation and the London School of Economics.

tions. In other words, they think Davidson offers the philosophical justification for what I below call the Standard View.

In this paper I challenge this orthodoxy. I aim to show that Davidson's alleged refutation of the Logical Connection Argument is not as decisive as it is generally taken to be. Indeed, *pace* Mele and Davidson's claims to the contrary, I do not think Davidson even presents a serious challenge for noncausalists to meet. Causalism may *de facto* be the biggest game in town, but Davidson should not be referred to as the philosopher authorising the rules of the game.

In what follows I first introduce the Standard View of seeing reasons as causes and intentional explanations as causal explanations. Second, I present the Logical Connection Argument against the Standard View. Finally, I discuss Davidson's alleged refutation of the Logical Connection Argument and the consensus of seeing Davidson as a (or indeed the) champion of the Standard View.

As I use the terms, scientific naturalism concerning intentional action and its explanation stands or falls with the Standard View.

2. The Standard View

The Standard View of intentional action and its explanation holds that the propositional attitudes of an agent – in particular, her desires and beliefs – that *rationalise* an action of the agent are also *causes* of the action. Moreover, the Standard View assumes that the reasons cause the action in virtue of being the *reasons* they are, i.e. in virtue of their propositional content. *What* I believe and *what* I desire is, according to the Standard View, the causally efficacious factor.

A clear and influential statement of the Standard View is Jon Elster's interpretation of Davidson's theory of action (e.g., Elster 1985). Accordingly, the Standard View is sometimes even referred to as the Elster-Davidson view. This is how Elster sets the stage:

> A successful intentional explanation establishes the behavior as an action and the performer as an agent. An explanation of this form amounts to demonstrating a three-place relation between the behavior (B), a set of cognitions (C) entertained by the individual, and a set of desires (D) that can also be imputed to him. (Elster 1985, p. 60.)

Elster then goes on to define the three-place relation in terms of three conditions that correspond to the features I have identified as he core of the Standard View, namely that desires and beliefs (i) rationalise actions, (ii) cause actions and (iii) do so in virtue of their contents. Or in terms of the quotation above, the following three claims state in Elster's (1985, pp. 60–61) mind the necessary and sufficient conditions for an acceptable intentional explanation of *B* in terms of *C* and *D*:

(i) Given *C*, *B* is the best means to realise *D*.
(ii) *C* and *D* caused *B*.
(iii) *C* and *D* caused *B qua* reasons.

The Standard View is a very ambitious view, and a great deal must be said before we can see why it must be rejected – and why Elster is mistaken in thinking that it is Davidson's view.

3. The Logical Connection Argument

Many philosophers have defended versions of the Logical Connection Argument. However, in this paper I concentrate only on the elegant version of the argument given by Georg Henrik von Wright (1971). Let us look at the standard form of an intentional explanation of action (cf. Rosenberg 1995, pp. 31–32).

1. *X* has a set of desires *D*.
2. *X* believes that the behaviour *B* is the best means to realise *D*.

(*L*) If any agent has the set of desires *D* and believes that *B* is the best means for realising *D*, then *X* does *B*.

Therefore: *X* does *B*.

The Standard View holds that (*L*) is a causal, descriptive *law* of nature applicable to rational beings and thus (*L*) expresses a causal connection between two self-sufficient sets of events: propositional attitudes (reasons) and meaningful behaviour (action). Von Wright's position, in contrast, is that (*L*) is a non-causal conceptual *norm* that interdefines desires, beliefs and actions such that (*L*) is at least partly *constitutive* of reasons and actions.

Hence, von Wright holds that reasons and actions cannot be separated from each other. What makes a behaviour an intentional action is that the behaviour *embodies* a meaning. According to von Wright, this requires that the action is essentially or conceptually (*logically*) connected to the reasons for the action. Such a connection *constitutes* the action *qua* action and not *qua* mere bodily behaviour, and similarly for reasons *qua* contentful reasons and not *qua* meaningless physical states. In other words, to have a belief *means* that if the belief is combined with certain desires, one is *rationally committed* to performing certain actions. (This explicit normativity sets the view apart from crude behaviourism.) Desires, in turn, have their content (exist *qua* desires) in virtue of their role in (normatively) connecting beliefs and actions. Finally, actions are actions only if reasons can be given and asked for them; otherwise they are mere bodily behaviours. In short, to identify a behaviour as an action is to see or redescribe the behaviour in terms of its reasons.

Thus, in von Wright's mind (*L*) does not connect independently existing events but rather explicates an aspect of a *normative framework*, which Wilfrid Sellars called the game of giving and asking for reasons, and which defines when certain states can count as contentful desires and beliefs and certain behaviours as meaningful actions. In other words, (*L*) expresses a conceptual (logical) connection that explicates a major constitutive principle of the normative framework relative to which we can meaningfully talk about desires, beliefs and actions. Thus, (*L*) cannot be a naturalistic, causal *law*, for it is, as one might say, a central *rule* in the game of giving and asking for reasons and (partly) constitutive of desires, beliefs and actions. Indeed, independently of (*L*), there *are* no desires, beliefs and actions that could stand in naturalistic, causal relations to one another. In sum, the Logical Connection Argument is based on a *holistic* view of reasons and actions. Reasons and actions (*qua* reasons and actions) exist only in virtue of their place in a normative pattern of *rational* relations such as (*L*).

4. Davidson's Alleged Refutation of the Logical Connection Argument

Davidson (1963) opens his famous attack on the Logical Connection Argument by pointing out that even if we grant that by explaining an action intentionally we redescribe a behaviour in terms of beliefs and desires (and thereby as an action), it does not follow that beliefs and desires cannot be causes of the action. As Davidson emphasises, we often do redescribe events precisely in terms of their causes. For example, when someone is injured we may redescribe the event by saying that he was burned (Davidson 1963, p. 10). In this case the man's injury is redescribed in terms of the man being burned, but surely the possibility of such a redescription does not prevent us from seeing burning as a cause of the man's injury.

However, an adherent of the Logical Connection Argument might think that this example is essentially different from a redescription of a behaviour in terms of the reasons for action. After all, as von Wright emphasises, the issue is the logical or conceptual connection between reasons and actions. We cannot make sense of the one without appealing to the other: a behaviour *is not* an action if it is not redescribable in terms of reasons. This does not hold for burnings and injuries. The connection between them is empirical and contingent, not conceptual or logical.

But Davidson has an answer readily available. Suppose "*A* caused *B*" is true. If so, we can redescribe "*A*" as "the cause of *B*". By substituting this coreferential redescription into the original claim, we have "the cause of *B* caused *B*". Here the connection between the *descriptions* "the cause of *B*" and "*B*" is conceptual, but by assumption the connection between the concrete particulars the descriptions pick out is nonetheless causal. Thus, as Davidson (1963, p. 14) puts it, "[t]he truth of a causal statement depends on *what* events are described; its status as analytic or synthetic [whether or not the connection is conceptual] depends on *how* the events are described".

In this manner Davidson thinks that the Logical Connection Argument is in fact *irrelevant* for causality, for the Logical Connection Argument applies only to *descriptions* at the abstract level of types of events, whereas causality is a material relation between *concrete particulars* in the world. However, intentional explanations of action are formulated precisely in terms of conceptually connected types. Thus Davidson's insistence that intentional explanations are causal explanations amounts to requiring that we add a causal claim about the relevant particulars to intentional explanations. Hence, we can represent Davidson's view of intentional explanation as follows:

1. *X* has a set of desires *D*.
2. *X* believes that the behaviour *B* is the best means to realise *D*.
3. This concrete particular instantiating *X*'s desires and *this concrete particular* instantiating *X*'s beliefs cause *this concrete particular* instantiating *X*'s action *B*.

(*L*) If any agent has the set of desires *D* and believes that *B* is the best means for realising *D*, then *X* does *B*.

Therefore: *X* does *B*.

How should we understand this? Has Davidson shown that reasons are causes and intentional explanations may be seen as causal explanations after all? Answering these questions requires profound understanding of the reasoning behind Davidson's view of intentional explanation.

Let us first observe that Davidson's famous doctrine of the Anomalism of the Mental (see, in particular, Davidson 1970 and 1974), namely, the doctrine that mental types are not covered by strict laws, is based on an explicit acceptance of the thesis of the *holistic* nature of the mental:

> There is no assigning beliefs to a person one by one on the basis of his verbal behaviour, his choices, or other local signs no matter ho plain and evident, for we make sense of particular beliefs only as they cohere with other beliefs, with preferences, with intentions, hopes, fears, expectations, and the rest. [...] [T]he content of a propositional attitude derives from its place in the pattern. (Davidson 1970, p. 221.)

> My point is that if we are intelligibly to attribute attitudes and beliefs, or usefully to describe motions as behaviour [*action* in my terminology], then we are committed

> to finding, in the pattern of behaviour, belief, and desire, a large degree of rationality and consistency. (Davidson 1974, p. 237.)

Further, "the satisfaction of conditions of consistency and rational coherence may be viewed as constitutive of the range of applications of such concepts as those of belief, desire, intention and action" (Davidson 1974, p. 237). Thus, for Davidson the norms of rationality, such as (*L*), *constitute* the mental realm, including the very possibility of intentional actions (Davidson 1970, p. 217; see also Kim 2003, p. 119).

Moreover, a corner stone of Davidson's philosophy of mind is the view that when we consider mental states *qua* mental, and actions *qua* actions (i.e., in terms of the reasons for action), "we necessarily impose conditions of coherence, rationality, and consistency. These conditions have no echo in physical theory, which is why we can look for no more than rough correlations between psychological and physical phenomena." (Davidson 1974, p. 231.) In other words, contentful mental states and intentional actions (so described) do not, in Davidson's view, fall into the scope of scientific naturalism (hence the title "Psychology as Philosophy" of Davidson 1974), and constitutive principles such as (*L*) cannot be causal laws, for they "have no echo" in the law-governed domain of natural science.

This, of course, is *precisely* what von Wright's (1971) Logical Connection Argument holds. So what should we say concerning Davidson's insistence that reasons are causes and intentional explanations of action are causal explanations? The crucial fact to take in is that Davidson in effect *accepts* the Logical Connection Argument when it comes to the paradigmatic model of intentional explanation of action, which consists of Premises 1, 2 and (*L*) and the conclusion (see Tuomela 1976 for a related reading of Davidson). The *explanans* has *nothing causal in it*, since the premises "have no echo in physical theory". This is why Davidson's insistence on causalism compels him to add (implicitly) the new Premise 3 that contains all the causal import there is in the *explanans*.

However, Premise 3 is merely a claim and hence in need of justification (Tuomela 1976, p. 203). In Davidson's mind this must be done in terms of laws, since "where there is causality, there must be a law: events related as cause and effect fall under strict deterministic laws" (Davidson 1970, p. 208). However, since the constitutive rules of the mental realm, such as (*L*), "have no echo in physical theory", the laws in question must be physical. This, as Davidson explains, amounts to redescribing the concrete particulars Premise 3 talks about in physical terms and subsuming these descriptions under a strict causal law:

> The Laws whose existence is required if reasons are causes of actions do not, we may be sure, deal in the concepts in which rationalizations [intentional explanations] must deal. If the causes of a class of events (actions) fall in a certain class (reasons) and there is no law to back each singular causal statement, it does not

> follow that there is any law connecting events classified as reasons with events classified as actions – the classifications may even be neurological, chemical, or physical. (Davidson 1963, p. 17.)

> Suppose *m*, a mental event, caused *p*, a physical event; then, under some description *m* and *p* instantiate a strict law. This law can only be physical [...]. But if *m* falls under a physical law, it has a physical description; which is to say it is a physical event. (Davidson 1970, p. 224.)

In other words, the Davidsonian possibility of seeing reasons as causes is based on the idea that we need *not* treat the desire and belief in question *qua* reasons, and the action *qua* action. Indeed, if we want to see the events in question as instantiating a causal process, we must think of the particulars in question in terms of non-intentional physical types, for the intentional types *desire*, *belief* and *action*, and the connection (*L*), "have no echo in physical theory". Thus, "[t]wo ideas are built into the concept on acting on a reason [...]: the idea of cause and the idea of rationality" (Davidson 1974, p. 233). But this simply means that in Davidson's model of intentional explanation there are in fact two *completely separate* (because of the "no echo" thesis) explanations that are artificially forced into one inference.

First, there is the paradigmatic intentional explanation of action in terms of Premises 1, 2 and (*L*), which is a *non-causal* explanation constituting the action *qua* action by embedding it into the normative framework *constituting* the mental realm. For Davidson, there *are* no mental states and action outside this framework. Second, there is a *causal* explanation of the behavioural component of the action in question (the action *qua* mere bodily movement) that picks out the particulars Premise 3 talks about as tokens of physical types (*qua* non-contentful physical events) subsumable under causal laws. In other words, in Davidson's scheme *there is no one causal, intentional explanation of action.* Rather there are two entirely distinct explanations (different *explananses* and different *explananda*). One is a causal, non-intentional explanation of a physical movement (behaviour) and the other is an intentional, non-causal explanation of an action. The Davidsonian obsession to include both explanations in a single model does not undermine the fact that there are two different issues at stake.

This, of course, does not tell against the Logical Connection Argument anymore. In fact, this *is* the point of the Logical Connection Argument. As von Wright puts it, the *explanandum* of a causal explanation of *behaviour* is "why *parts of his body move*, under the causal influence of stimulations of his nervous system, and not why *he moves parts of his body*" (von Wright 1971, p. 119), which is an *action* and consequently must be explained in terms of a non-causal intentional explanation. The Logical Connection Argument is not aimed to show that both explanations cannot be important and interesting, but only that one ought not to confuse the two, as they are genuinely different explanations. This is something Davidson's view does not – and cannot – challenge, for it is implied by Davidson's own anomalous monism.

So what should we say about Davidson's alleged refutation of the Logical Connection Argument? Clearly Davidson retains the unbridgeable gap between non-causal explanations of action and causal explanations of bodily behaviour. Hence, if the core of the Logical Connection Argument is taken to be the claim that intentional explanations as such are not predominantly causal explanations, I conclude that Davidson's alleged refutation fails. If the core is thought to be the idea that (*L*) does not express a causal law of nature but a logical or conceptual *normative* connection (that has "no echo in physical theory"), Davidson's refutation also fails. However, if one takes the Logical Connection Argument to be committed to the view that mental terms such "belief" and "desire" cannot be applied to particulars for which also a physical description may be given, then Davidson's refutation is successful. This, however, as the quotation from von Wright above makes clear, amounts to the *acceptance* of the Logical Connection Argument with an explicit emphasis on (i) the importance of the rejection of Cartesian dualism and (ii) the role of causal knowledge in modern natural science.

5. Davidson and the Standard View

So Davidson does not refute the Logical Connection Argument. What should we say about the other widely accepted claim, namely that Davidson provides the philosophical justification for the Standard View? As should by now be clear, in fact Davidson's line of reasoning renders the Standard View *incoherent* and unacceptable. To see this, let us again turn to Elster's (1985) formulation of the Standard View, for it is explicitly proposed as a Davidsonian account.

As we saw, Elster's first condition requires that reasons rationalise the action they explain (recall that *C* stands for the beliefs of the agent, *D* for her desires and *B* is the action):

(i) Given *C*, *B* is the best means to realise *D*.

In Elster's view (i) is a necessary but insufficient condition for *B* to be an intentional action of the agent. Elster's example is an actor who is asked to shudder as a part of a scene. The actor forms the relevant beliefs and desires and shudders, but even then we would not say he did it intentionally if his shuddering was caused by the sight of a snake. In such a case shuddering was not his action but rather something that happened to him.

Hence Elster adds his second condition:

(ii) *C* and *D* caused *B*.

If we consider (ii) to be a necessary condition for intentional action, we can rule out the case of the actor who shudders when he sees the snake. But are (i) and (ii) jointly sufficient? In the original version of his article "Actions, Reasons, and Causes" (1963, footnote 5), Davidson expresses his belief that (ii) "can be strengthened" to make the conditions jointly sufficient, although he does not explain how in the 1963 article.

However, Davidson returns to the problem a decade later (Davidson 1973) in terms of the following example:

> A climber might want to rid himself of the weight and danger of holding another man on a rope, and he might know that by loosening his hold on the rope he could rid himself of the weight and danger. This belief and want might so unnerve him as to cause him to loosen his hold, and yet it might be the case that he never chose to loosen his hold, nor did he do it intentionally. (Davidson 1973, p. 79.)

Here (i) and (ii) are satisfied, but as Davidson explicitly says, this is insufficient for intentional action. If the Standard View is to be defended, a further condition is required. Davidson suggests that the reasons must cause the action "in the *right* way" (Davidson 1973, p. 79) and that the "causal chain must follow the right sort of route" (Davidson 1979, p. 78). This is hardly satisfactory, which Davidson is well aware of: "What I despair of spelling out is the way in which attitudes must cause actions if they are to rationalize the action" (Davidson 1973, p. 79).

Elster, however, thinks he has a solution. For him the problems in the example are due to the fact that it is "a mere accident that in the case of the climber they [belief and desire] happen to cause the very same behavior for which they are reasons" (Elster 1985, p. 63). Hence, Elster adds his third condition:

(iii) *C* and *D* caused *B* qua reasons.

According to Elster, (iii) offers the kind of strengthening Davidson's original footnote in the 1963 article envisaged and the 1973 climber example requires. As we saw, together (i), (ii) and (iii) constitute the Standard View.

However, for Davidson it is clear that we *cannot* add anything like Elster's condition (iii). As we have seen, a crucial aspect of Davidson's thought is that we can treat intentional explanations as causal explanations *precisely to the extent* that the concrete particulars an explanation talks about are not considered *qua* reasons. In Davidson's picture the possibility of treating intentional explanations as causal explanations is based on an explicit denial of Elster's condition (iii). If Davidson is correct, then the standard view as formulated by Elster is inherently and inescapably incoherent.

Rather revealingly, Davidson *omits* the suggestion that the conditions (i) and (ii) could be strengthened to make them jointly sufficient for a causal theory of intentional action from the 1980 reprint of his 1963 article. In the 1980 version the footnote 5 seeks "to cancel any suggestion" that the conditions could be sufficient, whereas the original version implies the exact opposite.[2] Similarly, in his introduction to Davidson (1980), Davidson explains that the reasoning behind Davidson (1973) forced him to conclude that intentional action "cannot be analysed or defined" (Davidson 1980, p. xvii) in purely causal terms. A causalist theory cannot give "sufficient conditions of intentional (free) action" (Davidson 1980, p. xvii).

Indeed, the conclusion of Davidson (1973, p. 80), that "we cannot see how to complete the statement of the causal conditions of intentional action", is a direct rejection of both Davidson's own optimistic 1963 footnote and Elster's Standard View.

6. Conclusion

Contrary to the view held almost universally among contemporary action theorists, the upshot of Davidson's alleged refutation of the Logical Connection Argument is that the Logical Connection Argument remains intact. If Davidson's argumentation is sound, it is the Standard View that must be rejected and not the Logical Connection Argument. Elster shows what the Standard View requires, and Davidson argues that the requirements cannot be satisfied. Consequently, the common practice of referring to the Standard View as "Davidson's view", or indeed "the Elster-Davidson view", is highly misleading. In sum, Alfred Mele (2003, p. 82) may well be correct in saying that causalism is the biggest game in town. But, as we know, no "ought" from "is". In this paper I have shown that if causalism is to be defended, the defence cannot be Davidson's philosophy of action and mind.[3]

London School of Economics and Political Science

---

2 I thank Damien Fennell for drawing my attention to this interesting detail.

3 It seems to me that a main reason for insisting on the causalist Standard View is the conviction that if we want to hold on to (i) naturalism in the sense of *ontological materialism* and (ii) the thesis of *mental realism* in the sense of mental states making an irreducible difference, the Standard View is our best bet. This reasoning is based on a major confusion. As I have argued elsewhere (Saaristo 2004), the non-causalist view of intentional action and agency (and thus a non-causalist interpretation of mental realism) is the best way of remaining truly faithful to (i) and (ii).

References

Davidson, D. (1963), "Actions, Reasons, and Causes", in D. Davidson (1980), pp. 3-19. Originally in *Journal of Philosophy* 60, 1963.

Davidson, D. (1970), "Mental Events", in D. Davidson (1980), pp. 207-227.

Davidson, D. (1973), "Freedom to Act", in D. Davidson (1980), pp. 63-81.

Davidson, D. (1974), "Psychology as Philosophy", in D. Davidson (1980), 229-244.

Davidson, D. (1980), *Essays on Actions and Events*, Oxford, Clarendon Press.

Elster, J. (1985), "The Nature and Scope of Rational-Choice Explanation", in E. LePore and B. McLaughlin (eds.), *Actions and Events: Perspectives on Donald Davidson*, Oxford, Blackwell, pp. 60-72.

Kim, J. (2003), "Philosophy of Mind and Psychology", in K. Ludwig (ed.), *Donald Davidson*, Cambridge, Cambridge University Press, pp. 113-136.

Mele, A. R. (2003), "Philosophy of Action", in K. Ludwig (ed.), *Donald Davidson*, Cambridge, Cambridge University Press, pp. 64-84.

Rosenberg, A. (1995), *Philosophy of Social Science* (2nd edition), Boulder, Westview Press.

Saaristo, A. (2004), "Personhood as a Social Status", in H. Ikäheimo, J. Kotkavirta, A. Laitinen and P. Lyyra (eds.), *Personhood*, University of Jyväskylä, pp. 192-198.

Tuomela, R. (1976), "Explanation and Understanding of Human Behavior", in J. Manninen and R. Tuomela (eds.), *Essays on Explanation and Understanding*, Dordrecht, D. Reidel, pp. 183-205.

von Wright, G. H. (1971), *Explanation and Understanding*, Ithaca, NY, Cornell University Press.

Naturalized Epistemology Challenging Philosophy

Jarkko Vilkkilä

1. Introduction

Epistemology deals with questions concerning the nature, sources and scope of knowledge. In a word, epistemology is the theory of knowledge. My concern here is to explore the epistemological issues concerning belief, and I will try to clarify how these issues are dealt with in the debates of contemporary philosophy, especially in philosophy of mind. Even if there is an epistemological side to the debate about beliefs, the case of beliefs is not exclusively a part of epistemology, but it also concerns metaphysics. However, my point of view is more metaphilosophical than strictly epistemological or metaphysical. By "metaphilosophical" I mean that I will discuss how epistemological issues concerning beliefs affect our views about the nature of philosophy. I will restrict my discussion to contemporary philosophy that has its roots in Willard Van Orman Quine's naturalism. I will examine Quine's thinking in the light of theories of three philosophers. Two of them agree with the Quinean tradition of empiricism, but one is not so keen on naturalizing philosophy. These three philosophers are Daniel Dennett, Paul Churchland and David Chalmers.

2. Quine's Naturalized Epistemology

Quine had a project called "naturalized epistemology", which states that epistemology is closely connected to natural science. One of Quine's main arguments was that it is impossible to ground our knowledge of everyday life and science on the basis of our effort to derive statements about the world around us from statements about our own sensations. Quine was sure that we cannot strictly derive our beliefs about the world from our beliefs about sensations, and therefore we cannot be certain of the derived truths about the world.[1]

Abandoning the epistemological analysis of belief formation based on sensations, Quine stated that the stimulation of sensory receptors is all the evidence

1 Quine (1969b, pp. 72–75).

science needs in order to discover how we construct our picture of the world. In addition, Quine thought that this kind of discovery will be made in the field of psychology. Quine's idea was that we must study psychological processes that lead us from the stimulation of our sensory receptors to beliefs about the world. In his famous article "Epistemology Naturalized" (1969) the point is made clear:

> Epistemology, or something like it, simply falls into place as a chapter of psychology and hence of natural science. It studies a natural phenomenon, viz., a physical human subject.[2]

So Quine's suggestion was that epistemology is not a field of enquiry that is concerned with traditional epistemological questions about rationality and knowledge. Or at least these questions are not handled by means of traditional epistemology. The issues of epistemology according to Quine are issues of psychology, and psychology is a part of natural science.[3] In other words, Quine's ideas lead to a change of viewpoint, i.e., before Quine, it was thought that there is some epistemic relation between sensory experiences and beliefs about the world, but in Quine's conception this relation is a causal one which can be investigated by means of natural science.

Keeping in the mind the traditional epistemological problem of truth and falsity, one is tempted to ask what is the criterion of truth in Quine's conception. Recalling Quine's rejection of analyticity,[4] a sentence cannot be true in virtue of the mere meanings of words. Rather it is impossible to decide whether a sentence is true or false by appealing only to the meanings of words or to empirical facts. The truth or falsity of a sentence is related to what Quine called an observation sentence. As Quine defined it,

> [...] an observation sentence is one which all speakers of the language give the same verdict when given the same concurrent stimulation. To put the point negatively, an observation sentence is one that is not sensitive to differences in past experiences within the speech community.[5]

---

2 Quine (1969b, p. 82).
3 Quine (1969b, pp. 83–84).
4 See Quine (1951).
5 Quine (1969b, pp. 86–87).

Observation sentences play a dual role. First of all, the observation sentences constitute the foundation of the meanings of words. So we learn other languages on the basis of learned observation sentences. In other words, the learning of observation sentences seems to be linguistically prior to all other language learning.[6] Secondly, we use observation sentences to describe the world. So in Quine's conception observation sentences are the actual evidence for scientific enquiry. And finally, the observation sentences play a central role when we are evaluating scientific hypotheses intersubjectively.[7]

3. An Offspring of Epistemology Naturalized: Eliminative Materialism

Quine's naturalized epistemology has had a great influence upon philosophers of mind. In the philosophy of mind, Quine's naturalized epistemology leads to the view that the scientific study of consciousness does not deal with inner or private mental phenomena. Epistemologically we can't get a grasp on the nature of beliefs within the mind. Rather, any scientific study of consciousness has two alternatives. Scientists observe what people say or do, that is, they observe behaviour. Alternatively, scientists try to understand the facts about brain, and that is a matter of neuroscience. Both of these viewpoints can be labelled as "eliminative materialism", which holds that consciousness itself is just behaviour and brain-processes and the scientific study of consciousness describes them from a third-person perspective.

For example, Daniel Dennett is in favour of studying consciousness in terms of behavioural explanations. Dennett calls his method of studying consciousness "heterophenomenology".[8] As a matter of fact, Dennett's basic idea is a variant of the idea of the observation sentence. Heterophenomenology focuses on the behaviour of people, including both action and speech, and accepts that through action and speech people try to describe how things seem to them.[9] In order to perform this scientific study successfully one has to adopt the intentional stance. The intentional stance can be understood as treating the subject in question as a rational agent who exhibits beliefs, desires and intentionality, and stating that the subject's intentional behaviour is based on those beliefs and desires.[10] And as a

6 Quine (1969b, p. 89).
7 Quinc (1969b, pp. 88–87), see also Quine (1960, pp. 42, 226–227).
8 Dennett (1987, p. 153).
9 Dennett (1991, pp. 72–78).
10 Dennett (1987, p. 49).

result, data is collected by using the method of heterophenomenology, and the data can be represented as a set of sentences that creates the subject's "notional world" which should be thought of to be like a fictional novel.[11] Compared to Quine's naturalized epistemology we can say that one's notional world is a model that is based on observation sentences. And it seems to be the case that Dennett's way of doing science of consciousness is what Quine was after, for notional world is a source of evidence for enquiry of the science of consciousness.

Furthermore, if we think about Quine's argument that the stimulation of sensory receptors is all the evidence needed for psychological explanations, the field of study called "neurophilosophy" introduced by Patricia Smith Churchland seems to come very close to Quine's idea. Likewise, many ideas of Paul Churchland derive from Quine's ideas. For example, Paul Churchland is in favour of identifying the mind with physical substances or properties. He claims that all psychological theories are reducible to another scientific theory, i.e., neurophysiology. So it seems to be clear that Churchland's standpoint is a form of materialism. His argument states that if we can show that all of psychology is reducible to neurophysiology, then it is shown that the mind is nothing above and beyond the physical.[12] Of course, this kind of reduction requires that neurophysiology is reducible all the way down to physics. As a matter of fact, this reduction contains two levels. Firstly, languages are reduced to each other. Secondly, the ontology of psychology is claimed to be identical to the ontology of neurophysiology.[13]

The project of eliminative materialism is to show that consciousness does not exist at all and in more or less distant future the concepts of folk psychology are eliminated as neuroscience progresses. It is reasonable to think that Dennett is also an eliminative materialist, even though Dennett does not agree with Churchland who is trying to eliminate concepts of folk psychology. Nevertheless, Dennett is in the Quinean quest of getting rid of first-person methods that are supposed to give an insight into one's own mind. Eliminative materialists are assured that any scientific study of consciousness has to be conducted in terms of third-person science that leaves no room for speculative introspection. In conclusion, according to the theories of eliminative materialism all first-person data is considered as third-person data which is objective in nature.

11 Dennett (1987, p. 154).
12 Churchland (1981, pp. 210–212).
13 Churchland (1989, p. 49).

4. Rethinking the Nature of Epistemology

As is well known, some philosophers of mind do not agree with eliminative materialists. It is arguable that first-person data should be taken into consideration when examining the subjective character of experience. And this is the only way of formulating the true science of consciousness. One example of the non-reductionist theory of consciousness is introduced by David Chalmers. He has made a distinction between the easy and the hard problems of consciousness.[14] The former problems are susceptible to the standard methods of cognitive science, and they probably can be solved for example by understanding the computational or neural mechanisms involved, i.e., in terms of functionalism or neuroscience. The common feature of the easy problems is that they are about phenomena that are in some way associated with the notion of consciousness and they can be handled by empirical science. Instead, the hard problem of consciousness is experience. As Thomas Nagel has put it: the problem of what it is like to be a conscious being.[15] The hard problem concerns the question of how physical processes give rise to subjective experience.

Chalmers' distinction has not convinced the reductionists. Many of them suggest that the kind of science of consciousness that takes the hard problem seriously is not science at all. For example, Patricia Smith Churchland has stated that the hard problem is "a hornswoggle problem".[16] She asserts that there is no principled difference between hard and easy problems, and therefore she questions whether it is possible to make a systematic difference between easy and hard problems.

In any case, Chalmers' conception of an easy problem challenges the project of naturalized epistemology. Even if Chalmers is not denying the role of purely materialistic explanations, he still questions whether materialistic explanations can do the job just by themselves.[17] Chalmers asserts that pure materialism is a false theory, because he insists that consciousness is not logically supervenient on the physical, and the solution he offers is a kind of dualism.[18] Although dualism presents many problems that Chalmers is trying to solve, his theory of consciousness questions the capability of neuroscience alone to explain the nature of consciousness.

Moreover, Chalmers is in favor of first-person science, and thus he takes seriously the possibility of explaining beliefs without applying behavioural

14 Chalmers (2003, p. 103).
15 See Nagel (1974).
16 Churchland (1996, p. 37).
17 Chalmers (1996, pp. 46–50).
18 Chalmers (1996, pp. 123–124).

explanations. Chalmers is convinced that beliefs can be analysed in such a way that the analysis captures how things seem from the point of view of the subject whose beliefs are analysed. In other words, Chalmers is defending the possibility of conceptual analysis. Therefore, he has to handle the question of conceptual truths which are traditionally thought to be necessary and knowable a priori. Saul Kripke's book "Naming and Necessity" (1980) changed this picture. Kripke argued that meaning has its beginning in "the initial baptism", and from this point on begins a causal chain which continues through history to the present day. And theoretically the beginning of the use of every word can be traced to the point of "the initial baptism".[19] So Kripke is saying that a name is not identical to a description. Moreover, a name is "a rigid designator" whose reference is fixed by its definition. And Kripke also argues that there are statements that are metaphysical necessities known a posteriori, so this means that the identity of contents of thought are dependent on empirical issues, not conceptual.[20] Chalmers does not deny Kripke's theory of meaning as a whole, but according to Chalmers Kripke's theory does not lead to a conclusion that there is no sense in a priori conceptual analysis.[21]

Chalmers has presented his solution for the analysis of belief in a two-dimensional framework. This is actually a semantical framework but it can be applied to thought. For semantics deals with meanings, and it is plausible to think that meanings play a central role in thinking because our thoughts mean something. Hence if one analyses thoughts, one is concerned with meanings and semantics. In the two-dimensional picture of Chalmers the meaning of an expression is seen to consist of two intensions: the primary and the secondary. The secondary intension of a sentence can be thought to be identical with the proposition expressed by the sentence in question. And the proposition is understood as a set of possible worlds that the sentence is true of. This aspect of meaning is dependent on non-semantic factors of context of utterance. On the other hand, the primary intension is a function from worlds to extensions. And the world in question is thought of as actual. The primary intension is independent of empirical factors, and it specifies how reference depends on the way the external world turns out to be.[22]

Now, going back to Quine's ideas, we can say that Chalmers is an anti-Quinean, although his two-dimensional framework is an idea derived from Quine. For if we consider Quine's idea of eternal sentences, we can see that there is some resemblance with the two-dimensional framework. According to Quine, eternal sentences are standing sentences whose truth-values remain constant over time and

---

19 Kripke (1980, p. 96).

20 Kripke (1980 pp. 34–39) and Kripke (1980, pp. 48–53).

21 Chalmers (1996, p. 59).

22 Chalmers (1996, pp. 57–65).

from speaker to speaker.[23] And the possible worlds of primary intensions are seen as centered worlds that are marked with a center, which is an ordered pair of an individual and a time. The center of the world can be thought of as representing the perspective of a speaker in the world.[24] So in a sense, it can be said that the primary intensions deal with cognitive and rational relations between thoughts.

5. Philosophy Naturalized Entirely?

Though Quine has had a great influence on contemporary philosopical issues concerning beliefs and human rationale, and Quine's naturalized epistemology has been bridging a gap between philosophy and natural science, there still remain questions and problems that cannot be solved by natural science alone. The vivid dispute over epistemological issues, such as whether there is a priori knowledge, goes on among contemporary philosophers who are trying to deal with problems that were raised by Quine. And even though Quine's naturalism suggests that there is no strict demarcation between scientific work and philosophical analysis, many philosophers have been debating whether naturalism is acceptable.

By scrutinizing the *status quo* of contermporary philosophy, it seems that there is currently some discussion going on about whether naturalism rules out certain areas of philosophy altogether, such as semantics and ethics, or excludes the use of mentalistic vocabulary in the philosophy of mind. Whatever the outcome of the debate will be, the debate itself suggests that there are some areas of philosophy that cannot be naturalized entirely. On the other hand, the followers and objectors of Quine's naturalism suggest that philosophical analysis is needed alongside empirical knowledge in order to gain scientific knowledge. Consequently, naturalized epistemology can be seen to lead us to the field of cognitive science that involves both philosophical and scientific thinking.

University of Tampere

References

Chalmers, D. (1996), *The Conscious Mind*, Oxford, Oxford University Press.
Chalmers, D. (2003), "Consciousness and its Place in Nature", in Stich and Warfield (2003), pp. 102-142.

23 Quine (1960, p. 193) and Quine (1969c, pp. 139-144).
24 Chalmers (1996, pp. 60-61).

Churchland, P. (1981), "Eliminative Materialism and the Propositional Attitudes", in Lycan (1990), pp. 206-223.
Churcland, P. (1989), *A Neurocomputational Perspective*, Cambridge, MA, The MIT Press.
Churchland, P. S. (1996), "The Hornswoggle Problem", in Shear (1997), pp. 37-44.
Dennett, D. (1987), *The Intentional Stance*, Cambridge, MA, The MIT Press.
Dennett, D. (1991), *Consciousness Explained*, Little, Brown, Boston.
Kripke, S. (1980), *Naming and Necessity*, Oxford, Basil Blackwell.
Lycan, W. (ed.) (1990), *Mind and Cognition – A Reader*, Guildford, Basil Blackwell.
Nagel, T. (1974), "What Is It Like To Be a Bat", *The Philosophical Review* 83, 435-450.
Quine, W. V. O. (1951), "Two Dogmas of Empiricism", *The Philosophical Review* 60, 20-43.
Quine, W. V. O. (1960), *Word and Object*, Cambridge, MA, The MIT Press.
Quine, W. V. O. (1969a), *Ontological Relativity and Other Essays*, New York, Columbia University Press.
Quine, W. V. O. (1969b), "Epistemology Naturalized", in Quine (1969a), pp. 69-90.
Quine, W. V. O. (1969c), "Propositional Objects", in Quine (1969a), pp. 139-160.
Shear, J. (ed.) (1997), *Explaining Consciousness – The Hard Problem*, Cambridge, MA, The MIT Press.
Stich, S. and T. Warfield (eds.) (2003), *The Blackwell Guide to Philosophy of Mind*, Oxford, Blackwell.

III

Science, Values, and Literature

Iris Murdoch's Naturalism

Kate Larson

1.

Iris Murdoch (1919–1999) was a prolific novelist, the author of twenty-six novels. As a philosopher she published less, and her reputation is mainly due to a slim volume published in 1970, *The Sovereignty of Good*, which consists of three essays. In these she displays her originality as a thinker; against the *techne* of her time, she fuses her interests, art and philosophy, under the considerations of truth and life. Her main influence has been in moral philosophy, the narrowness of which she has inspired to expand. Also, she is a major figure among those who have tried to explore the interdisciplinary connections and possibilities between philosophy and literature.

In this paper I will initially focus on her early writings, which are mainly polemical in tone and more circumspect than the later texts. When Iris Murdoch, in the texts that make up *The Sovereignty of Good*, articulates her own position, it reads as follows:

> I offer frankly a sketch of a metaphysical theory, a kind of inconclusive non-dogmatic naturalism, which has the circularity of definition characteristic of such theories.[1]

In its context this develops into a Platonic position. Although original, it refrained several of her colleagues from considering the thesis. My aim is to show how the early writings, through an unbiased criticism, pave the way for the naturalism Murdoch speaks about in the above-quoted passage.

2.

Iris Murdoch voiced in the fifties a critique of the contemporary and prevalent view in British moral philosophy. Her major criticism was that this view or rather philosophical attitude was putatively neutral. She gave it a name – "the liberal standpoint" – and thus situated it in history, and as more or less general examples

---

1 Murdoch (1997b, p. 335).

of its expounders she choose the at the time influential thinkers Stuart Hampshire and R. M. Hare.

As historically situated in the philosophical tradition this position is the result of the elimination of metaphysics from ethics. Its method is linguistic analysis. When Murdoch scrutinizes it she finds at its base a fusion between an anti-metaphysical and an anti-naturalistic argument, both of which supports its supposed neutral or logical stance. The linguistic analysis applied to moral statements is capable of strictly separating the descriptive from the evaluative not just considering propositions, but regarding moral words as well. Given the supposition of the rationality of moral judgements, their claim to universality, the judgement involves an objective description of the facts of the situation in which it takes place. The moral words acquire their meaning through the functions of recommendation and specification. And a moral concept, Murdoch summarizes, would be "roughly an objective definition of a certain area of activity plus a recommendation or prohibition" (1997d, p. 77).

The current philosophical view is then, in Murdoch's own words "simple, behaviouristic, anti-metaphysical, and leaves no place for commerce with 'the transcendent'" (1997d, p. 80).

When Iris Murdoch in 1961 surveys her contemporary philosophical scene her judgement is: "we have suffered a general loss of concepts" (1997a, p. 290) and because of this also a drainage of experience. The linguistic analyst shows mistrust in language, she provokingly writes. But she does not urge her colleagues[2] to abandon the linguistic method, only to take it seriously and by using language creatively to extend "as poets may extend, the limits of language, and enable it to illuminate regions of reality which were formerly dark" (1997d, p. 90).

What Murdoch challenges is the limitation of morality and moral life to the question of "a series of overt choices, which takes place in a series of specifiable situations" (1997d, p. 77). This way of defining the moral is a disregard both for the transcendent – something restricting the agent's choice from outside – and the inner life. Murdoch defends both elements as morally important when considering our understanding of both our own conduct and others:

> Moral differences can be differences of concepts as well as differences of choice. A moral change shows in our vocabulary. How we see and describe the world is morals too – and the relation of this to our conduct may be complicated.[3]

---

2 The critique is pressing also concerning non-analytical thinkers as Saussure and Derrida, whom Murdoch is about to discuss later on, as contemporaries, in Murdoch (1993, chapter 7).

3 Murdoch (1997c, p. 73).

In her analysis of the "current view" she was able to name it "the liberal standpoint", because she together with the arguments against metaphysical suppositions and naturalism detected another less neutral argument. The anti-metaphysical and anti-naturalistic arguments do have the character of exposing a logical fallacy, showing the impossibility of philosophically establishing a metaphysical belief and that a moral term cannot be defined by a non-moral term. That is, you cannot, as Hume once put it, derive 'ought' from 'is'. The principal argument is that if you do there is a suppressed evaluative premise in the judgement. But attached to these two arguments is a view on what morality *should be*; a conception of freedom, rationality and the necessity of reflection, which makes it a moral attitude, among others, and not the instrument surveying all possible attitudes which the method claims to be.

By limiting the area of the moral to action and choices and the meaning of moral words to recommendations and specifications, this view, according to Murdoch, narrows the position of the ethical in the field of philosophy as well as makes more shallow the picture of what it is to be a human being.

She argues from everyday experience:

> When we apprehend and assess other people we do not consider only their solutions to specifiable practical problems, we consider something more elusive which may be called their total vision of life.[4]

This vision can be displayed and communicated or inwardly elaborated and just guessed at from the outside, but it cannot be named totally private and superfluous, as it makes up the framework for the individual's conduct in large.

She also argues with reference to existing viewpoints, which has a completely different conception of the moral agent's task. Marxism and Thomism are examples of a "Natural Law-view", in contrast to the Liberal, which understand the individual as interacting with and moving towards a reality or framework that transcends him. Morality in this view is an increasing integration with that transcendent reality; the individual's freedom is more of an ongoing task, an achievement, than the singular freedom of choice.

The difference between these two positions is conceptual; it is a difference of conceptual frameworks, of vision.

4 Murdoch (1997d, p. 81).

3.

> A moral concept seems less like a moveable and extensible ring laid down to cover a certain area of fact, and more like a total difference of *Gestalt*. We differ not only because we select different objects out of the same world but also because we see different worlds.[5]

A moral concept is not always the description of a situation, open to discussion, but could show itself as the form or colour of reality. This brings with it that the area for the moral will include understanding and interpretation, and its activity that of reflection and contemplation, attitudes that do not necessarily have to be communicated overtly.

So if moral differences are to be regarded as conceptual rather than pertaining to specific and actual situations, what happens to the supposed universality of rationality that for example R. M. Hare proposed? And what about naturalism? Is it still a fallacy?

Iris Murdoch would not dispute that there in a given situation are facts that we would all agree on; still, she advocates the idea of "moral facts". That is: an interpretation of the situation in moral terms where the moral concept decides the nature of the situation. If it where removed, we would face another situation, other facts.

> If we interest ourselves in the conceptual background of choice, and the 'vision' and 'moral being' of the chooser, we shall see naturalism not as a fallacy but as a different system of concepts. The current model, so far from refuting naturalism, merely summarises a non-naturalistic moral attitude.[6]

There is no neutral area when it comes to morality; there is no definite separation between fact and value, because if moral concepts, as Murdoch writes, are regarded as "deep moral configurations of the world" rather than as moveable rings over factual areas, then there is "no facts 'behind them' for them to be erroneously defined in terms of" (1997d, p. 95).

I do not know if Murdoch at the time had read Simone Weil's essay on reading ("Essai sur la notion de lecture", 1946), which is a direct comment on the relation between fact and value but put in the broader terms of our different readings – interpretations and/or perceptions – of a situation, of the world as it presents itself to us. However, Murdoch was reading Weil's notebooks during the fifties and would certainly have been stirred by their discussions of the impossibility, in an

5 Ibid., p. 82.
6 Ibid., p. 96.

academic and philosophical way, to solve or even elucidate the problems. Rather, Weil shows that there is work to be done, painstakingly so, by way of attention, the ever richer perception of the otherness of the world.

Murdoch has acknowledged her debt to Simone Weil and her adoption of the concept of attention as the key moral word in her thought. In Simone Weil's thinking attention is a non-prejudiced reception of the world and the other. Due to the ego's self-centred perspective this reception or attention is also the creative act of love. In Murdoch's terminology attention is the "just and loving gaze directed upon an individual reality" (1997b, p. 327). With this she qualifies the moral sense of seeing which is the inventive result of moral imagination and effort; that is love. "I would like on the whole to use the word 'attention' as a good word and use some more general term like 'looking' as the neutral word" (1997b, p. 329).

The concept of attention has not, in my view, been fully explored in the reception of Murdoch's ideas. It has been reduced to the, not uninteresting but more narrow, emotional and personal receptivity or sensibility to particular situations, which for the more sensible is morally coloured, for others not. The outcome is an ethics of particularity, which to some degree echoes the ethics of universality that Murdoch criticized in Hare.

As Charles Taylor puts it when in the essay "Moral Philosophy" he tries to situate Iris Murdoch in the wider context of moral philosophy and ethics:

> Iris Murdoch was criticizing the narrowness of moral philosophy well before the present counterwave – and not entirely on the same grounds. Indeed, her reflections would lead us to make a further extension of the ethical, beyond those proposed by neo-Nietzscheans and neo-Aristotelians. Put briefly, while these take us beyond morality to issues about the good life, they stop there. Murdoch seems to me to take the first step, but then one further. Beyond obligatory action, she opens up the question of what it is good to be.[7]

4.

The question of what it is good to be, as Charles Taylor summarizes Iris Murdoch's contribution to moral philosophy or the equation between the cognitive and the moral, as Cora Diamond rephrases it, will lead us to Murdoch's pronounced Platonism. She read Plato as one of the few realistic thinkers considering human existence, its limits and potentiality. Plato was a pessimist, but allowed and embraced the concepts of development, education and enlightenment in respect to each individual endeavour. Murdoch observes that Plato, contrary to his diverse

---

7 Taylor (1996, p. 5).

critics, "saves metaphysics by showing how the noumenal and the phenomenal exist *inside* each human life" (1993, p. 182). This reading of Plato, and the appreciation that he regards individuals not as solitary subjects but as persons interacting in and with a multifaceted world, is foundational for Murdoch's philosophy, which places the individual consciousness at the centre as a *value-bearing continuum* (1993, p. 148). Values cannot be derived from facts because of the ubiquity of value – moral discrimination, she writes, is not a natural instinct among others. It is for human beings something *sui generis*: unique and special, forming and transforming us.[8]

When Murdoch criticized "the current philosophical view" in her essay "Vision and Choice in Morality" (1956) she said that it left no room for our "commerce with the transcendent" (1997d, p. 80). This commerce is Murdoch's main interest because it also is the possibility of our own transcendence. The transcendent is in Murdoch's thought ranging from our fellow human beings, the other in his or her individual reality or freedom, to the postulated and transcendent value of the Good, the Platonic form which could be understood as an image of our truth-seeking relationship to reality.

This relationship can of course philosophically only be studied through the nature of the human being, something that, as she writes, occurs at "a point of great conceptual sensibility". And further: "Man is a creature who makes pictures of himself and then comes to resemble the picture. This is the process which moral philosophy must try to describe" (1997c, p. 75).

Murdoch sketched her philosophical ambition in *The Sovereignty of Good* as an ambition to offer a metaphysical theory, an inconclusive, non-dogmatic naturalism. Perhaps the non-dogmatic element will appear to you if I quote at length from the summary in *Metaphysics as a Guide to Morals*, Murdoch's last philosophical work from 1992. Something which will also, I hope, show you the continuity of her thought.

> Good is also something clearly seen and indubitably discovered in our ordinary unmysterious experience of transcendence, the progressive illuminating and inspiring discovery of *other*, the positive *experience* of truth, which comes to us all the time in a weak form and comes to most of us sometimes in a strong form (in art or love or work or looking at nature) and which remains with us as a standard or vision, an *orientation*, a *proof* of what is possible and a vista of what might be. I see here no need for a dramatic voluntarist 'way' to be pressed upon us by theologians. The ordinary way is the way. It is not in that sense theology, and the 'mysticism' involved is an accessible experience. The 'proof' may be said to resemble the

8 See the discussion in chapter 2 of Murdoch (1993).

> Cartesian *cogito* in that it is an intuitive movement which we may prompt ourselves to make, a proof by morality and by love, not by logic.[9]

To conclude I would like to return to the beginning of this paper and Murdoch's analysis of the current view, the anti-naturalistic and anti-metaphysical philosophical stance as an attitude and not the impartial and logical position it posed as. The philosophical attitude to be preferred, she seems to say, should be the one of love. This pertains to her discussion of moral differences as conceptual differences. "Great philosophers coin new moral concepts and communicate new moral visions and modes of understanding", she writes (1997d, p. 83). This is an act of love, or rather of attention, the term she borrowed from Simone Weil which portrays the individual's interest in the world other than herself, thus experiencing and communicating an ever richer and complex reality trough imaginative and precise use of language.

Uppsala University

References

Antonaccio, M. and W. Schweiker (1996), *Iris Murdoch and the Search for Human Goodness*, London, The University of Chicago Press.
Murdoch, I. (1993), *Metaphysics as a Guide to Morals*, London, Penguin Books .
Murdoch, I. (1997a), "Against Dryness", *Existentialists and Mystics*, London, Chatto & Windus, pp. 287-295.
Murdoch, I. (1997b), "The Idea of Perfection", *Existentialists and Mystics*, London, Chatto & Windus, pp. 299-336.
Murdoch, I. (1997c), "Metaphysics and Ethics", *Existentialists and Mystics*, London, Chatto & Windus, pp. 59-75.
Murdoch, I. (1997d), "Vision and Choice in Morality", *Existentialists and Mystics*, London, Chatto & Windus, pp. 76-98.
Taylor, C. (1996), "Moral Philosophy", in Antonaccio and Schweiker (1996), pp. 3-28.

9 Ibid., p. 508.

Truth in Science and Ethics

Cheryl Misak

1. Introduction

My suggestion in this paper will be that the concept of truth is applicable in both science and ethics. If this suggestion is to be made plausible, it is clear that not just any old concept of truth will do. I am going to assume that everyone sees that a view of truth which takes truth to be solidarity – or what is believed by this or that community – is not a view of truth that is suited for science. I am also going to assume that everyone sees that a view of truth which takes truth to be correspondence to a believer-independent world is not a view of truth that is suited for ethics. Indeed, I think that there is just one view of truth on which it makes sense to think that beliefs in both science and ethics might be true. That account of truth is the one given to us by the founder of pragmatism: C. S. Peirce. He argued that a true belief is a belief which would be indefeasible or which would stand up to the rigors of inquiry. (CP 5.569, 6.485.[1]) A true belief is one which is "unassailable by doubt"; it is a belief which would meet every demand we were to place on it. (CP 5.416.)

This account of truth is entirely general – that is, it is applicable in principle to any kind of discourse or domain of inquiry.[2] Of course, different kinds of inquiry (moral, mathematical, scientific, etc.) will appeal to different standards – standards of indefeasibility will vary from kind of inquiry to kind of inquiry. Nonetheless, we have just one concept of truth here: truth is indefeasibility. And the property this concept picks out is a stable property – a belief is either true (indefeasible) or not. It is not a property that fluctuates from community to community – if a belief is really indefeasible, it would stand up to whatever could be thrown at it, by any community of inquirers.

I will be happy if I succeed in this paper in doing two modest things. First, I will try to motivate the suggestion that beliefs in science and in ethics are both aspirants to truth. Second, I will gesture at what we might gain if we can make good sense of this idea.

---

1 "CP n.m" refers to the *Collected Papers*, where n is volume number and m is paragraph number.

2 See Misak (2000; 2004) for details about how it is at home in science and in mathematics.

2. The First Motivation

The first way of trying to motivate a unified view of truth and inquiry turns on a familiar point about the phenomenology of inquiry. Our practices in both science and ethics betray the fact that I don't aim at what my own standards point to (what is justified by my lights) or at what my community's standards point to (what is justified by our lights). We distinguish between thinking that one is right and being right; we criticize the beliefs and actions of others; we think that we can improve our judgements; and we take ourselves to be able to *learn* by gathering more evidence, by paying attention to the views of others, by putting ourselves in another's shoes, by examining the arguments of the other side, by broadening our horizons, etc. We think that 'rational' persuasion, not brow-beating or force, is the appropriate way to get people to agree with us. Indeed, we want others to *agree* with us, not to merely mouth what we say or fall in line with it.

That is, our practices – our ways of inquiring and deliberating in science and in ethics – rest upon or presuppose a notion of truth that goes beyond what is justified by my lights or by the lights of my local community. Now clearly this kind of argument from the phenomenology of inquiry is a point of which the pragmatist is fond. The pragmatist thinks that in order to fully understand *any* concept, we need to explore its relationship with practical endeavors. In the case of the concept of truth, we need to explore its relationship with the practices of belief, assertion, and inquiry. As Peirce put it: "We must not begin by talking of pure ideas, – vagabond thoughts that tramp the public roads without any human habitation, – but must begin with men and their conversation" (CP 8.112). As David Wiggins (2002, p. 316) puts it, when a concept is "already fundamental to human thought and long since possessed of an autonomous interest", it is pointless to try to define it. Rather, we ought to attempt to get leverage on the concept, or a fix on it, by exploring its connections with practice.

But of course, those who don't already think that there is something right about pragmatism will be less enamored of an argument that starts straightforwardly from practice or the phenomenology of inquiry. They will ask: what if we are mistaken in our reliance upon a robust notion of truth? What if not everyone relies upon such a notion of truth? What closes the gap between what we find ourselves doing (what our practices are) and what we *ought* to do or *must* do? And they may say that even if you accept the legitimacy of starting from practice, you don't get to the conclusion that we aim at truth in ethics. There is only disagreement and variance in ethics, nothing like agreement and truth.

So we need a stronger motivating argument. We need an argument that concludes that it makes little sense to suggest that we might be mistaken in relying upon a robust notion or truth or that not everyone might rely on such a notion of truth. As Bernard Williams has put it, we need to show that that truth will be valued in any human society. I will offer two versions of such an argument – one I put forward in *Truth and the End of Inquiry*, another recently put forward by Huw Price.

3. The Second Motivation: Truth and its Connection to Belief and Assertion

Reflect on the difference between the phrases "I suspect that *p*" or "it seems to me that *p*", on the one hand, and "I believe that *p*" or "I assert that *p*", on the other. What I do when I use the first two phrases is distance myself from the obligations which come with belief and assertion. (From here on I will stick to belief, although everything I say applies equally to assertion.)

Those obligations can be summarised by saying that if I believe that *p*, I commit myself to saying what could speak for or against *p* and to giving up *p* in the face of sustained evidence and argument against it. A belief, in order to be a belief, is such that it is responsive to or answerable to reasons and evidence. That is a very part of what is to have a belief – it is a *constitutive norm* of belief. Part of what it is to be a belief, as opposed to some other mental state, such as an entertaining of an interesting but idle thought, a lie about what one believes, or a dogmatic opinion, is that there must be something that can speak for or against a belief and that belief must be sensitive to what can speak for or against it.[3] Some cognitive states – those not appropriately connected to reasons – are not deserving of the label 'belief'.

A believer thinks that her belief fits best with the evidence and argument. I cannot get myself to believe that *p* by deciding that if the coin I am about to flip lands heads, I will believe it, and if it lands tails, I will not. In order to believe *p* I have to be convinced that I have good reason to believe it. A corollary of this point is that, as Bernard Williams (1973) has argued, we cannot decide to believe – we cannot get ourselves, by an act of will alone, to believe that something is the case. Raz (1999, p. 10f) has rightly pointed out that much of the discussion about whether we can choose to believe would have been better conducted if attention had been paid to the fact that beliefs belong to the active side of our lives, not the passive. Beliefs do not just come upon us – we reach for them. We respond to reasons. We can decide to believe something only in the sense that we can decide to deliberate. We might then conclude that we have good reason to believe *p*, but

3 Of course, the reasons standing behind a belief need not be explicitly acknowledged – I might believe and act habitually, but for reasons nonetheless. When I stop at the red light, I need not articulate to myself the reasons for stopping, but the reasons are there. I would articulate them if asked.

we cannot pick a belief and decide to believe it, not caring where the reasons may fall.

So this motivating argument is that what it is to assert, to make a claim, to believe, to judge, is to be engaged in a process of justification aimed at getting things right. If (to use a well-known Peircean contrast) we are to retain the distinction between mere tenacity (holding on to our opinions come what may) and belief, then our beliefs must in principle be responsive to evidence and argument and they must be aimed at being such that they would be account for that experience and argument. They must be aimed at truth. This holds in any domain of inquiry in which we have beliefs and we have genuine beliefs in both science and ethics.

4. The Third Motivation: Price on Truth and Disagreement

Huw Price has recently put forward a very nice variation of the second motivating argument. The pragmatist of Rorty's stripe, Price argues, fails to see that when you examine practice, you find that truth is not just solidarity or what this or that community happens to find best to believe. When you examine practice, you find that the goal of inquiry is truth as we have always thought of it – something stable and independent of what this or that person or community might think. Speakers take themselves to be subject to the norm of truth – not just the norms of sincerity, justification, warrant, or keeping in step with the beliefs of the community. Price thinks that "in practice we find it impossible to stop caring about truth". (Price 2004, p. 187.)

His thought is that the behavior pattern which betrays the commitment to truth is so central to what we presently regard as a worthwhile human life that no reasonable person would experiment with trying to do without the norm of truth. (Price 2004, p. 168.) No reasonable person would try to rest with the aim of getting beliefs that are warranted or justified by current community standards. For the norm of truth plays an essential, indeed a constitutive, role in assertion and dialogue as we know them. To do without truth is to silence our conversations – both our conversations with others and our own internal conversations.

And Price gives us something important to say to our nay-sayer, who points to the disagreement prevalent in ethics. "The very essence of the norm of truth", argues Price, is to give disagreement "its immediate normative character". (Price 2004, p. 168.) It is to make disagreement matter: "Without truth, the wheels of argument do not engage; disagreements slide past one another." (Price 2004, p. 185.) In order to really *engage* others in conversation or dialogue, we have to see their disagreement as implying a mistake on someone's part. Otherwise, we are merely talking past each other.

The flip side of this point is that without truth, we could not find the conceptual space to formulate the idea of improving upon our beliefs. Indeed, the norm of truth encourages improvement by motivating speakers who disagree to try to resolve the disagreement. For disagreement, again, implies a mistake on someone's part.[4]

Price brings the norm of truth into focus (and shows that it plays the above distinct and critical roles in linguistic practice) by asking us to imagine a community which uses language primarily for expressing preferences – a community of speakers which obeys only the weaker norms of subjective assertibility (sincerity) and personal warranted assertibility (justification). The norm of sincerity says that to assert that *p* without believing that *p* is a ground for censure or disapprobation. The norm of justification says that to assert that *p* without having adequate personal grounds for believing that *p* is a ground for censure or disapprobation. A person who meets both of these norms may be said to have done as much as possible, by her own current lights, to ensure that her assertion that *p* is in good order. These speakers can criticize each other for being insincere or for making personally unjustified statements, but not for failing to tell the truth. They have merely-opinionated assertions (MOAs) – hence the name 'Mo'ans' for the members of this community. (Price 2004, p. 177.)

Mo'ans are like dedicated lunchers whose language atrophies to the bare essentials (Price 2004, p. 177). They might criticize expressed preferences on the ground that they were not sincere or on the ground that they were not well-founded by the speaker's own lights. ("You always say that you want the goat, but when it arrives, you don't eat it!") Mo'ans, that is, use linguistic utterances to express their 'beliefs', preferences, and desires. They differ from us in that they do not take disagreement to indicate that one or the other speaker is at fault or mistaken. They do not say "goat tastes disgusting and if you disagree, you are mistaken". Disagreements are of a no-fault kind.

A Mo'an, that is, cannot be criticized for getting things wrong, even when she is not talking about what she would like to eat, but about whether there is e-coli in the petri dish or whether there is milk in the refrigerator. For Mo'ans do not treat a disagreement between two speakers as an indication that one speaker has made a mistake.

---

4 See Misak (2000, pp. 77, 94f.) for a similar thought – the notions of improvement and mistake require a robust concept of truth. Price notes that Rorty can identify improvements and mistakes by contrasting a belief with the beliefs of the community. But, given that we also want to be able to say that a community's beliefs can be mistaken and can be improved, we are directed to something more like Peirce's account of truth, on which the community's belief can be contrasted with the belief that would stand up to inquiry, were inquiry to proceed as far as it fruitfully could go.

We can now see how our linguistic practice differs from that of Mo'ans. In our practice, we can distinguish between telling the truth and being sincere. I can be sincere (say what I really believe) without telling the truth. The converse is also possible: I can tell the truth accidentally – i.e. without being sincere. We can also distinguish between being justified and telling the truth. I can have the best possible (personal) justification for *p* and yet *p* can be false. The converse is again possible: *p* can be true and I can lack justification for it. The Mo'ans are unable to make these distinctions.

By noting such differences, we can see what the third norm – the norm of truth – adds to our practice. The main difference, Price tells us, is that in the Mo'ans linguistic practice we cannot find dialogue and argumentation as we know them. We cannot, that is, find the *engagement* of individual opinions, as opposed to the mere roll call of individual opinions. (Price 2004, p. 169.) The third norm adds something new to a practice governed merely by the first two norms: it adds "the grit that makes our individual opinions engage with one another" (Price 2004, p. 169). He thinks that the difficulty we have in holding on to the idea of such a community as the Mo'ans

> stems from our almost irresistible urge to see the situation in terms of our own normative standards. There really is a third norm, we are inclined to think, even if these simple creatures don't know it. If two of them make incompatible assertions then one of them must be objectively incorrect, even if by their lights they both meet the only norms they themselves recognize. (Price 2004, p. 178.)

In the following statement, the affinities between Price's argument and the argument I presented as the Second Motivation should be clear:

> The crucial point is thus that assertoric dialogue requires an intolerance of disagreement. This needs to be present already in the background, a pragmatic presupposition of judgement itself. I am not a maker of assertions, a judger, at all, unless I am already playing the game to win, in the sense defined by the third norm. (Price 2004, p. 186.)

Assertion, judgement, and belief presuppose or require that one aims at getting things right, where 'getting things right' means getting to the answer which is not merely justified by me, or by my community, but justified *tout court*. And, although this is not explicitly a part of Price's position, it is at least plausible that we engage others in moral deliberation – that here too, disagreement *matters*.

5. Ethics

Enough said about the motivation for the search for a concept of truth that can do duty for ethics as well as for science. Let us turn now to my suggestion that Peirce's account fits the bill and to what it might give us. Peirce, we have seen, holds that a true belief is one that would satisfy all of our aims in the inquiry to which it belongs – it would be indefeasible or would stand up to whatever experience and argument were put to it. Peirce gave careful consideration to mathematics, arguing that it is an inquiry which is responsive to experience and argument. Indeed, he carried out significant first-order inquiry in logic and mathematics, experimenting upon diagrams, just as his conception of experience for that kind of inquiry requires.

He was less clear about ethics, but the upshot of his discussion is that here too we find an inquiry which is responsive to experience and argument. We are, as Price suggests, bothered by disagreement and we think that we can improve our judgements. When we deliberate about what we ought to do, we take ourselves to be sensitive to reasons, argument, thought experiments, and first-person experience. We try to put ourselves in the shoes of others, to broaden our horizons, to listen to the arguments of the other side. That is part of what it is to make a moral decision and part of what it is to try to live a moral life. It would not be a moral life – it would not be engaged with the complexities of moral requirements – if we simply made our decisions about how to treat others by following a oracle, or an astrologer, or the toss of the dice.

Ethics, Peirce argues, is like metaphysics in that it rests on observations "and the only reason that this is not universally recognized is that it rests upon kinds of phenomena with which every man's experience is so saturated that he usually pays no particular attention to them" (CP 6.2, 1898). The observation of phenomena in the special sciences requires special instruments, precautions and skill. (CP 1.242, 1902.) The observation of phenomena in metaphysics and ethics does not, but metaphysical and ethical phenomena are "harder to see, simply because they surround us on every hand; we are immersed in them and have no background against which to view them" (CP 6.562, 1905).

Given that the place of experts and equipment in moral deliberation is at best marginal, this kind of commonplace and non-technical observation does indeed seem appropriate to moral beliefs. We often feel the force of experience in moral matters and, as in mathematics, this can take two forms. One is what often gets called by philosophers 'intuition' or 'felt response'. Upon observing a group of teenagers bully and torment a younger child, we simply 'see' that it is cruel. The other kind of experience in moral deliberation is that we find some reasons, arguments, imagined situations, and thought experiments compelling and may, in light of them, revise our moral judgements. We can learn from debate and discussion, from 'morals' we find in novels, films, Aesop's fables, the parables of the

New Testament, and from the philosopher's hypothetical situations and counter-examples to general moral principles.

We need not, and must not, suppose here that we have a special faculty for perceiving right and wrong or that our intuitions are self-evident or indefeasible. We need only suppose that we find ourselves compelled in moral matters, as we find ourselves compelled when considering whether Toronto is north of Buffalo or whether 2+2=4. Observation in the moral domain is of course fallible. What is intuitively obvious in any kind of inquiry can be discredited – think of initially plausible stereotypes in the social sciences and the fact that Einstein's relativity theory was taken at the outset by many to be intuitively false. Indeed, our very ability to recognize mistakes like these turns on the idea that what strikes us as obvious is subject to further experience. We can, that is, be critical of our 'intuitions'. Those intuitions are just our deeply held, but revisable, background beliefs and habits. Society, Peirce argues (with James) may be seen as a long-running experiment aimed at identifying the best kind of conduct. Its conventions thus deserve respect, for, while remaining fallible, they capture the experience of generations. Peirce, and every pragmatist after him, insists that those beliefs which inquiry has not thrown into doubt in such ways are such that it is pointless to try to whip up a paper or a Cartesian doubt about them. What this means for epistemologists and for moral philosophers is that we have no choice but to begin where we find ourselves, with our current beliefs and with our thoughts about what counts as evidence, about what the standards for justified beliefs are, etc. There is

> but one state of mind from which you can 'set out', namely the very state of mind in which you actually find yourself at the time you do 'set out' – a state in which you are laden with an immense mass of cognition already formed, of which you cannot divest yourself if you would. (CP 5.416, 1905.)

In both scientific and moral matters, we have cherished beliefs which are nonetheless responsive to experience. In ethics, as in science, we act on our experience-driven background beliefs, while realising that they might yet be overthrown by further experience.

Clearly, we can mistakenly preserve the status quo; we can be too reluctant to revise our background beliefs. The only way of guarding against this – in any area of inquiry – is to recognise the possibility and then to try to stay well away from it by listening to challenges to deeply held beliefs and practices. Those challenges can come from within, when my own judgements or principles conflict and I feel a pull towards revising them. And they can come from without – from the senses or when I see that the judgements of others, from within my circle or from afar, conflict with my own judgements and I feel a pull towards reconsidering them. I can then revise those beliefs while holding enough stable against which the revision can take place.

There are of course tensions involved in thinking that our human, contingent, historically rooted judgements about how to treat others are such that they might be objectively correct or incorrect. If we are to get right the thought that our moral judgements aim at truth, we need to preserve these tensions. Failure to do so will result in a view which does not reflect the difficult, problematic, and perhaps tragic nature of moral deliberation.

At one point Peirce distinguishes disagreement in moral matters from disagreement about taste: "However it may be about taste, in regard to morals, we can see ground for hope that debate will ultimately cause one party or both to modify their sentiments up to complete accord." (CP 2.151, 1902.) Then he delivers what seems to be an inconsistent thought:

> Should it turn out otherwise, what can be said except that some men have one aim and some another? It would be monstrous for either party to pronounce the moral judgments of the other to be *bad*. That would imply an appeal to some other tribunal. (CP 2.151, 1902.)

At first glance, this looks like a straightforward contradiction. But Peirce here is articulating his rather subtle position on bivalence – a position especially suited to the nature of moral judgement. Should it turn out that there is no possibility of accord – should it turn out that this hope or regulative assumption of inquiry would in some case not be fulfilled – then there is no truth of the matter at stake. We frequently find Peirce saying, in an entirely general fashion:

> It is true that we cannot know for certain that experience, however long and full, ever would bring all men to the same way of thinking concerning the subject of inquiry. But that is the only result that can satisfy us, so that we must forever continue in the hope that it will come, at last.[5]

We must hope, for any question into which we inquire, that bivalence will hold, but we cannot know that it will. And here is the point to which any cognitivist must be alert: we may expect that bivalence will fail more often in moral inquiry than in chemistry and less often than in deliberation about matters of taste.

A particularly helpful text here is the 1901 "Truth and Reality and Error". Peirce rehearses his general stance on bivalence and says that perhaps the question of whether there is free will is such that no answer would be forthcoming, no matter how long the discussion were to go on and no matter how advanced our methods of inquiry were to become:

---

5 MS 1342, p. 2 of variants, undated. See also MS 408, p. 147, 1893–5.

> Then in regard to that question, there certainly is no *truth*. But whether or not there would be perhaps any *reality* is a question for the metaphysician ... Even if the metaphysician decides that where there is no truth there is no reality, still the distinction between the character of truth and the character of reality is plain and definable. (CP 5.565, 1901.)

Chris Hookway (2000) has drawn our attention to this distinction in an especially helpful way. Truth, for Peirce, may be indeterminate and the underlying reality determinate. 'True' applies to propositions, statements, or beliefs if they would survive all attempts at refutation – if they would never lead to disappointment. (CP 5.569, 1901.) 'Real' applies to things as they are irrespective of what any mind or collection of minds takes them to be (CP 5.565, 1901).

After distinguishing truth from reality, Peirce very clearly says that the distinction holds not just for science, but also for ethics (CP 5.566, 1901) and for mathematics (CP 5.567, 1901). Elsewhere he says that all of these inquiries aim at the truth:

> Now the different sciences deal with different kinds of truth; mathematical truth is one thing, ethical truth is another, the actually existing state of the universe is a third; but all those different conceptions have in common something very marked and clear. We all hope that the different scientific inquiries in which we are severally engaged are going ultimately to lead to some definitely established conclusion, which conclusion we endeavour to anticipate in some measure. Agreement with that ultimate proposition that we look forward to, – agreement with that, whatever it may turn out to be, is the scientific truth. (CP 7.187, 1901.)

There will be differences between kinds of inquiry: the mathematician, the chemist, and the inquirer into what is the morally right thing to do will not all find themselves engaged in identical methods. Nor will they find that their aspirations have identical prospects. Nor will they all be talking about the same sort of reality. As Hookway puts it:

> We might agree that mathematical propositions, ethical propositions, propositions from the more theoretical reaches of science can all be assessed as true or false. Each, we might suppose, can be tested or 'compared with reality'. This might involve looking for a proof, considering how the ethical proposition would appeal to someone who took up a distinctive disinterested viewpoint on things, or making explanatory inferences about what best systematises our other theoretical beliefs and experimental results. (Hookway 2000, p. 97.)

That is, comparing hypotheses with 'reality' is bound to take different forms in different inquiries.

Peirce goes on to explicitly make the point that I have been pressing: what is central and common to all these various inquiries is the surprise of experience, against a background of stable expectations or beliefs. He says: "Thus it is that all knowledge begins by the discovery that there has been an erroneous expectation ... Each branch of science begins with a new phenomenon which violates a[n] ... expectation." (CP 7.188, 1901.)

Let us look at the mathematical case, about which Peirce is exceptionally clear. In "Truth and Falsity and Error", he says:

> [t]he pure mathematician deals exclusively with hypotheses. Whether or not there is any corresponding real thing, he does not care. His hypotheses are creatures of his own imagination; but he discovers in them relations which surprise him sometimes. A metaphysician may hold that this very forcing upon the mathematician's acceptance of propositions for which he was not prepared, proves, or even constitutes, a mode of being independent of the mathematician's thought, and so a *reality*. But whether there is any reality or not, the truth of the pure mathematical proposition is constituted by the impossibility of ever finding a case in which it fails. (CP 5.567, 1901.)

Peirce himself argued that mathematics does not answer to a *physical* reality – it is not concerned with physical objects, but with possibilities (CP 4.234, 1902, 3.527, 1896) or the forms of relations (CP 4.530, 1905). And he was a realist about possibilities. But his point here is that, whatever your metaphysics might be, mathematics aims at the truth in the same way other kinds of inquiry aim at the truth. A true belief would be the best belief, were we to pursue our inquiries as far as they could fruitfully go and what makes for a best belief in mathematics might differ from what makes a belief best in science, or in morals. As Hookway puts it, Peirce held that the idea of truth is metaphysically neutral.

"Truth and Reality and Error" is thus a wonderful text for the pragmatist who wants to think of moral belief as being truth-apt. We have Peirce saying, very clearly, that a belief can be sensitive to experience even if there is no underlying physical reality. Perhaps a domain of inquiry which rests on an underlying physical reality will contain more statements which are bivalent. But some kinds of inquiry, such as mathematics, will be full of bivalent statements and yet they are such that there is no underlying physical reality.

Moral deliberation presents us with especially interesting phenomena. We might, for instance, think that there are situations in our moral lives where no decision can be right. Such tragic contexts are exemplified by those horrors from Nazi Germany, where a concentration camp guard tells a mother that she must choose one of her two children for the gas chamber. If she does so, one child will live; if she refuses to choose, both will die. There seems to be no decent solution to this kind of problem. Not even a random choice, a flipping of a coin, will be right, for the mother will surely feel that in making the decision in any way, or in refusing to make it, she betrays something that is valuable and fundamental.

We should not want to set up our moral epistemology so that we do away with the possibility that some issues will prove to be impossible – so that we do away with situations in which wretched compromise is the best that we can manage. To return to the point at which I started, ruling out such possibilities would be untrue to the phenomenology or practice of morals. It would be untrue to all those occasions where we feel at a loss, where we feel that no matter how hard we persevere, there is no right answer to be had, or where we feel regret that, even though we did something acceptable, we were unable to act on the opposing considerations. To aim for precision where there may be none would be to do a disservice to the kind of inquiry we are trying to characterise. The Peircean view of truth, with its explicit accommodation of the possibility of the failure of bivalence, is well-suited to capture this kind of phenomenon.

Another kind of problematic situation which arises in moral deliberation is that in which there seems to be a number of equally good ways of answering a moral question. It can seem, for instance, that those with one cultural and religious background will have one answer to a question about the right thing to do and those with a different cultural and religious background will have an equally good, but different, answer.

Here the Peircean can think of the potential agreement not as agreement about which one way of life is best, but as agreement about which ways of life are reasonable or permissible. What we might agree upon is that a number of conceptions of how best to live are acceptable and these conceptions will produce different, equally acceptable, answers to some moral questions. There is nothing in the Peircean view which suggests we must hope for an agreement which will level all difference.

Peirce's view, that is, is not committed to a melting-pot metaphor. It is not that we need lots of perspectives in order for the best one or the best amalgam to emerge. Rather, one would expect that plurality of belief would sometimes be preserved, were moral deliberation to continue. (*P or Q*) may be true for the whole community of inquirers. This is very different from saying that *P* is true for one subset of the community and *Q* is true for another.

This is also very different from saying that 'anything goes'. The aim of inquiry is to get true belief, which, on Peirce's view, is belief which would forever fit with experience and argument. The best means to this aim is clearly a method by which we test our hypotheses against experience and argument. If truth is that which would best fit with experience and argument, then if we want true beliefs, we should expose them to experience and argument which might overturn them. Better to find out now that a belief is defeated, rather than down the line.

Not only are we justified in exposing our beliefs to experience, but we are justified in exposing them to as broad a variety of experience as we can. A scientist who refused to take into account any of the experimental results of, say the English or the Italians, would be adopting a very bad method for getting beliefs that would stand up to experience. Similarly, those engaged in moral deliberation who denigrate or ignore the experiences of those of a certain skin colour, gender, or religion are also adopting a method unlikely to reach the truth. Their methods and their beliefs can be criticised in this second-order way (as failing to be sensitive to all the experience and argument or as failing to properly aim at truth), as well as in the usual first-order kind of way (as being disrespectful or as failing to treat others as equals, etc.).

So, while *(P or Q)* might be equally acceptable, if *R* turns on denigrating some group or other, then it is unacceptable. And the judgements in which it issues are not likely to be true. Peirce's position, while leaving plenty of room for difference and plurality, has the wherewithal to rule out some beliefs and some methods (intolerant ones, for instance) as being on the wrong track.

The upshot is that the fact that the sort of experience relevant to morality is liable to vary from culture to culture need not be destructive for Peirce's position. For that position allows for some failures of bivalence and it allows us to criticise cultures which do not take the moral experience of others seriously. It brings moral deliberation within the scope of an experience-driven and reason-driven inquiry which is aimed at the truth, yet it does not pretend that ethics is identical to science or has the same prospects as science.

Indeed, we are not short of explanations for why we find things as we do in moral deliberation. It might be that moral judgements require more collateral information. The judgement that the bullies are cruel requires, for instance, that we know that they intend to cause distress.[6] Or perhaps science has a more precise theory of error; perhaps disagreement there can often be explained by citing malfunctioning or misuse of equipment, for instance. Perhaps the scarcity of goods and resources explains some kinds of irresolvable conflict in the moral and political domain. All these things said, we must be careful also to notice the tremendous amount of underlying agreement in moral matters as well as the diversity of background theory and belief in science and mathematics. As Putnam (1987) has argued, nothing very uniform can be said of the sciences over and above a few thin methodological points – points like Peirce's argument about the need to take experience seriously.

We need to remind ourselves here of the importance of the distinction between belief (which is aimed at the truth) and tenacity. If we hold this distinction steady when we think of moral deliberation, we will want to retain, against the obstacles,

---

6 See Wiggins (1996, pp. 278– 279).

the thought that moral judgements are responsive to experience. What it is to assert, to make a claim, to believe, to judge, is to be engaged in a process of justification. It is to commit oneself to giving reasons – to be prepared, in the appropriate circumstances, to justify one's assertion to others, and to oneself. Those claiming to hold a belief, something which has a truth-value, or those who are engaged in dialogue and argument with others, commit themselves to being open to evidence and argument. Those who engage in moral *deliberation*, as opposed to holding on to our moral opinions come what may, must take their beliefs to be responsive to new reasons, arguments and experiences about what is cruel, kind, odious, just, etc. If one is committed to never revising an opinion, no matter what evidence against it is brought to light, then that is something like blind dogmatism and not a genuine judgement about what is true. Moral deliberation, for those who engage in it, is experience-driven, example-driven, and argument-driven. And we do sometimes, if not always, engage or try to engage in moral deliberation rather than in mere tenacity.

University of Toronto

References

Crisp, R. (ed.) (1996), *How Should One Live?*, Oxford, Clarendon Press.

Hookway, C. (2000), *Truth, Rationality, and Pragmatism: Themes From Peirce*, Oxford, Oxford University Press.

Misak, C. (2000), *Truth, Politics, Morality: Pragmatism and Deliberation*, London, Routledge.

Misak, C. (2004), *Truth and the End of Inquiry: A Pragmatist Account of Truth*, 2nd, expanded ed., Oxford, Clarendon Press.

Peirce, C. S. (1931–35), *Collected Papers of Charles Sanders Peirce*, vols. 1–6 ed. by C. Hartshorne and P. Weiss, Cambridge, MA., Harvard University Press.

Peirce, C. S. (1958), *Collected Papers of Charles Sanders Peirce*, vols. 7–8 ed. by A. Burks, Cambridge, MA., Harvard University Press.

Peirce, C. S. *The Charles S. Peirce Papers*, microfilm, Harvard University.

Price, H. (2004), "Truth as Convenient Friction", *The Journal of Philosophy* 100:4.

Putnam, H. (1987), "The Diversity of the Sciences", in P. Petit, R. Sylvan, and J. Norman (eds.), *Metaphysics and Morality: Essays in Honour of J. J. C. Smart*, Oxford, Blackwell.

Raz, J. (1999), *Engaging Reason: On the Theory of Value and Action*, Oxford, Oxford University Press.

Schantz, R. (ed.) (2002), *What is Truth?*, Berlin, de Gruyter.

Wiggins, D. (1996), "Natural and Artificial Virtues: A Vindication of Hume's Scheme", in R. Crisp (ed.) (1996), pp. 131–140.

Wiggins, D. (2002), "An Indefinibilist cum Normative View of Truth and the Marks of Truth", in R. Schantz (ed.) (2002).

Williams, B. (1973), "Deciding to Believe", in Williams, *Problems of the Self*, Cambridge, Cambridge University Press.

Philosophy and the Science of Literature

Mogens Pahuus

1. Introduction

There can be no doubt that there is a connection or affiliation between poetry or literature on the one hand and philosophy, or rather philosophy concerning the human being, on the other. For most poems – including novels, short stories, dramas – at least, it may be claimed that they provide a picture of life or of a part of life, which can be so conclusive that it may be called a philosophy of life or a perception of life. "Poetry orders life for us in clear and perceivable images", as the Danish literary scholar Svend Møller Kristensen writes in his book *Digtningens teori (The Theory of Poetry)*.[1] Accordingly, we may say that philosophy of life seeks an understanding of human existence in all its possibilities and limitations with special focus upon the question of successful self-expression, i.e. the good life. In this sense, both poetry and philosophy of life are attempts at formulating a contribution to our understanding of life.

Philosophy has not always valued this connection to poetry. There have been many types of philosophy that have not considered it a philosophical issue to investigate human self-expression and to address the question of the good life. Since antiquity, these questions have been suppressed. First of all, the question of goodness has often been limited to the field of ethics or morality in its narrow sense. Secondly, within philosophy it has been taken for granted that there is only one form of knowledge or cognition, and that is conceptual-scientific cognition, whereby art, hereunder literature, is excluded from the field of epistemology.

Today, however, there are many philosophers who both see philosophy of life as a legitimate philosophical discipline and who are aware of the connection between this type of philosophy and poetry. The theme of my discussions in the following is this: given that there is a close relationship between poetry and philosophy, how should we perceive the relationship between the science of literature and philosophy? (Here I use the term 'science' in its broadest sense. In some traditions literary studies have not been regarded as a science. In others it definitely has, especially when the paradigms for research have been inspired by the sciences of linguistics, psychology or sociology.)

1 Kristensen (1973, p. 126).

When it comes to dealing with such a conclusive and multi-faceted question, it can prove useful to link these general considerations to a limited exemplary area. I will therefore focus on one field: the phenomenon of love in its various forms – a field that is analysed in philosophy and articulated in poetry. Following this, I will introduce another exemplary area – an example from poetry, as well as the literary theory connected to this example. My example is Hans Christian Andersen's poetry, to be more specific his fairy tales, and to be even more specific, one of the fairy tales where he articulates different forms of love: "The story of a mother" – as well as research into Hans Christian Andersen's writing.

Let me briefly introduce these two exemplary areas, which I shall repeatedly return to throughout this article.

Love is a theme for philosophy in several ways. In philosophy of life, erotically defined love is considered an important ideal. In ethics, loving one's neighbour – in the form of sympathy and empathy – is considered important in terms of the ethical notion of care. In pedagogic philosophy (and the philosophy of psychology), the significance of parents' love towards their children is a theme. There are a number of philosophical theories on the nature and forms of love – from Plato's notion of Eros to Schopenhauer's theory of the love of the sexes to Kierkegaard's theory of the relationship between selfishness and selflessness in the various types of love, and to theories of the relationship between natural love and loving one's neighbour, and of the relationship between body and spirit in love.

As for love as a theme within poetry, there can be no doubt that we are dealing with a central theme. There are whole genres defined by this phenomenon: love poetry, love novels. The Danish term "roman", meaning novel, is in fact derived from this theme. We have oeuvres which are to a large extent concerned with the relationship between specific forms of love and other ideals for life – as for instance those of Ibsen and Pontoppidan, both of whom are concerned with the relationship between – and tension between – self-expression and self-devotion. Another example is Hans Christian Andersen's oeuvre, wherein love is a main theme, not only in many of his fairy tales, but also in his five novels, and where we can find an idea of love as something making itself manifest in a number of forms, ranging from the lower to the higher.

I have thus come to the presentation of my second exemplary area, the fairy tales of Hans Christian Andersen. Although Hans Christian Andersen quickly gained fame throughout Europe, for his novels in particular, then as H. C. Ørsted pointed out at an early stage, his fairy tales would make him even more famous – and in fact immortalise him. And as we know, that is what happened. Consequently, there can be no doubt that, at least as far as his most significant fairy tales are concerned, we are dealing with poetry at its finest. Having said that, I could still imagine many people would question, whether these fairy tales give an important contribution to an understanding of life, and therein an explication of love, for the value of the fairy tale could be of a different nature. Indeed it is the case for many fairy tales – let me briefly mention for instance "The Tinder Box" and "Numskull Jack" – that they are first and foremost read for the sake of their humour and language. But I do also believe that most Andersen readers and scholars have always considered several fairy tales or stories – for instance "The Little Mermaid", "The Marsh King's Daughter", "The Shadow", "The Story of a Mother" – as deep interpretations of our existence. Of these, I choose the simplest and shortest, "The Story of a Mother", as my specific example.

Let me then briefly remind you of its content: the protagonist is a mother who watches over her very sick child. A stranger knocks on the door one winter's night, and is given permission to come in and be warmed. While he is there, the watching mother slumbers for a brief moment, and when she comes around, the stranger – death – has left with her child. The mother rushes after him, comes across a woman clad in black – night – whom she is to sing for (just as she sang for her child) in order to be told which direction death has gone with her child; she then comes across a thorn-bush, which she has to warm by her heart for further information; and finally a large lake to which she has to give her eyes in order to cross and thereby reach the garden of death, where she has to give the woman gardener her black hair to be told among which of the plants she may find her child's tree or flower of life. When death then comes to take away the sick flower and replant it in Paradise, the mother out of desperation threatens to pull up some of the plants, but when death tells her that she would thereby make other mothers unhappy, she refrains from doing so, and when death then shows her the future that awaits those two children whose life plants she put under threat, the one as unhappy as the other is happy, and at the same time tells her that one of these paths of fate lies in store for her own child, the mother goes down on her knees and prays to God: "Do not head me when I pray against your will, which is best." The story ends with these words: "and death went with her child into the unknown country."

As is already clear from this short paraphrase, the first half of the tale depicts maternal love in its strength and perseverance, and this type of love is then confronted with both love of your neighbour and love of God, seen as a surrendering to God's will.

Despite its short and simple nature, this tale has always made a deep impression on both the lay reader and scholars. There is something moving in this tragic tale. But at the same time, significant differences appear in the way in which the ordinary reader and the scholar deal with this story (and the other great fairy tales).

2. Existential Reading and the Science of Literature

The reading which this story encourages or the natural way of reading this story is a reading which seeks to grasp its existential content, wherein the artistic form may be understood – in this case fairy tale elements such as the night appearing as a woman dressed in black and the lake talking to the mother – as formal elements whose aim is to mediate the existential understanding of love. At the same time, this reading immediately presumes that the author has written this tale in order to communicate an understanding or an interpretation which has been urgent to him, and where we as readers expect and are prepared to have our own understanding of life explicated, and where we thus read for our own sake.

The description I have given here of the engaged reader's experience of "The Story of a Mother" is based on the classical model[2] of the four main elements through which reader, interpretation, experience may be understood:

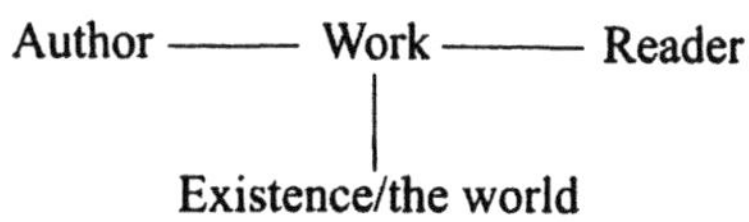

There is an author who produces a work which provides an image of existence, or the world, and it is this picture the reader or the public are concerned with.

This model covers the literary mode of communication – and it is therefore not surprising that it may be rediscovered in Habermas' general model of communication.

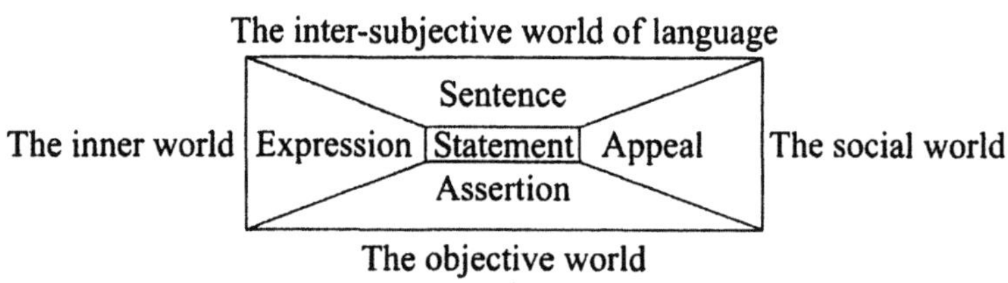

If we now turn from the engaged reading to research, i.e. reflections on and the scientific investigation into literature, we then discover that there has been a tendency to suppress the engaged reading. This development can be seen in both the aesthetic theories (which are highly influenced by philosophy) and in the science of literature (which has also been influenced by philosophy).

2 Abrams (1953, chapter 1).

Following Abrams, it is possible to present the development within aesthetic theory as running through four stages. The first stage focuses on the work of art as an image of the world or of existence. Here we find the theory of *mimesis* in Plato and Aristotle, a perception which continues a long way forward in time, and which especially influences the 1700s (with among others theorists such as Batteux and Lessing). The second phase, which is founded in the romantic poetics with Horats as its pivotal figure, is more concerned with the audience or the reader. Here the pragmatic aesthetic theory is dominant, whose main focus is the effect of the work of art, i.e. that it is to entertain and instruct. This theory is likewise continued and is also present in the 1700s, namely in Beattie and Samuel Johnson. With the Romantic period there then comes a breach where the third stage begins, in which the theory is formulated of art as an expressive form, thereby focusing on the author, who is seen to express himself and his individuality in his work of art. And finally at the end of the 1800s, the *l'art pour l'art* theory – art for art's sake – emerges, whose focus is the work itself as a distinctive aesthetic object.

The development within literary science (and especially within literary interpretation and criticism) can be described as being somewhat parallel to the development within aesthetic theory sketched above.

In the first phase – the philological stage – the work is primarily seen as a statement concerning existence or the world. It is actually taken for granted that the reader can understand this statement, as long as you make sure – and this is where the task for literary science is to be found – to provide reliable versions of the works (textual criticism) as well as ensure a linguistically and factually correct perception of the text (through text commentary of linguistic and factual nature). The next phase, from ca. 1800, thus becomes the historical phase, which is actually divided into two – first a romantic historicism emerges, where the work is seen as an expression of the individuality of the author or of the soul of the people. Then follows positivist historicism, which is a study into the problems of causality concerning the individual authorships and the individual works of poetry. These causalities are either sought within the temperance of the poet, his course of life and social affiliations – the biographic method – or within his literary formation (the comparative method). The third phase is then the emergence of New Criticism in the beginning of the 20th century, in which the work as an autonomous aesthetic entity becomes the central issue.

So we see a course that to some extent corresponds to the course for aesthetic thinking. Only it is not to the same extent possible to recognise a phase – positioned after the first phase – in which the reader is the focal point, although it could be said that this reader-orientation appears in literary interpretation in its latest contemporary phase.

The period after new criticism – in Denmark the latest thirty to fourty years – has been dominated by several theories that conceptualize the linguistic, psychological and historical-sociological aspects of literature (cf. the Habermas-model), i.e. all other aspects then the text taken as a statement about the world or human existence. The linguistic aspect has been investigated in structuralist theories (both the Saussurian and the Peircean type). And it has been continued in poststructuralist deconstruction, which problematizes the idea of a coherent and unambigous text (Derrida). The psychological aspect is investigated in different forms of psychoanalytically inspired theories, later developed by Lacan, in the socialization-theories of Foucault and in the feminist approaches to literature. The historical-sociological aspect has been investigated in different types of Marxist theories of literature and later in new historicism and cultural studies generally. And lastly we have – as mentioned – research into reception.

If we now turn to Hans Christian Andersen research, we can see a recurrence of the types of literary science mentioned. Only it should be mentioned that the phasal separation becomes less clear, although it is still present. At this point I will merely mention the types of interpretation (and here I stick to Danish research). By far the most dominant type is the biographical and comparative research. Early significant names are Brandes and Brix. Later, psychoanalytically inspired research appears, as well as a Jungian inspired biographical research (Nyborg). Most recently, a peculiar type of biographical research has appeared, inspired by the Danish scholar Åge Henriksen, based in among other things Eastern thinking, but also Jung (Klaus P. Mortensen). Next I should mention a more philologically oriented research, which lends particular attention to genre specification and to specification of elements of style (Rubow – to some extent also Kofoed and Bo Grønbech), and which at a later stage becomes even more work-oriented as in New Criticism, where focus is given to language, especially imagery (Baggesen). Finally there are a number of approaches closer to ideological criticism and general social history – Peer E. Sørensen, Jørgen Holmgaard, Finn Hauberg Mortensen – which is most recently affiliated with receptions theory. Chronologically we can see it as a movement from a focus on the author via a focus on the work, and first and foremost its formal aspects, to a focussing on the reader or the reception.

This means that what I have described as the natural mode of reading, which the deeper tales imply, cannot be said to have influenced scholarship. If each of the elements in the natural mode of reading – the author and his intention, the text in which understanding of existence is the central aspect, and the reader as someone who is existentially involved – are considered individually, then we may recognise the three elements within the approach of the science of literature, but the three elements are no longer perceived through the perspective of existential description. When it comes to explaining how things have gone this way in literary research, we should refer in particular to the influence from more established and prestigious sciences in connection with influences more in key with the theory of science, i.e. from the field of philosophy.

The biographical research was stimulated by the method thought to be found in the natural sciences and which in philosophy took the shape of positivism. New Criticism was developed through an impression of the development partly within psychology and partly within linguistics, since both fields were developing totalising or structural theories (Gestalt psychology and structural linguistics) – i.e. in the direction of nomothetic sciences. The latter is also what lay behind French structuralism, which in the same way was elevated to form a comprehensive frame of understanding in parts of philosophy and theory of science. The psychological approaches were stimulated by the development of psychoanalysis, which again in parts of philosophy and theory of science was elevated to become a general model for the understanding of humanity. And just as psychoanalysis was elevated to become an overall framework for the understanding of the human being (a type of psychologism), the same thing happened with the Marxist oriented field of social science (a type of sociologism), which after the student revolt spread to all humanist subjects – not least literary science. As mentioned before it has then been followed in the latest decades by other reductive perspectives such as feminism, constructivism, deconstructivism, which again are frameworks established in fields within philosophy and the theory of science – the whole thing often referred to as "theory", not "literary theory", because "theory" is a sort of mix between philosophy and theory of literature.

Considering Abrams' model and Habermas' communications model, it is natural that the development in linguistics as well as psychology and social science plays a major role in the development of the science of literature. Literature – as a specific type of communication – has a linguistic aspect, a psychological aspect (sender) and a societal or social aspect (recipient or audience). What is strange, though, is that these aspects can imply a partial overlooking of the aspect that is central for the engaged reader: the work as a statement about existence or about the world.

Philosophy's role for the science of literature is also worth noting. In all the types of literary research mentioned, philosophy has played a large role. If we take a look at the founders of positivist literary science – Sainte Beuve and Taine – we can see that Comte was significant to both of them. And when the Freudian and Jungian type of psychoanalysis is later to play such a large role for some scholars within literary science, this is not least because Freud himself devised a philosophy of culture, as did Jung. Lacan, who at an even later date came to play a significant role, was one of the philosophising psychoanalysts. As far as the sociological inspiration is concerned, which in the science of literature is expressed in ideological criticism and in a broader socio-historical scholarship, it goes back especially to Marx, Lukacs and Adorno. The semiological tradition goes back to Peirce. And the tradition that was inaugurated by Saussure has been developed in literary studies by the philosopher Greimas. And finally for poststructuralism in the broadest sense of the term, it is the case that it is most significantly inspired by Derrida's deconstruction and Lyotard's postmodernism.

A third factor, which has been decisive for the development of the science of literature, is the development within literature itself. The most evident example is New Criticism, which developed in America through the influence of especially Eliot's modernist poetry. When New Criticism is then revised and reformed or dismissed by the deconstructivists, this is under the influence of the postmodern retaliation against modernism – especially within poetry, for instance that of William Carlos Williams. At the same time, this deconstructivist retaliation towards New Criticism is also linked to a reestablishment of romantic poetry (Wordsworth and Shelley), from whom the modernists dissociated themselves. It is of course fine that the science of literature is developed concurrently with literature, but the problem is that this produces a generalisation of a specific mode of reflection, so that what is legitimate in relation to a specific type of literature is considered to apply to all literature.

My outlining of the history of aesthetics and the science of literature – hereunder Hans Christian Andersen scholarship – has of course pushed the situation to its extreme in its description of this development as being misguided in relation to the natural or actual state. Most literary scholars have of course always been concerned with the existential cognition articulated in poetry. But many have not felt that it could be of scholarly interest to be directly concerned with this aspect. They have certainly meant their own research – regardless of its orientation, be it biographical, comparative, stylistic, sociological, etc. – to be a contribution serving the elucidation of a work, and today, many are of the opinion that it is precisely through the juxtaposition of various approaches to a work that it is best elucidated. It should also be mentioned that there have been literary scholars who have been less influenced by the theoretical currents and who have clearly been primarily concerned with poetry as expressions of cognition. This also applies to Hans Christian Andersen research. But generally speaking, I would maintain that to a certain degree, there has been a misguidance. The fact that this has also been the case for Hans Christian Andersen scholarship may be confirmed by quotes from some of the Danish Hans Christian Andersen scholars. In 1975, Johan de Mylius wrote that "Considering how widely known and highly esteemed Hans Christian Andersen's oeuvre is, then even in this the centennial year of his death, one is amazed of how little scholarship is actually concerned with interpreting and positioning his works and defining the key tendency of his oeuvre".[3] And in 1968, Aage Jørgensen wrote: "Hans Christian Andersen scholarship is already conclusive – but only rarely does the scholar reach beyond an interest in the degree to which the author's life is reflected in his works and into a significant occupation with

---

3 de Mylius (1976).

these works as artistic expressions of an interpretation of existence and its conditions – as formulations of cognition."[4]

In my opinion, this state in Hans Christian Andersen scholarship as well as in the science of literature – which has not changed to any noteworthy degree in the past 30 years – presents a challenge for philosophy. Philosophy has to great extent played its role in shaping the science of literature in such a way that this has neglected much of its genuine task. The question then is whether there are other aspects of philosophy – other views and theorists – that can contribute to a change of direction for the science of literature. It is my belief that there are.

If the science of literature is to focus on what the reader focuses on in her or his reading of classics such as the more profound of Hans Christian Andersen's fairy tales, then there are a number of theoretical schools within philosophy that can be utilised. I will mention two types of these: 1. Various theories concerning the artistic mode of expression as an articulation and communication of a specific type of cognition. 2. Various suggestions for a method for uncovering the insight (the picture of life) articulated by a specific literary text.

3. Theories Concerning the Artistic Mode of Expression

As far as these theories are concerned, we may note the consensus that the two key elements in the poetic or literary mode of expression are narration and imagery or metaphors. In terms of the theory of cognition (both within and outside philosophy), discussions of these elements have been concerned, first of all, with the fact that narration and imagery are applied far more frequently and in more central places in a number of sciences, not least in the science of history (Hayden White). This has led some people to render these as relative forms of scientific cognition. But it may also be argued – as other scholars have done – that the elements of narrative and imagery discovered in the sciences are there because they contain an insight, an understanding, which is indispensable. This is the case for instance for Ricoeur as it is in narrative psychology, just as a related view is to be found in cognitive semantics (Lakoff and Johnson).

Here I shall merely outline – and apply – a single theoretical school – namely Hans Lipps and K. E. Løgstrup's description of the breadth of meaning to be found in the majority of everyday words.[5] This theory, which among other things assigns to verbs a breadth that is grounded in the fact that we are dealing with an aim that

4 Jørgensen (1968, p. 126).
5 Løgstrup (1976).

cross-matches factually diverse fields, not least a crossing of the sensory and historical-existential fields, which also has to do with the fact that we use our bodies to gesture and gesticulate, can shed light on imagery in poetry – for instance the imagery we find in "The Story of a Mother".

When a person is struck by an accident, sickness or death, we may say that he or she has been stalled, that the person has been enclosed, that they have lost their grip, that they cannot go on, that they are no longer standing on firm ground, that they cannot touch bottom, that they do not have a clue, that all possibilities feel exhausted. It is this breadth of meaning that enables us to understand the image of the mother and the lake. In the story it goes: "Then she came to a large lake, where there was neither ship nor boat. The lake was not frozen solid enough to bear her, and not open and shallow enough for her to wade through, and cross it she must if she was to find her child. Then she lay down to drink the lake dry, which is of course impossible for a human being, but the sad mother thought that perhaps a miracle could happen" – so we see that the mother cannot go on, she doesn't have a clue what to do, she loses support, she cannot touch bottom. She is confronted with the impossible – the insurmountable, the inexhaustible – but love challenges the impossible, refuses to see the impossible where everyone else sees something to be impossible. When the lake charges her the price of her eyes to carry her across, this should be understood – first and foremost – by way of the expression "love is blind", which is to be understood here in its positive sense: not because it blinds you or because it leads to gullibility and naivety, but because love does not stick to the superficial level because it believes in the other person despite the apparent – the more superficial – characteristics. Secondly, the expression that "the eyes are the mirror of the soul" – compare the fact that the gaze may be soulful, i.e. concerned with what is most sincerely and deeply felt (compare the fact that the gaze may be warm or tender).

4. Methods for Analysis

As far as new methods are concerned, it is clear that of the methods hitherto formulated for the analysis of works of poetry, it is the New Critics and the phenomenologists and existentialists who are most concerned with the picture of life. There is actually a connection between the two methods. As pointed out by Bo Hakon Jørgensen in his book *Intentionality*, concerning literary analysis on a phenomenological basis, Wellek and Warren base their major new criticist work *Theory of Literature* on the phenomenological theory concerning the layers of the literary work.

In the context of Danish literary science, it is Jørgen Vosmar who has presented the hitherto most convincing outlining of a phenomenological picture of life – an analysis which can disclose the world and the attitude of the work. In particular, Vosmar bases his theory on Heidegger's analysis in *Sein und Zeit* and he applies the existentials, defined through analysis by Heidegger and subsequent existential phenomenologists, as categories for analysis – both in his well-known article, "Værkets verden, værkets holdning" ("The World of the Work – the Attitude of the Work"),[6] and in his comprehensive analysis of J. P. Jacobsen.[7] The latter greatly applies Binswanger's contribution to the further development and revision of the Heideggerian analysis of existentials.

Although Heidegger's primary goal in *Sein und Zeit* is not to formulate a philosophical anthropology, it is my belief that he has actually presented the elements for one, which in particular has inspired new forms of psychology and psychiatry – and thereby also a new form of literary analysis (this mediation of Heidegger through existential psychology applies to Vosmar at least). This has meant that the Heideggerian philosophical psychology, seen as a comprehensive view of the human being, has been extended and revised in various types of existential and phenomenological psychology. Naturally, there are other types of philosophical psychology (in Anglo-Saxon philosophy, too), but these have not to the same extent concerned themselves with presenting a complete set of fundamental concepts for the understanding of human self-expression, and are therefore not so clearly suited as a foundation for a literary method of analysis. It is therefore my belief that if philosophical psychology was to be inspired by the aims of existential and phenomenological psychology – and interacted with literature and the science of literature – it could develop a conclusive vision of human self-expression – a complete set of existentials or fundamental concepts – that would be a fitting analytical tool for the science of literature. In a couple of books on the philosophy of life and on various authors, I have attempted to contribute to such a further development, and will now mention a few concepts that are relevant in terms of our exemplary area – the phenomenon of love.

Taking as our starting point the classical division of the psychological field into the emotive, the conative and the cognitive, we may say that Heidegger is mostly concerned with the conative – aspiring and taking action – and the cognitive – an openness towards the world. I have subsequently – inspired by the very field of existential psychology, but also by revising it – added an analysis of the emotive aspect. In this context – following Maslow – we can differentiate between two fundamental types of motivation – need caused by lack and the need for expansion or self-realisation. In order to understand love, it has to be rooted, not merely in need (as for Maslow), but to an even further extent in the yearning for self-realisation, which may be seen as a yearning to acquire an adoption of the world.

6 Vosmar (1968).

7 Vosmar (1972).

As the name implies, this is linked to a special type of openness towards the world (a forgetting of the self in an openness towards and concern with), just as it implies an endeavour or a yearning – and action – which has the disposition of something with a mind of its own, free and without force (as opposed to the openness towards the world defined by need, which is selective or limited and split, and which implies a yearning and an action which is formed by the will, in which one has a grip of oneself and the situation). Furthermore, I therefore believe that the need for self-realisation should be divided into two types – self-realisation or acquisition of the world and self-submission or the bonding of oneself.

Based in these fundamental concepts, the phenomenon of love may be explained. Common for all forms of love is the caring for someone or something, the fact that someone or something means a lot to you. This is the same as being, in thought and feeling, in the place of that someone or something – that one is therefore in the presence of something in an unmediated fashion, that one is open and concerned in a self-forgetting fashion. When we speak of love both in connection with natural love and love of ones neighbour, this is because there is a degree of self-forgetting concern in both cases. What then enables us to speak of the love of God is its similarity to natural love – especially erotically defined love – the giving of oneself to another person, to something else, which is the epitome of self-sacrifice. In the love of God, this comes in the shape of what we term belief, meaning trust – the feeling that one has of something supporting us – life itself, which is greater than the individual and which the individual does not control – which one may rely on and therefore also feel a need to sacrifice oneself to. And which we can still feel is there even if we are hit hard by accident or loss.

This is merely to suggest that a fully composed philosophical anthropology would be a good foundation upon which to analyse literature – in this particular case Hans Christian Andersen's perception of love in "The Story of a Mother". The analysis should of course be elaborated if we furthermore wished to draw in the essential element of Andersen's perception of love, which is a longing or a yearning for – in for instance "The Little Mermaid". But the difference between need and the urge for self-realisation is exactly what explains the difference between lacking and yearning for or longing for.

Inspiration may also be gathered for a more adequate literary approach or method in the development of existential philosophical hermeneutics – just as aesthetic theory since the beginning of the 1800s has been characterised by the theory of expression, which has emphasised the author and her or his inner life as the central aspect, and just as literary theory from the beginning of the 1800s has been correspondingly characterised by a biographical interest, focussing on the interplay of life and work, the same thing is the case for the hermeneutics of Schleiermacher and his followers that interpretation is understood as an empathising with the author's thoughts and feelings. And just as biographical science of literature was later expanded by a more socially oriented perspective, in which there was an interest in the institutions into which each and every work is written and is organised through – currents, genre, etc. – in this way, the hermeneutic psychological conception of understanding and interpretation was added and partly replaced by a more sociological concept of interpretation based in given sets of rules and forms of community (Dilthey). We can therefore go on to say that the development of existential hermeneutics by Gadamer and Ricoeur was concerned with retaining the perception of classical hermeneutics that understanding is primarily concerned with the truth within the text.

I shall now finish off by attempting to show the productive results of an interaction between literature, philosophy and the theory of literature. This interaction would enable for us a multi-faceted elucidation of the type of phenomena articulated in poetry, which is a significant aspect of poetry's *raison d'etre* and is just as much the central issue of the philosophy of life.

5. The Systematic Exposition of the Phenomena of Human Life

Let us consider our area of exemplification, the phenomena of love. Our poetic example, Hans Christian Andersen's "The Story of a Mother", and in a broader sense the fairy tales and oeuvre in general, present us with an account of love as a phenomenon, which at one and the same time is of many shapes and yet linked by something of a unifying nature. I have already mentioned this unifying nature of love. As far as understanding natural love is concerned, philosophy can learn from poetry. In philosophy, a spectrum of views on this phenomenon is to be found. There is the notion that natural love may be perceived as a type of empirical *Neigung* of no particular ethical value (Kant), as a feeling whose partiality at the end of the day means it is of a selfish nature (Kierkegaard) and on the other hand as something which plays a part in defining human nature and the that which is of fundamental value in a human being's life – love as the culmination of I-you relationship (Buber), as *Über-die-Welt-hinaus-Sein* (Binswanger). If we were to juxtapose these theories with Hans Christian Andersen's artistic expression of motherly love, it becomes clear that while *Neigung* for Kant is not linked to the actual essence of a person, and is something passive, the motherly love portrayed by the poet is of a particularly powerful nature, and it is exactly in the most essential part of the mother that it is rooted, and is the opposite of passivity. In the same sense, it is evident that we are not dealing with selfishness. There may be partiality, but the forgetting and sacrifice of the self mean that we are dealing with the very opposite of selfishness.

As far as the hierarchy of the various forms of love is concerned, the philosophical discussion may also learn from poetry. The relationship between natural love and the love of one's neighbour is central in connection with the question of the relationship between philosophy of life and ethics. The question is whether the ethical aspect always outweighs considerations regarding the success of one's own life. Within a Christian ethics, some would interpret it as being so (Løgstrup). Within a more secular context, some people interpret Habermas as believing that considerations concerning the just always carry more weight than the considerations about the good life. It is my opinion that both interpretations are wrong. What is right about this perspective – and what Habermas means – is exactly what is expressed in this tale: the fact that one can never obtain a life of fulfilment by sacrificing another person.

So far, we have been dealing with how philosophical analysis can be inspired by poetic articulation. But philosophy can also shed light on and adjust poetic articulation. As far as the latter is concerned in "The Story of a Mother" – the love of God – most contemporary readers would probably feel there are flaws. First of all, God is dealt with as a personified entity that stands behind everything, deciding what happens. Secondly, the thought is articulated that the fate of the individual is predetermined from birth. The latter in particular is unreliable, but many people would also experience the former as an image of God that can no longer be maintained. But this does not mean that we cannot link anything to the phenomenon of giving oneself of love to God in the sense that we may say "thy will be done".

There lies a kind of resignation in saying "thy will be done". But there are two forms of resignation – one passive, the other active. When you are hit hard, in an unrecognisable, irreparable way, then you can gradually lapse into a state of inner resignation, where for instance you try to forget the one you have lost, or even deny the value of what has been lost so as to lessen the pain, but you could also maintain that it was of value that it was your happiness in life, and precisely being concerned with the value of the thing lost means that you still see existence as such as something rich and inexhaustible. Then you are dealing with fate in such a way that you can say "thy will be done".

When, like Hans Christian Andersen, you are faced with the problem of how to gestalt this opinion in a conceivable form, you come to conceive things too demonstrably. Hans Christian Andersen does actually think too demonstrably and naïvely about religion in other places too. He takes God and immortality all too literally. Here the interplay between literature and philosophy is such that philosophy can correct poetry – through a kind of demythologising.

We can now finally see what the science of literature can offer philosophy. When philosophy applies literature in its analysis of phenomena, there is often the risk of misunderstanding or overlooking something, because one is not sufficiently familiar with the context, which may be the oeuvre in its entirety, the period in its entirety, or the genre in its historical development. When for instance, it comes to applying some of Hans Christian Andersen's tales in an analysis of the various types of love, then it would be a great advantage to be able to rely on monographs from the science of literature concerning Hans Christian Andersen's philosophy of life – interpreted in terms of his entire oeuvre, Hans Christian Andersen's view of love, Hans Christian Andersen's view of love in relation to the view of the entire period (the period of romanticism) or in relation to Oehenschläger, Ingemann, Paludan-Müller, Kierkegaard, and others. These literary science contributions are only partially available.

Another aspect is that when in philosophy it comes to considering the relationship between the ahistorical or universal and the historically-changing (for instance in connection with an understanding of the types of love), then help can be found, not only in the literature itself, but also in the presentations within the science of literature of the historical development.

University of Aalborg

References

Abrams, M. H. (1953), *The Mirror and the Lamp: Romantic Theory and the Critical Tradition*, New York.

Jørgensen, A. (1968), *Kundskaben på godt og ondt*, Aarhus, Akademisk Bokhandel.

Kristensen, S. M. (1973), *Digtningens teori*, København, Gyldendal.

Løgstrup, K. E. (1976), *Vidde og prægnans*, København, Gyldendal.

de Mylius, J. (1976), "Problemer omkring H. C. Andersens realisme", in J. Breitenstein (ed.), *H. C. Andersen og hans kunst i nyt lys*, Odense.

Vosmar, J. (1968), "Værkets verden, værkets holdning", *Kritik* 8, København.

Vosmar, J. (1972), *J. P. Jacobsen*, København, Gyldendal.

Death, Rationality, and Irrationality: Challenging Science and Conceptual Categorization through Horror Fiction*

Sami Pihlström

1. Introduction

This paper consists of three main parts. First, I will (in section 2) do some groundwork for a study of the relation between horror and irrationality (or madness) through an examination of some conceptual and metaphysical issues surrounding the notion of *death*. It is, after all, death, in particular our own mortality, that is the source of the darkest horror there can be in human life; death is something we find especially hard to think rationally about, something the contemplation of which can, indeed, drive us into madness. It hardly helps to know that science can deal with death, too. The sense in which death can be scientifically studied – or, more generally, rationally discussed, "clearly" thought about (cf. Rosenberg 1998) – may not be the sense of death as something mysterious that evokes horror in us. Moreover, the concept of *the undead*, invoked in Gothic fiction both in its seriously artistic and in its more popular forms, can be used to illuminate this horror. This concept, which will be my central example in what follows, is connected in interesting ways with the Christian concept of resurrection, although these concepts are by no means the same.

In the second part of the paper (section 3), the concepts introduced in the first part will be put to work in the realm of the philosophy of art, in order to consider the aesthetic functions of horror fiction operating with the category of the undead. The more specific example to be used in highlighting the aesthetic and epistemological role of this category-breaking category is Bram Stoker's novel *Dracula*. I will distinguish between different forms of horror and briefly examine the concepts of the natural and the supernatural, in close relation to the concept(s) of (ir)-rationality.

---

* This paper was not presented at the XV Internordic Philosophical Symposium, but it was presented at another inter-Nordic meeting, the NOS-H workshop, *Madness and Irrationality*, at Copenhagen University, October 8–9, 2004. I am grateful to the participants of the workshop, especially Bjørn Ramberg and Dan Zahavi, for stimulating questions and criticism.

Finally, the third part of the paper (section 4) will pull some of the threads together by offering some "reflexive reflections" on the nature of the experience of being horrified by something deeply different or unnatural. It turns out that our concepts of the natural and the unnatural (or supernatural), just like our concepts of the rational and the irrational, must be relativized to the natural world we take ourselves to be living in; it is precisely this world, as well as its describability and/or comprehensibility, that is challenged by the conceptual framework of the undead. Thus, my investigation, while partly about Gothic fiction as a threat to scientific (or, perhaps, scientistic) attempts to get rid of mystery and unintelligibility in our relations to the world, is not confined to, or even primarily part of, aesthetics or philosophy of art. Some of my key results concern the notion of the natural, which brings my approach close to epistemology and philosophy of science. Indeed, the kind of horror and irrationality we find associated with the concept of (human) death may be considered a serious challenge to science – a challenge to its characteristic attempt to provide a "full coverage", a complete, adequate, or even "absolute" picture of reality. Human nature, including our human capacity to describe and categorize reality, is more complex, and this is something that our mortality, as thematized by Gothic art, may help us realize.

2. Death, Irrationality, and Horror: Conceptual and Metaphysical Issues

In this section, the concept of the undead will be explicated in order to throw some light on the connections between the concepts of death, the undead, horror, and madness or irrationality. Some analytic philosophers dealing with the metaphysics of death and dying – e.g., Antony Flew (1987) and Jay F. Rosenberg (1998) – have argued that there are severe conceptual difficulties involved in the very notion of surviving one's own death, in particular, in the idea of immaterial soul-like existence, and that it is, therefore, deeply irrational to believe in disembodied spiritual existence or, for that matter, in any kind of life after death. Their arguments, which I am not going to deal with in any detail, contain important insights that may, however, be problematized by invoking the concept of the "undead".

The undead are, by definition, corporeal and thus material beings; yet, they have died and have "come back", finding no "rest" in their death. Vampires and other monsters, whom presumably no reasonable scientifically educated person in the twenty-first century believes to actually exist, are of course paradigmatic examples of such beings, but similar or closely related conceptual issues concern, in principle, the Christian belief in resurrection. According to the Christian view, or one well-known and widespread version of it, resurrected humans (including Christ as resurrected, even after his ascension to Heaven) are material beings who have died but live again; both the soul and the body will have an eternal life. Yet, unlike the undead, the resurrected will not "come back" to the Earth (though Christ supposedly did that, for a while, and is believed by Christians to do that again) but will rather go into a transcendent Heaven, or possibly Hell. On the other hand, the undead do not exactly "come back", either; that is, they do not come back to their normal life as earthly creatures but to something quite different. A living dead, a living corpse, such as Count Dracula, the famous vampire, is a perverted version of a resurrected human person – or indeed of Christ as resurrected.

The Christian background is, obviously, ineliminable from Gothic fiction operating with the concept of the undead – which is one reason for realizing that serious Gothic fiction (literature, movie, etc.) is very different from the shallow and ridiculous horror fiction one may consume by watching "grade C" video movies. Sophisticated Gothic works of art are typically characterized by philosophical and psychological problems, particularly by the *epistemological uncertainty* resulting from the mixing of conceptual categories – something that we will reflect on in the following.

In the paradigmatic cases, the undead, characterized in the above terms, are, or have been, *humans* (though some examples might be drawn from popular fiction in which this is not true). Thus, their being undead, or perversely resurrected, is a mark of their utmost *inhumanity*. Given the important analogy (though by no means identity) between the undead and human beings after resurrection, it is evident that the relevance of the philosophical issues related to the notion of the undead is wider than one might initially think. In any event, it is clear that no simple anti-Cartesian argument against the possibility of un- or disembodied post-mortal existence is efficient against views in which the category of the undead figures centrally. The undead could be as inseparable mixtures of "soul" and "body" as we are, although their "life of the soul" would undoubtedly be very different from ours.

A crucial issue is whether the undead are so much as *possible* entities. If they are, this result should affect our metaphysical theories of death and of the mind. We should bear in mind that resurrected humans are not immortal souls, because they are, after all, corporeal; nor is the category of the undead the same as the category of ghosts. The latter are usually thought to be immaterial, though in some cases vaguely materialized. Ghosts might actually be seen as occupying an intermediary category between immaterial soul-like entities, e.g., Cartesian substantial souls that might continue to exist after the destruction of the body, and the fully corporeal undead; ghosts have also, in a way, "come back" to, or have never entirely "gone away" from, this world.

What does this have to do with the concepts of irrationality and madness? First, obviously, we twenty-first century educated Western people, within the scientifically advanced culture we live in, think that it is irrational to believe in the existence of the undead. Science has rightly taught us that there are no such things.

If we "see a ghost", we will immediately start looking for some natural explanation for our experience. It must be admitted, to be sure, that many people whom we take to be relatively "rational" in their daily lives do believe in extraordinary things. According to even relatively recent statistics, a significant number of, say, Americans actually believe in the existence of angels, demons, or even ghosts. Let us leave this matter aside, however, because the main issue of this paper is not the irrationality of believing in ghosts, or in the undead in general, or in some other supernatural beings – these are beliefs that I find obviously irrational and in no need of further discussion. Secondly, more importantly from our present point of view, the very conceptual framework of the undead – a framework which can be used for the purpose of evoking horror in a work of art, for instance – challenges some of our cherised assumptions of rational thought. If the undead are possible entities, we may have to rethink our ways of classifying reality. And I can think of no convincing argument against their *possibility*, although I am strongly (quite unshakeably) convinced that such things do not actually exist. At least, I do think that the undead are *imaginable* entities, though I may be mistaken here – I would be the first to admit that, for all I know, my intuitions about what is imaginable and what is not are fallible.

Be that as it may, our problems belong to a grey zone between philosophy of science and philosophy of art. We may examine whether, and how, we should revise our standard view of science as something that rationally describes and explains reality, when challenged by the sheer possibility of there being things not dreamt of in our science. Such an examination, motivated by the concept of the undead, should be based on more specific insights into the kind of horror we may imaginatively experience upon meeting such creatures.

3. The Aesthetics of Horror and the Undead

These thoughts lead us to the *aesthetics of horror*, in which the notions of irrationality and madness play a significant role. The metaphysical and conceptual issues briefly dealt with above have applications in the field of aesthetics. Thus, this section will take a look at the problem of the undead from the point of view of the philosophy of art. As is well known, the undead – vampires, zombies, and other horrible monsters – busily wander around in the world of popular fiction, though in some cases the boundary between "high" and "popular" art is vague – just think about, for instance, Bram Stoker's novel *Dracula* (1897) and Francis Ford Coppola's movie (1992) based upon it. While the importance of fictional characters, including vampires, for our understanding of *evil* has been rightly emphasized by Morton (2004), among others, it should be noted that also non-fictional evil – whose many forms we encounter simply by opening the television and watching the latest news – may be a challenge to our categories of understanding, something that makes us think about concepts such as "madness" or "insanity". My emphasis on fiction here by no means precludes our need to cope with these non-fictional challenges as well.

According to an influential epistemologically focused view on the aesthetic significance of horror, the horribility of such fictional figures as vampires is based on the breaking down of our ordinary categories of intelligibility. (See Carroll 1987; 1990. Let me note that I am only using Carroll's theory as a point of departure for my own reflections, i.e., I am not subscribing to his views on the genre of horror in general.) There are, however, different varieties of such a breakdown and thus of horror:

- the *naturally horrible* or terrifying (e.g., a tiger, an earthquake, a car accident) – i.e., something *actually* existing and potentially harmful, which may destroy one's life but does *not* challenge our ordinary categories of intelligibility and which is, therefore, rationally and, in principle, scientifically explainable as a natural object, event, or process in the world;
- the *supernaturally horrible* or terrifying (e.g., a vampire, a ghost, a rising mummy) – i.e., something that presumably does not actually exist but is, for all we know, logically or conceptually, or even metaphysically, *possible* and, as such, is a challenge to our standard way of categorizing reality in terms of, e.g., the concepts of life and death (viz., the undead are "in between" these two categories, which are usually thought to be mutually exclusive – though, as we saw, vampires and ghosts can be said to be "in between" them in somewhat different ways);
- the *totally unintelligible or unconceptualizable* and therefore horrible (e.g., a situation in which time or temporality, or other Kantian-like preconditions of meaningfully arranged experience, break down completely, as in some of H.P. Lovecraft's horror stories; cf. Airaksinen 1998) – i.e., something totally "mad" or *impossible*, at least for us, as the kind of world-experiencing and -conceptualizing beings we in fact are.

In terms of this threefold classification, the humanly natural and almost universal fear or horror of death does not constitute a category of its own; rather, the horror of death can be and usually is involved in all of the three forms of the horrible. I do not mean to provide any exact definition of "horror", or of related notions such as fear, but I find it natural to speak about horror in cases in which one is threatened by death, the loss of one's life. In horror, no rational discussion or calculation is possible. Horror may be a "natural" feeling or state in some given situation, even in a situation in which something supernatural – really or imaginatively – happens; it would, however, be odd to describe it as "rational", even though it may be rational for us to be afraid of various kinds of things.

Moreover, the mixing of the categories of life and death in the concept of the undead is more serious than, say, the mixing of the categories of the human and the machine, or of the human and the animal, although these may be extremely terrifying, too, and may even be partly related to the notion of the undead. It seems that we can incorporate in our scientific world-view new concepts referring to entities that are thought to be some sort of mixtures between the human and the machine, for example. To some extent, the mixing of the categories of life and death may also be scientifically acceptable – consider, say, viruses, which are thought to be neither fully living nor "dead", or the debates over the most plausible criteria of death, e.g., in terms of total termination of brain activity. But in a more profound sense, life and death cannot be mixed, e.g., in the sense of claiming that a certain person – the whole person, not some of her/his cells or some parts of her/his brain – is both dead and alive. Such claims do not, at the present moment, at least, belong to our rational classifications of things, either in science or in ordinary life. To claim such a thing is to be committed to utter madness, to step out of the language-game(s) in which words like "life", "to live", "death", and "to die" have their meaningful use.

Let us take a look at the case of the vampire, Dracula in particular, in some more detail (with the help of Carroll 1992; see also Summers 1928; Roth 1982; Reed 1988; and Botting 1996, ch. 7). Note, first, that Dracula (as pointed out to me by Dan Zahavi in conversation) is not the best possible example of an "undead" figure, as defined above, because, at least according to the introductory story of Coppola's 1992 movie (a story not to be found in Stoker's novel), Dracula, unlike other vampires, never died but became an immortal "living corpse" after having cursed God. Yet, I use Dracula as my main example here, because this vampire is obviously one of the best known among all demonic characters in Gothic literature. Dracula, as depicted in Stoker's novel, is physically threatening, dangerous, fearsome, but also dirty, disgusting; one feels both fear or horror and disgust when confronting him. As the narrator of the novel tells us: "His face was not a good face; it was hard, and cruel, and sensual, and his big white teeth, that looked all the whiter because his lips were so red, were pointed like an animal's." (Stoker 1897, p. 170.) Moreover, the disgust is not just physical but also, more deeply, conceptual. Dracula is "in between" the ordinary classes of things, breaks the normal "forms" of things. As Carroll's (1987, 1990) theory well explains, horror as an artistic way of expression can be compared to postmodernist ideas about the inconsistency and unstability of our conceptual and classificatory frameworks. If this is so, then an undead figure like Dracula is a most philosophical construction. As a being that breaks the dead vs. alive classification, in particular, he is an epistemically threatening being, something that simply does not fit our scientific or everyday categories. He threatens us with his "polymorphic" way of being: he lies not only between death and life but also between the past and the present, East and West, barbary and civilization, irrationality and rationality, the uncanny and the homely (cf. Botting 1996, p. 150). The threat he poses is therefore not only physical but also, perhaps primarily, epistemic and conceptual. Or, one might say that the threat is physical from the point of view of the characters within the novel but epistemic-conceptual from *our* point of view, for us readers, who do not or at least need not believe that Dracula is real.

One of the central symbols in *Dracula*, blood, is also similarly conceptually confusing and ambiguous: it is the symbol of both death and life, and Dracula's pathological taste for blood keeps him "alive" as an "undead", as a living corpse. Without blood, he would have to return to the normal state of dead people, non-existence. Sucking blood and "living" through it is perhaps the most unnatural thing about him. Dracula is, then, dirty or filthy in a mythical, archetypal way. As is well known, the New Testament teaches us that the blood of Christ saves the humankind, liberating us from the grip of sin and death; again, this "saving" is depicted in an ironic way in *Dracula*. In Dracula's case, blood – the object of his unholy appetite and the condition of the continuation of his "life" as an undead – is as ambiguous and conceptually challenging as Dracula himself.

On the basis of this reading of Stoker's novel, we may say that the key philosophical issue of horror fiction lies in concepts such as uncanniness, otherness, unknowability, incomprehensibility, indescribability, and epistemic fragility – in the confrontation with something that we cannot see as anything else but evil, filthy, or deeply irrational (in a word, *inhuman*). The ways of understanding and conceptualizing reality that are built into our ordinary practices and habits of thought, whether scientific or commonsensical, are violated when we face such an imagined reality. Our own humanity may, through such a violation, break down. Of course, in some cases we may be even more upset and horrified when something definitely *human* – a real person, such as Adolf Hitler – turns out to be monstrous, worse than any of the evil fictional characters we could think of. Such an experience may even be strengthened if we experience a partly fictional work of art based on the life of such a person, or on the historically real monstrosities that really took place, because fictional and non-fictional evils may be – again challengingly – mixed up in such an artwork.

In its celebration of the unknowable, the indescribable, and the uncanny, Gothic fiction can, in any case, be said to be characterized by features such as excess and transgression (see Botting 1996, ch. 1). Furthermore, as Airaksinen (1998) notes, Lovecraft's horror stories are usually realistic until something inexplainable takes place, something which indicates that the world we thought to be familiar and safe is not at all like that. Reality is transformed into something... well, mad. Our sense of reality is fundamentally distracted. In particular, our trust in the scientific worldview collapses within the projected world of such stories. This is also a central theme in *Dracula*: something as pre-modern as vampires can exist in the modern, scientifically and technologically advanced England of the late 19th century. As soon as something like this must be admitted by the characters in the story, their categorization of the world breaks into pieces. Small wonder, then, that the motive of *getting lost* has also been popular in horror fiction: showing how our lives are epistemically fragile, such fiction shows that we may easily get lost in our daily lives and lose sight of how to get along with the conceptual categories we habitually employ. A popular movie from a few years back, *The Blair Witch Project* (1999), brings this out quite explicitly by developing the motive of getting lost in the woods: as the protagonists get lost, they become increasingly challenged, physically as well as conceptually and epistemically, by the evidence about witchcraft they originally took to be mere make-believe. Yet, unlike the characters in *Dracula,* they are surrounded by a threat that never becomes fully visible; the evil that haunts them remains unknown to the very end.

It may be noted that even in the case of the naturally horrible, some moderate version of the breaking down of conceptual categories and thus of "getting lost" may be involved. For example, many people are afraid of flying; it may be thought that the very idea of human beings as flying is a conceptual confusion – humans just don't fly, do they? However, as airplanes are part of our daily life, we have become used to this kind of "mixing up" of categories; even many of those who do fear flying usually think that, while their fear may be naturally explainable in terms of psychology, with reference to, say, the loss of control involved in the passenger cabin, there is "really" nothing to be afraid of in normal passenger air traffic, at least compared to other much more risky forms of traffic.

In any case, the one who fears flying – just like the one who is afraid of vampires or other undead – is presumably above all afraid of dying. The horror of death is the context within which all different forms of horror can be distinguished. Heidegger (1927) may have got something essentially right in arguing that we humans are living toward death in the sense that the expectation and inevitability of death colors our entire existence. In this sense, the horror of death has no specific object and is thus to be distinguished from "normal" kinds of fear; whatever the immediate object encountered or imagined (a tiger, an airplane crash, a vampire, an unconceptualizable "something"), the "mood" of the horror is provided by the expected, imagined, or acknowledged, but in any case inevitable, loss of one's life. Heidegger could, indeed, have used horror as an example of the kind of anxiety characterizing human temporal being-in-the-world.

4. Reflexive Reflections

These reflections are, thus, intimately related to the Heideggerian view that our existence is irreducibly temporal. Temporality as a condition of intelligibility (in Kantian terms, as a form of pure intuition) is highly central in the philosophy of horror, insofar as the latter – as in the case of the undead – is conceptualized in terms of the breaking down of our usual conceptual categories. However, we should immediately notice that the modalities involved in the distinctions between the three main types of horror given above, as related to the different types of "breakdown" and/or unintelligibility, cannot be construed in the ancient temporal sense, according to which what is possible exists at some moment of time and what is necessary exists at every moment, i.e., always, but should be treated in the modern terms of possible worlds. Now, the concepts of the *natural* and the *supernatural* turn out to be crucial, and they obviously require further scrutiny. What is natural – and what defines our notion of the naturally horrible – is relative to our world; this notion can be construed with reference to the set of possible worlds in which the natural laws of the actual world (or of "sufficiently similar" ones) hold. The undead are un- or supernatural *for us*, in our world, because they do not obey the natural laws governing all known living and dying organisms; the aesthetic effect of such entities is to a great extent based on their being *partly* natural (corporeal, acting) and partly supernatural – and thus "in between", mixing up conceptual categories, as we have seen.

The typology of different forms of horror must, then, be relativized to the world we actually live in and whose objects, events, and processes our science reliably explains and predicts. For different (possible but not actual – for us) creatures, the challenging of *their* conceptual categories might result in forms of horror quite different from ours. Aesthetically, a most interesting challenge would be to apply, reflexively, these concepts to themselves. How could someone whose very existence breaks down our conceptual categories her- or himself experience a similar breakdown? That is, what would it be like for, say, a vampire to experience something truly horrible and/or unintelligible? Can *we* imagine anything like that? Similarly, can we really imagine what it would be like for a person we consider mad, or deeply mentally disturbed, to arrive at a worldview very different from the one s/he actually has, e.g., a more "normal" or "cured" one, something that we take ourselves to have? These questions may lack clear answers. The undead, such as Count Dracula, are, we may assume, capable of feeling horrified; that is also part of their aesthetic effect upon us. However, we must again be careful here: for example, zombies, lacking mental life altogether, can hardly experience such feelings, and even the "emotional life" of a vampire may be so inhuman that it might be difficult for us to ascribe our notions of fear and horror to such an entity. Yet, a vampire can, presumably, be afraid of, e.g., a crucifix, can't s/he? S/he might even be afraid of something that we can hardly understand or even imagine. (Dan Zahavi reminded me of "Dayblood", a story by Roger Zelazny, included in his collection, *Frost and Fire* – a story in which vampires themselves are haunted by other monsters.)

Let us return to Dracula for a moment. His dirtiness and irrationality, the fact that he cannot be accommodated by either scientific or commonsensical classifications, is not restricted to his evil and violent character. There is a deeper paradox lying in this figure. We might be led to think that Dracula, like many other monsters, is such an inhuman figure that we cannot adopt to him the kind of attitude we adopt to other human beings. Our attitude to him, presumably, cannot be an "attitude towards a soul" (cf. Wittgenstein 1953, II, iv) – or can it? It seems that the profound way in which Dracula challenges our conceptual categories lies in his being also a suffering creature whom we must pity, someone for whom getting rid of the nightmare of eternal life would even amount to a kind of ironic Christian atonement (see also Roth 1982, pp. 96, 126). Before the crucial battle in *Dracula*, Mina Harker tells the others to feel pity for Dracula, too: "That poor soul who has wrought all this misery is the saddest case of all. [...] You must be pitiful to him, too, though it may not hold your hands from his destruction." (Stoker 1897, p. 306.) As Dracula is in the end destroyed, he finds peace and his sufferings are over. From the point of view this Christ-like figure and the motive of atonement, Dracula's ambiguity is highlighted. He breaks our categories, once again, by being, somewhat metonymically, both Satan and Christ. On the one hand, he is evil through and through, almost an Antichrist, the one who produces unspeakable suffering and misery; on the other hand, he suffers himself, and through his suffering, he also releases and "saves" his victims, including Mina Harker. As Botting (1996, pp. 150–151) points out, Coppola's movie, *Bram Stoker's Dracula* (1992), makes it particularly clear that the vampire is no longer an "absolute Other"; on Botting's interpretation, Coppola moves Stokerian horror into the 1990s, an age which longs for love, tolerance, and understanding – even for a vampire and for otherness in general. Thus, the movie is "emptied" from its Gothic elements and signifies the "end of Gothic" (ibid.).

The closing of *Dracula* is not unambiguous, either, however. Van Helsing's final words, "We want no proofs; we ask none to believe us!" (Stoker 1897, p. 376), can also be read ironically. A "happy end" would be too easy a resolution. The fundamentally problematic nature of human life, the fact that surprising, un- or supernatural events and the distractions of reality they produce cannot be *a priori* rejected, is again highlighted toward the end of the novel. The possibility of evil and the fundamental inexplainability, incomprehensibility, or intractability of this possibility (cf. Bernstein 2002; Neiman 2002; Morton 2004) remain ineliminable features of our human world. The inexplainability of reality, the fact that the world – both in its good and evil aspects – is an irresolvable mystery for us humans, can be seen as a key theme of *Dracula*. This is a feature of the world itself, the world as we humans know and experience it, that Stoker's great work describes – and it seems to be a feature of our world *not* adequately accounted for by science. "There are always mysteries in life", Van Helsing says and asks his friends: "Do you know all the mysteries of life and death?" (Stoker 1897, p. 189.) And further: "Do you not think that there are things which you cannot understand; and yet which are; that some people see things that others cannot? But there are things old and new which must not be contemplate [*sic*] by men's eyes, because they know – or think they know – some things which other men have told them. Ah, it is the fault of our science that it wants to explain all; and if it explain not, then it says there is nothing to explain." (Ibid.)

Van Helsing, a scientist and "metaphysician" himself, thus abandons the pursuit of a "total" or "absolute" scientific conception of the world. This need not be interpreted as the thesis of the novel itself, but *Dracula* might be seen as a novel attempting to open our eyes, a text whose author – or, rather, implied author, as narratologists like to put it (cf. Booth 1983) – tries to maintain something from the original human openness to the mystical realms of being. Without the religious context of the book, nothing like this could be achieved.

On a meta-level, *Dracula* can be read not only as a piece of Gothic fiction raising feelings of horror in us by its symbolic descriptions of our conflicting ways of being in the world, but also as a thesis about the very nature of human ways of describing and categorizing the world. It can be read as claiming that our categories are never fully stabilized, never metaphysically secured – that they are, indeed, ineliminably human ones. They are not Nature's or the World's "own" categories, which (as it were) could open up for us reality as it is in itself. On the contrary, our engagements with the world we live in constantly transform the categories within which we find such engagements possible. Our own way of being human may thereby also be transformed. Our categorizations of the world and of ourselves are, then, never immutable or eternal; they must dynamically change in the course of the changes of our lives themselves. Insofar as Stoker's novel (and, possibly, many other Gothic works of fiction) can be read as a reminder of this way in which the categorization of reality is tied to our limited and changing perspectives, it is not merely a novel dealing with our epistemological uncertainty and the limits of human knowledge, that is, about the possibility that something quite unexpected may happen, but also a literary treatment of the *metaepistemological* issue of categorization itself.

Let us draw some more general conclusions based on our case study of *Dracula*. It may, as the example of Dracula testifies, be impossible for us to truly understand the imagined "inner life" of (for us) merely possible creatures, such as the undead, because the very classification of things into natural and unnatural that we rely on and use in classifying the undead themselves as unnatural would be quite different from their point of view, were such a point of view ever actualized. We may find an analogy here: it has been recently argued by Rupert Read (2004) that there is no such thing as "understanding" severe schizophrenia or other grave mental illnesses, because there is no sense, content, or claim "in" there, in the life or words of a person suffering from such an illness. If this is correct, we only experience an illusion in "understanding" what such a person thinks or does, just as we experience an illusion of understanding the words of a metaphysical philosopher. (Read's argument is based on his "neo-Wittgensteinian", "resolute" conception of nonsense, according to which there is no profound, somehow meaningful nonsense that "gestures at" something that cannot literally be said, be put to words. Cf. the essays in Crary & Read 2000.) Mina Harker's request to feel pity for Dracula might then itself be ironic: for all she knows, she might be mistaken in her assumption that there is something to be understood in Dracula. If Dracula, as the category-breaking being he is, is truly, deeply, different, then there might be nothing "in" him that we could ever understand or, *a fortiori*, pity. Perhaps he is, once again, in between categories, this time between the categories of the comprehensible and the incomprehensible; and it might be entirely incomprehensible for us how anyone or anything could lie between *those* categories.

The deepest horror of all might indeed lie in our fear of becoming incomprehensible or unintelligible – to others, to ourselves, or both. We may fear that the challenge to our conceptual categories and classifications launched by the aesthetic use of the concept of the undead, for instance, takes away the very natural vs. supernatural classification we have relied on and agreed on with our fellow humans. Or, obviously, we might be afraid of becoming mentally ill. In these cases, as varied as they may be, our world would clash into pieces, lose its rational integrity. We might even, as a philosophical thought experiment, imagine, or rather, desperately try to imagine, a world in which a "transcendental chaos" would be a rule: a world, that is, in which the objects and events of our environment would not posses a sufficient, to us recognizable degree of stability and variety, so as to render self-conscious human cognitive experience of determinate objects and events so much as possible (cf. Westphal 2003; 2004, as well as Westphal's paper in this collection). If we are afraid of transcendental chaos, thus defined, or are afraid of madness or utter irrationality interpreted as a transcendental chaos in this sense, we are afraid of losing the integrity of experience itself – or, even worse, of losing self-consciousness, which, of course, we do lose in death. All of these cases are, unsurprisingly, entangled with our horror of death, although mere horror of death as such is too simple a concept to cover these different conceivable situations.

5. Conclusion

Generally speaking, we should notice the occasionally ingeniously self-conscious *reflexivity* characteristic of Gothic works of art. When describing this human – and inhuman – world, e.g., extreme evil, Gothic art often describes the very difficulties involved in the project of describing the world, or even the impossibility of describing it accurately, as it is "in itself". Admittedly, I have not shown, merely by considering horror fiction and the possibility of the undead, that (say) science, or the philosophical interpretation of science known as scientific realism, would really be in trouble in the project of representing and categorizing the world. But I have argued that world-categorization is a more complex matter than we might initially think – a reflexively complex one, that is. We should also note that Gothic works of art may even question *their own ability* to describe or represent anything, including the difficulty of describing or representing anything. Reflexively, they thus question not only the absoluteness of the scientific perspective on reality but also *their own status as works of art*. It is a commonplace to note that the problematization of the project of narrating about or linguistically describing the world is a crucial feature of particularly twentieth century Gothic (cf. Botting 1996, p. 14); it may, however, be more interesting to note that this conception of horror fiction can be connected with the ways in which the notions of irrationality and madness are used in our accounts of our own difficulties with describing the world we live in, and ourselves. We may, simply, feel like living in an irrational world, a world that can only be described as "crazy" or "mad" – and in that sense indescribable. The dialectics of description, or representation, runs deep. Probing the limits of conceivability, experienceability, and conceptualizability in this manner may even lead us to appreciate the condition that Leszek Kolakowski (2001) labels "metaphysical horror".

Through the reflexive reflections provided in the previous section, the metaphysical and aesthetic issues of this paper can be combined. The artistic functions of the category of the undead are largely based on the conceptual strangeness, uncanniness, of this category and on the resulting epistemological problematizing of our normal human abilities to categorize, scientifically or otherwise, a meaningfully structured reality. The undead – or even the possibility of imagining them – are, as we have seen, a challenge to our conceptual capacities, to our "rational" being in the world, even if we are convinced that there actually are no such things. Realizing this may lead us to appreciate the philosophical value of Gothic fiction, but also the value of thought experiments such as Westphal's (2003; 2004) "transcendental chaos". I have, in other words, tried to point out the philosophical relevance of art (in particular, horror fiction, but in principle other forms of art as well) in our on-going reflections on what it is to describe a world that we occasionally experience as undescribable or even inexperienceable. I have, that is, tried to show that our understanding of ourselves may be enriched by attempts to imagine "mad", category-breaking entities and situations apparently very different from what human life looks like in the light of science and common sense.

University of Helsinki

References

Airaksinen, T. (1998), *The Philosophy of H. P. Lovecraft*, New York, Peter Lang.

Bernstein, R. J. (2002), *Radical Evil: A Philosophical Interrogation*, Cambridge, Polity Press.

Booth, W. C. (1983), *The Rhetoric of Fiction*, 2nd ed., Chicago and London, The University of Chicago Press (1st ed. 1961).

Botting, F. (1996), *Gothic*, London and New York, Routledge.

Carroll, N. (1987), "The Nature of Horror", *The Journal of Aesthetics and Art Criticism* 46, 51-59.

Carroll, N. (1990), *Philosophy of Horror or Paradoxes of the Heart*, London and New York, Routledge.

Crary, A. and R. Read (eds.) (2000), *The New Wittgenstein*, London and New York, Routledge.

Flew, A. (1987), *The Logic of Mortality*, Oxford, Blackwell.

Heidegger, M. (1927), *Sein und Zeit*, Tübingen, Max Niemeyer, 1961.

Kolakowski, L. (2001), *Metaphysical Horror*, rev. ed., ed. by A. Kolakowska, Harmondsworth, Penguin Books (1st ed. 1988).

Morton, A. (2004), *On Evil*, London and New York, Routledge.

Neiman, S. (2002), *Evil in Modern Thought: An Alternative History of Philosophy*, Princeton, Princeton University Press, 2004.

Read, R. (2004), "Wittgenstein and Faulkner's Benjy", in J. Gibson and W. Huemer (eds.), *The Literary Wittgenstein*, London and New York, Routledge.

Reed, T. (1988), *Demon-Lovers and Their Victims in British Fiction*, The University Press of Kentucky.

Rosenberg, J. F. (1998), *Thinking Clearly about Death*, rev. ed., Indianapolis, Hackett (1st ed. 1983).

Roth, P. A. (1982), *Bram Stoker*, London, Twayne Publishers.

Stoker, B. (1897), *Dracula*, New York, Permabooks, 1957.

Summers, M. (1928), *The Vampire*, London, Studio Editions, 1995.

Westphal, K. R. (2003), *Hegel's Epistemology: An Introduction to the Phenomenology of Spirit*, Indianapolis, Hackett.

Westphal, K. R. (2004), *Kant's Transcendental Proof of Realism*, Cambridge, Cambridge University Press.

Wittgenstein, L. (1953), *Philosophical Investigations*, trans. G. E. M. Anscombe, Oxford, Basil Blackwell, 1958.

The Scope and Limits of Value-Freedom in Science

Panu Raatikainen

1. Introduction

The issue of whether science is, or can be, value-free has been debated for more than a century. The idea of value-free science is of course as old as science itself, and so are the arguments against this idea. Plato defended it against Protagorean relativism and skepticism. Bacon warned about "idols" that could divert scientific inquiry. Nineteenth century writers often used the term "ideology" as a sort of code word to underline the threat to the objectivity of science posed by the intrusion of values alien to science proper (cf. Proctor 1991).[1]

However, the dispute truly began with Max Weber's classical defense of value-freedom and the neutrality of social science in particular (see Weber 1904; 1917; 1919; 1949). For Weber, the fundamental issue was that objective knowledge was possible in social science. Accordingly, the arguments against value-freedom have often included the claim that scientific research and its results are so thoroughly value-laden that objective knowledge is not at all possible in science.

Weber wanted to defend the autonomy of science with respect to religion and politics. As is often noted, value-freedom for Weber was, at least in part, a shield against the accusations of socialism in social science. But it is important to note that Weber also wanted to protect values from *scientism*, that is, he opposed the view that one can derive all values from science – that science can answer all value questions. According to him, scientists should not say categorically what one ought to do.

In Weber's view, the scientist *qua* scientist should not make categorical value judgments. He was reacting to the "ethical economics" – popular at the time – which was quick to give all sorts of recommendations on what one should do. Weber, on the other hand, emphasized that one should keep science and politics separated, so that one would not disguise politics in science's clothing. He compared science to a map: it cannot tell us where to go, but it can tell us how to get

---

1 Proctor (1991) is a comprehensive historical survey of the history of the controversy over the value-freedom of science.

there. One cannot derive ultimate values from scientific facts. Weber, however, admitted that values may affect the choice of a research problem.

The logical positivists defended a strong form of the value-freedom thesis. They held that moral statements are radically different from factual statements, as the former do not even have cognitive content – they endorsed emotivism in ethics. But of course, acceptance of the idea of the value-freedom of science does not require this radical view. In any case, logical positivism thus differed from classical French positivism (e.g. Comte), which rather adhered to scientism, which is incompatible with the idea of value-free science.

Nazism, on the other hand, emphasized that science is not, and should not be, value-free. According to it, there was Jewish science, and there was Aryan science. The ideal of value-freedom was also denied by the leftist Frankfurt critical school, by e.g. Horkheimer, Adorno, Marcuse – and later Habermas. So it was also in the Soviet Union, where one had to take a side between "bourgeois science" and "proletarian science". In 1948, Lysenko's totally unscientific theory of the inheritance of acquired characteristics was endorsed as the official state biology of the Soviet Union, and Mendelian genetics, as well as the relativity theory and quantum mechanics, were denounced.

Richard Rudner (1953) famously argued against Weber that the scientist *qua* scientist does make value-judgments in a typically ethical sense. For example, if the hypothesis under consideration were to the effect that a toxic ingredient of a drug was not present in a lethal quantity, we would require a high degree of confirmation, whereas in some less critical case we may accept a hypothesis on the basis of a lower degree of confirmation. In general, Rudner stated, before a scientist can accept any hypothesis, the value decision must be made in the light of how serious it would be to make a mistake, and whether the evidence is strong enough to warrant its acceptance. How great a risk of being wrong one is willing to take will depend upon how seriously in the typically ethical sense one views the consequences of making a mistake, Rudner concluded.

The possibility of value-freedom was also questioned by Charles Taylor (1967). He argued that political science necessarily requires theoretical frameworks which always implicitly support some values. In his view, to say that something fulfills human needs and wants always constitutes a reason for calling it good. According to Taylor, political science cannot be morally neutral.

Sociologist Alvin Gouldner (1962) complained that "value-freedom" had become a hollow catechism, an excuse to not think seriously about social problems – that "value-freedom" had come to mean a willingness to sell one's skills to the highest bidder. According to Gouldner, "value-freedom" allowed one to believe that it is no worse to work to spread disease than to cure it; value-freedom meant that there was nothing wrong with doing market research for tobacco companies to help them sell cigarettes.

More generally, the extreme interpretation of "value-freedom" (it is hard to find anyone to defend it explicitly, but certainly the view has been out there) holds that, first, it is valuable to pursue any research problem – that all knowledge is positive – so it does not matter which research problem you choose, and, second, that a scientist is not responsible for the possible applications of his or her research, but that the responsibility lies with those who apply it in practice. In any case, this is not what Weber and the others following him defended as the value-freedom of science.

2. Some Distinctions

The value-freedom of science has clearly meant different things to different people. I submit that one can only make progress in answering the question of whether science is value-free by making several clarifying distinctions.

First, it is important to distinguish *epistemic values* (such as explanatory power, simplicity, and empirical support) from *non-epistemic values* (e.g. moral and political values). Obviously, scientists are committed to such values as the scientific ethos (in Merton's sense), honesty and regard for facts. More generally, science is characterized by the values of objectivity, openness, criticality and publicity. And certainly, epistemic values such as probability, simplicity, explanatory power, or support from observations are effective in the acceptance of theories and hypotheses.

None of this is really controversial, although such trivialities are sometimes offered as evidence for the conclusion that science is not value-free. In particular, such values in no way threaten the objectivity of science. The real and interesting question of the value-freedom of science therefore concerns the possible effect of non-epistemic values on science (and *vice versa*).

What exactly are the epistemic values of science and how they function is, of course, controversial, and the issue is debated in the philosophy of science. But it is clear that they are essentially different from such non-epistemic values as moral or political values. (See e.g. Levi 1967; Kuhn 1977; McMullin 1983; Laudan 1984; Niiniluoto 1984; 1993.)

Second, one should keep in mind that "science" is an ambiguous term; by it, one may mean, for example: (1) science as an institution (scientific workers, the organization of research), (2) scientific research activity, or the research process, (3) results generally accepted in the scientific community at a particular moment; scientific knowledge, and (4) the scientific method as a critical and intersubjective method for accepting beliefs. (cf. Niiniluoto 1984, ch. 1).

Science understood as "the scientific method" is a strongly normative concept (although not in any moral sense), and not everything that claims to be science or is even widely taken as science really counts as science in this sense. Consequently, one can also understand the other senses of "science" in a normative way, e.g. delimit "the scientific institution" to only those instances that truly follow the scientific method. In that case, a community of investigators ceases to be a *scientific* community if it gives up, or is forced to give up, its autonomy and neutrality (cf. Niiniluoto 1984, ch. 1).

Alternatively, more external criteria can be used to limit the definition, so that workers whose work does not really satisfy the requirements of good science still count, if they are natural parts of the same institution of science as a whole. The same applies, *mutatis mutandis*, to the notions of "science" as research activity and as scientific knowledge, or to the results of science. But especially in the case of the results of science, it may be more natural to think that only the outcomes of a critical discussion within the scientific community may be tentatively accepted as the results of inquiry.[2] In this sense, the real "subjects" of scientific knowledge are the scientific communities rather than individual scientists, as has been emphasized by e.g. Peirce, Weber and Popper (cf. Niiniluoto 1984, ch. 1).

Third, it is important to understand the difference between *basic research* and *applied research*. The traditional philosophy of science often totally ignores applied sciences. Various critical thinkers, on the other hand, assume that all science is applied, and thus at least implicitly deny the very existence of basic research.

The classical characterization of the distinction was presented by the OECD in 1966: Basic research is defined as "the systematic pursuit of new scientific knowledge without the aim of specific practical application", and applied research as "the pursuit of knowledge with the aim of obtaining a specific goal"; further, "development" uses the results of research to develop "new products, results, and means of production" (OECD 1966). This is certainly quite vague, for it is unclear whose aims are relevant here. Nevertheless, there are certainly clear enough examples of each; in Kitcher's words, "we can contrast extreme cases .... When any links to practical projects are buried in a distant past, with no bearing on contemporary applications, and when it's very hard to forecast how results from this inquiry could be adapted to novel practical problems, the researchers can quite legitimately declare their intentions to be thoroughly epistemic" (Kitcher 2000, p. 89).

Niiniluoto (1993) has presented a less vague approach. He suggests that the old basic-applied distinction is less fundamental than the distinction between what he

---

2 Note, by the way, that also human sociobiology and orthodox evolutionary psychology, fields of study now often discussed in connection with the value-freedom thesis, do not at the moment, and in my mind probably never will, count as scientific knowledge in this sense. More generally, I think that many examples that the critics of science and the idea of value-freedom use do not count as scientific knowledge in this sense.

calls *descriptive sciences* and *design sciences*. Descriptive sciences primarily aim to describe, explain and understand the reality surrounding us. Design sciences, on the other hand, aim at knowledge that is useful for the activity of design, i.e. aim to enhance human art and skill. The nature of design sciences can be schematically illuminated with "technical norms" (this notion is from G. H. von Wright, 1963):

> If you want *A*, and you believe that you are in a situation *B*,
> then you ought to do *X*.

Technical norms express the typical structure of the knowledge provided by design science. Such norms are factual, true or false statements about the relation between means and ends, and can be objective scientific results just like descriptive statements. One can accept or reject such norms through normal scientific methods.

Technical norms often follow from the corresponding causal claims:

> *X* causes *A* in circumstances *B*.

Statements of this form are typically results of ordinary descriptive science. In such a case, a design science really is applied science. In other cases, however, no general theory is at hand, and a technical norm is obtained by more practical trial-and-error procedures and experimental tests. Thus, not all design science is applied science. Design sciences are based upon manipulable causal relations in which we can do or bring about *X*. That many causal relations are not humanly manipulable – one can think of astronomy – makes it clear that purely descriptive science also exists. Note, however, that the distinction between descriptive science and design science cuts across disciplines (see Niiniluoto 1993).

And fourth, the scientific research process has different phases. We can roughly divide it into three possible different phases:

(1) the choice of a research problem;
(2) the acceptance of results; and
(3) the possible practical applications.

One can attribute value-freedom or value-ladenness to any or each of these. Originally, Weber only meant his thesis of value-freedom to apply to phase (2), and defended the objectivity requirement concerning results. Sometimes, however, both scientists willing to avoid any responsibility for the applications of their work and critics of the idea of value-freedom have assumed that the thesis of value-freedom applies to all these phases.

In consequence of all the above, the question of the value-freedom of science is multiply ambiguous.

3. Conclusions

With these distinctions at hand, let us now return to our main questions about the possibility and limits of the value-freedom of science. One can recap the content of Weber's original idea of value-freedom with the following two theses:

Objectivity Requirement: In science, one should not, in accepting or refuting hypotheses, appeal to any non-epistemic (such as religious, political or moral) values.

Anti-Scientism: Ultimate values (more exactly, unconditional value judgments) cannot be derived from the results of science.

I think that once the issue is clearly understood, the implausibility of scientism should be quite obvious. It is simply inconceivable that, say, human rights would some day be considered to be unconditional facts of science. Likely assumptions of the contrary are usually only implicit and based on confusions – I doubt that very few nowadays really want to argue for scientism. Unlike conditional technical norms, the categorical unconditional imperative "You ought to do *X*" does not even have a truth value. In other words, such unconditional recommendations are qualitatively different from the results of science.

How about the requirement of objectivity in accessing hypotheses? If we identify "science" by external criteria, and count e.g. Lysenko's theory or "ethical economy" (which was Weber's target) as science, it is rather trivial that science is *not* value-free – plenty of bad science and pseudo-science around violate that requirement. If, on the other hand, we understand science, and scientific knowledge, in a more normative way (see above), science is objective and value-free by definition.[3] The question then is: Is there any science in this more demanding, normative sense? The extremely strong skeptical claim that there has never been, and could never be, good science is quite implausible and totally unjustified. It is rarely stated explicitly but is implicitly assumed in the many radical arguments against the value-freedom of science. The spectacular practical success of modern science strongly suggests that science has not done at all badly.

A further question is whether there is scientific research that is value-free also with respect to the choice of the research problem and applications. This turns to the question of whether there is non-applied basic research or descriptive science.

Behavioralism (e.g. Carnap, Jeffrey, Rudner) holds that scientists do not accept or reject propositions at all, unless such acceptance or rejection is related to

3 Many disputes between philosophers and sociologists of science seem to me only verbal, for the participants interpret "science" very differently.

practical objectives. Scientists are considered to be decision-makers or guidance counselors. From behavioralism's perspective, all the results of science should be viewed as mere recommendations for action in practical decision-making situations. More generally, *Instrumentalism* (some pragmatists and Marxists, e.g. Dewey, Habermas) treats all science as applied science, and denies that scientific knowledge of reality could be valued for its own sake. Science always has and must have instrumental value. Opposed to the both above views is *Cognitivism* (see Levi 1967), which recognizes the quest for truth as a legitimate human activity whose aims and products are not directly relevant to practical application – that at least *some* scientific inquiries may be concerned with accepting propositions as true, rejecting them as false, or suspending judgment on them (this acceptance may, of course, be tentative). Cognitivism appears to give a much more accurate picture of science than its opponents (see e.g. Levi 1967; Niiniluoto 1984). Consequently, there is also basic, descriptive science (cf. above).

For example, Rudner's much-cited argument against value-freedom is based on his assumption of behavioralism, and ignores the difference between basic and applied research. It shows, if anything, at best that in the practical application of scientific knowledge one has to make value-judgments. In basic descriptive science there are no such applications (cf. Levi 1960, 1967). More generally, many attacks on the idea of the value-freedom of science assume a behavioralist or instrumentalist view. But once we see the shortcomings of these views, many such critiques lose their power.

Every scientist must somehow select some specific topic of study; no one can study everything. Does this mean that all science must be value-laden at the point of selection? Not at all! That would be the case only if one wholly denied the existence of basic, descriptive science – but we have seen good reasons to rebut such instrumentalist views. In basic research, research problems are chosen primarily because of their potential scientific relevance – that is, only epistemic values are involved. Such research is thus also value-free with respect to the choice of the research problem. And because phase (3) of the research process, applications, is missing by definition, one can say that such research is value-free in all three phases. In applied research, or design science, the situation is of course quite different.

At the level of results and their acceptance, all good science, whether basic or applied, descriptive or design science, is value-free and objective. On the other hand, in applied research, or design science, the question of which issues are studied is by no means morally neutral. The choice of research problem may be significantly value-laden. Further, in many situations it is reasonable to consider scientists morally responsible for the applications of their results. This is especially obvious if the morally questionable application has been clear from the beginning, e.g. in developing weapons of mass destruction, effective methods of torture, or marketing morally problematic products.

Further, as Kitcher points out, mere ignorance of morally problematic applications does not always free the scientist from moral responsibility: "when only little curiosity is needed to see the propensity of others to engage in morally consequential applications, the researcher who proclaims solely epistemic intent is guilty of self-deception (at the very least)" (Kitcher 2001, p. 89). Part of the moral duties of a scientists is to actively examine the question of potential ethically problematic applications of one's research. Only if one has done one's best in this respect, one may be considered not to be responsible for some unpredictable immoral applications of one's research.

University of Helsinki

References

Gouldner, A. (1962), "Anti-Minotaur: The Myth of a Value-Free Sociology", *Social Problems* 9, 199–213.

Kitcher, P. (2001), *Science, Truth, and Democracy*, New York, Oxford University Press.

Kuhn, T. S. (1977), "Objectivity, Value Judgments, and Theory Choice", in Kuhn, *The Essential Tension*, Chicago, University of Chicago Press, pp. 320–339.

Laudan, L. (1984), *Science and Values*, Berkeley, University of California Press.

Levi, I. (1960), "Must the Scientist Make Value Judgments?", *Journal of Philosophy* 57, 345–357.

Levi, I. (1967), *Gambling with Truth*, Cambridge, MA., The MIT Press.

McMullin, E. (1983), "Values in Science", *PSA 1982*, Vol. 2, East Lansing, Philosophy of Science Association, pp. 3–28.

Niiniluoto, I. (1984), *Is Science Progressive?*, Dordrecht, Reidel.

Niiniluoto, I. (1993), "The Aim and Structure of Applied Research", *Erkenntnis* 38, 1–21.

OECD (1966), *Ministerial Meeting on Science: Fundamental Research and the Policies of Government*, Paris, 1966.

Proctor, R. N. (1991), *Value-Free Science? Purity and Power in Modern Knowledge*, Cambridge, MA & London, Harvard University Press.

Rudner, R. (1953), "The Scientist qua Scientist Makes Value Judgments", *Philosophy of Science* 20, 1–6.

Taylor, C. (1964), "Neutrality in Political Science", in P. Lanslett and W.G. Runciman (eds.), *Philosophy, Politics and Society*, third series, Oxford, Basil Blackwell, pp. 25–57.

Weber, M. (1904), "Die 'Objektivität' sozialwissenschaftlicher und sozialpolitischer Erkenntnis", in Weber, *Gesammelte Aufsätze zur Wissenschaftslehre*, 3rd ed., Tübingen, 1968. Translated as "'Objectivity' in Social Science and Social Policy" in Weber (1949).

Weber, M. (1917), "Der Sinn der 'Wertfreheit' der soziologischen und ökonomischen Wissenschaften", in Weber, *Gesammelte Aufsätze zur Wissenschaftslehre*, 3rd ed., Tübingen, 1968. Translated as "The Meaning of 'Ethical Neutrality' in Sociology and Economics", in Weber (1949).

Weber, M. (1919), "Wissenschaft als Beruf", in Weber, *Gesammelte Aufsätze zur Wissenschaftslehre*, 3rd ed., Tübingen, 1968. Translated as "Science as a Vocation" in Hans Gerth and C. Wright Mills (eds.), *From Max Weber*, New York, 1958.

Weber, M. (1949), *The Methodology of the Social Sciences*, New York, The Free Press.

von Wright, G. H. (1963), *Norm and Action*, London, Routledge and Kegan Paul.

Social Epistemology of Scientific Inquiry[1]

Kristina Rolin

1. Introduction

The last three decades have witnessed an emergence of social studies of science which aim to show that social values play a more significant role in scientific inquiry than what philosophers of science have acknowledged. Philosophers' responses to these studies have been divergent. One response has been to design ways of strengthening the methods of science in the hope of eliminating the influence of social values in scientific inquiry. Another response has been to argue that social values do not inevitably undermine the epistemic integrity of science; instead, they can contribute to the epistemic success of science. These developments have given rise to a subfield in philosophy of science, *social epistemology of scientific inquiry*, which aims to understand why social processes are of epistemic importance in science and why some social processes are epistemologically better than others. Social epistemologists acknowledge the social and value-laden nature of scientific inquiry and are concerned with normative and prescriptive questions which have been the traditional preoccupations of philosophers (Longino 1994, p. 135).

Miriam Solomon's *Social Empiricism* (2001) is one of the most ambitious attempts to develop an account of scientific inquiry which is both social and normative. Solomon argues that social values can play a positive role in scientific inquiry by generating and maintaining an efficient distribution of research effort among those theories that have some empirical successes. Solomon develops this argument into a thoroughgoing criticism of individualism in philosophy of science. Instead of evaluating the reasoning and decision-making of individual scientists, we should evaluate scientific communities on the basis of how well they succeed in distributing research effort, she claims.

---

1 I am grateful to K. Brad Wray for an opportunity to discuss Solomon's social empiricism in detail as part of our collaboration in social epistemology. I also wish to thank Panu Raatikainen and Petri Ylikoski for their insightful comments on an earlier version of this paper presented in the XV Internordic Philosophical Symposium in Helsinki, 2004.

In this paper, I argue that Solomon's social empiricism is inadequate as a normative theory of scientific communities. Paradoxically, it is both too social and not social enough. It is too social because it emphasizes the rationality of communities at the expense of the rationality of individuals. It is not social enough because it operates with a narrow understanding of the social dimensions of scientific inquiry. The "social" in science is understood to refer to scientists' *social values*. I wish to draw attention to the epistemic importance of other social dimensions of scientific inquiry. The "social" in science can also refer to *social relations* among scientists.

2. What is Social Empiricism?

Miriam Solomon invites us to take a fresh look at the traditional distinction between epistemic and non-epistemic values as it is used, for example, by Thomas Kuhn (1977). By epistemic values, Kuhn refers to such values as accuracy, consistency, broad scope, simplicity, and fruitfulness. These values are to be understood as desirable features of scientific theories. By non-epistemic values, Kuhn refers to scientists' moral and social values. In order to replace the concepts of epistemic and non-epistemic value, Solomon introduces the epistemologically neutral concept of "decision vector." A decision vector is any factor that influences the direction of research (2001, p. 53). According to Solomon, "scientific rationality – conduciveness to scientific success – is not an intrinsic property of most decision vectors" (p. 63). Therefore, we should not prematurely judge any decision vector as either irrational or arational since it can function in many ways in scientific inquiry. Whether or not a decision vector is conducive to scientific success will depend on circumstances (pp. 53, 63). For example, desire for fame and fortune can motivate scientists either to aspire towards higher standards of research or to succumb to fraud. In the former case, desire for fame and fortune would be conducive to scientific success. In the latter case, it would be an obstacle to scientific success. This is why Solomon rejects the widespread practice of calling some decision vectors as "biasing factors" (p. 53).

However, Solomon makes a distinction between two types of decision vectors, *empirical* and *non-empirical*. According to her definition, "empirical decision vectors are causes of preference for theories with empirical success, either success in general or one success in particular" (2001, p. 56), and "non-empirical decision vectors are other reasons or causes for choice" (p. 56). Solomon emphasizes that "only the empirical decision vectors are always conducive to some scientific success, and even then, they do not typically maximize attainable empirical success" (p. 63). She cites the salience of data and the ability to generate novel predictions as examples of empirical decision vectors (p. 57). Her examples of non-empirical decisions vectors include conservativeness, competitiveness, peer pressure, deference to authority, elegance, and simplicity (pp. 57–58).

By *scientific rationality*, Solomon refers not merely to scientific reasoning but to any scientific practice that leads to empirical success or truth (2001, p. 52). Her definition of rationality is externalist in the sense that scientific rationality depends on the consequences of scientific practices (p. 16). It does not matter whether scientific practices are "logical," "clear," "algorithmic" or "objective" (p. 52). It matters merely whether they are conducive of empirical success or truth. Empirical success can come in many forms. For example, successful predictions of new phenomena, new explanations of already known phenomena, successful control and manipulation of natural processes count as empirical successes (p. 27). Solomon argues that empirical successes are the primary goals of scientific inquiry because they are "contingent on the world outside the inquirers" (p. 17). In contrast, theoretical successes, such as simplicity and elegance, are not (p. 17).

These definitions are preliminary steps towards what I think is Solomon's major contribution to contemporary philosophy of science. The major contribution in *Social Empiricism* (2001) is that it aims to provide a novel solution to the problem of under-determination in philosophy of science. The problem of under-determination is the question of what criteria ought to guide theory choice when theory choice is under-determined by empirical evidence. Solomon's solution includes two parts. First, she suggests that a normative theory of scientific inquiry should not ban non-empirical decision vectors in the individual level, in determining an individual scientist's choice of one theory over another. Second, she suggests that a normative theory of scientific inquiry ought to address the role of non-empirical decision vectors in the community level, in determining a rational distribution of research effort in the community. Let me explain each part in more detail.

When different theories have different empirical successes and none of the theories has all available empirical successes in a domain of inquiry, scientists are faced with the under-determination of theory choice by empirical evidence. Solomon argues that in this situation non-empirical decision vectors are allowed to influence an individual scientist's theory choice since they can play a *rational* role in scientific inquiry. They play a rational role insofar as they distribute research effort among theories that have empirical successes. As Solomon explains, "distribution of research effort is generally conducive to scientific success because it is not possible to tell in advance which theory or theories will be empirically successful or true" (2001, p. 53). And she adds that "scientists often have the opportunity to explore a number of potentially successful directions of research, and it is most efficient to accomplish this when different individuals, or groups, pursue different theories or strategies at the same time" (p. 65).

However, not just any distribution of research effort will do. As a normative theory of scientific inquiry, social empiricism defines conditions for a *rational* distribution of research effort. According to Solomon, a rational distribution of research effort requires two things. It requires first that empirical decision vectors be *equitably* distributed in proportion to the empirical successes of the various theories under consideration so that each theory will receive its fair share of attention, and second that non-empirical decision vectors be *equally* distributed among those theories that have some empirical successes (2001, pp. 77, 95, 117–118). Solomon emphasizes that a rational distribution of research effort is not always an equal distribution of research effort since one theory may have more empirical successes than others and hence deserve more attention than others (p. 76). On the other hand, not all scientists should abandon a theory which has less empirical successes than another theory. Solomon believes that every theory that has some empirical successes deserves some attention. Scientists should pursue only those theories that have empirical successes; theories without any empirical success do not have scientific merit (p. 95). Besides defining conditions for a rational distribution of research effort, social empiricism defines conditions for a *normatively appropriate consensus*. According to Solomon, a consensus on a theory is normatively appropriate when the theory has all the empirical successes available in a domain of inquiry (p. 119).

Finally, Solomon develops social empiricism into a thoroughgoing critique of individualist assumptions underlying most philosophy of science. As she explains, "social empiricism is social because what matters, normatively speaking, is the distribution of empirical and non-empirical decision vectors across a community of investigators" (2001, p. 120). "Normative judgments are not made of the thoughts and decisions of individual scientists," she says (p. 120). For an individual scientist, social empiricism gives merely one guideline. An individual scientist should aim for and work with empirically successful theories (pp. 150–151). Perhaps the most striking feature of social empiricism is that it imposes no constraints on the influence of non-empirical decision vectors on the reasoning and decision-making of individual scientists. As Solomon puts it, social empiricism demands less than most normative accounts of scientific inquiry (p. 135). "In particular, it does not require individual scientists to make overall impartial assessments" (p. 135). As a normative account of scientific inquiry, social empiricism is meant to be applied in science policy and funding decisions rather than in the everyday work of individual scientists.

3. Limitations of Social Empiricism

I would argue that social empiricism is inadequate as a normative account of scientific inquiry because it does not pose sufficiently stringent norms on the reasoning and decision-making of individual scientists. It is not sufficient to require merely, as Solomon does, that individual scientists work with theories with some empirical successes (2001, pp. 150–151). Given this norm, it is difficult to understand how scientific change could take place at all. Obviously, what counts as a rational distribution of research effort in a community changes over a period of time if the distribution of empirical successes among competing theories changes over the same period of time. As a normative theory of scientific inquiry, social empiricism recommends that research effort is focused to a larger extent on those theories that have been successful in accumulating empirical successes. However, it is difficult to perceive how this shift in the distribution of research effort can take place unless we require that individual scientists are responsive to a change in the distribution of empirical successes. For example, those scientists who are hired in permanent positions and who are equipped with some research resources by their institutions cannot be forced to respond to a change in the distribution of empirical successes merely by an outside pressure coming from science policy makers and external funding institutions. Within the confines of social empiricism, these established scientists could continue research on a theory which has not been successful in accumulating new empirical successes. To say the least, relying merely on science policy makers and funding institutions as agents of scientific change is too slow a means to implement a normatively required shift in the distribution of research effort.

To understand how scientific change can take place, a normative theory of scientific inquiry would have to include norms that are applicable to individual scientists' reasoning and decision-making. In fact, there is no shortage of such norms since the reasoning and decision-making of individual scientists is the major object of inquiry in traditional philosophy of science which Solomon wishes to leave behind. As K. Brad Wray (1997) points out, we should expect at least that the effects of non-empirical decision vectors on individual scientists' reasoning and decision-making diminish as empirical evidence accumulates. Therefore, it is a mistake to suggest, as Solomon does, that *all* that matters is the distribution of empirical and non-empirical decision vectors across a community (2001, p. 120). It matters also whether individual scientists are willing to let empirical decision vectors overweigh non-empirical ones in their decision-making. And it is a mistake to claim, as Solomon does, that a normative theory of scientific inquiry does not need to make normative judgments of the thoughts and decisions of individual scientists (p. 120).

An unfortunate consequence of Solomon's anti-individualism is that social empiricism gives an idealized account of non-empirical decision vectors in science. It is one thing to say that non-empirical decision vectors are allowed to influence an individual scientist's choice of one theory over others when different theories have different empirical successes and none of the theories has all available empirical successes in a domain of inquiry. It is quite another thing to say that non-empirical decision vectors are allowed to influence an individual scientist's choice of one theory over others *even after* the other theories have succeeded in accumulating more empirical successes than what they have had before. Solomon is right in suggesting that an adequate account of non-empirical decision vectors would recognize that they sometimes play a rational role in science by generating and maintaining a rational distribution of research effort. However, an adequate account of non-empirical decision vectors would also recognize that a non-empirical decision vector which plays a rational role at a particular time t_1 can play an irrational role at another time t_2.

These problems in social empiricism derive not only from Solomon's commitment to anti-individualism but also from a conceptual confusion between scientists' motivations and the standards of argumentation that make scientists accountable for each other. Recall that Solomon defines a decision vector as any factor that influences the direction of research (2001, p. 53). By decision vectors, she refers to motivations which drive scientists in their scientific work. Yet, motivations are not the same as standards of argumentation which guide conversations among scientists. A scientist can pursue research for almost any motivation but when she presents her results to her colleagues she has an epistemic responsibility to argue in accordance with the standards of argumentation that are shared in the community (or to present arguments in favor of a revised set of standards). I have argued that Solomon's normative theory of scientific inquiry is inadequate because it does not include such standards.

Paying attention to such standards would require that one pays attention to other social dimensions of scientific inquiry besides the distribution of research effort in a scientific community. One would have to pay attention to social relations among members of scientific communities. If a normative account of scientific inquiry aims to be social, it should include norms which guarantee that members of scientific communities are accountable for each other. Solomon's social empiricism does not offer such norms.

4. Conclusion

As a conclusion, I would argue that Solomon's social empiricism is *not social enough* because it focuses on a rational distribution of research effort at the expense of developing a theory of the norms that guide social interactions among members of scientific communities. I would argue also that Solomon's social empiricism is *too social* because it focuses on a rational distribution of research effort at the expense of developing a theory of the reasoning and decision-making of individual scientists. It is possible to appreciate the insight that non-empirical decision vectors sometimes play a rational role in science without committing oneself to Solomon's thoroughgoing anti-individualism.

Academy of Finland
Helsinki School of Economics

References

Kuhn, T. S. (1977), *The Essential Tension*, Chicago, The University of Chicago Press.
Longino, H. (1994), "The Fate of Knowledge in Social Theories of Science", in F. Schmitt (ed.), *Socializing Epistemology*, Lanham, Rowman & Littlefield Publishers.
Solomon, M. (2001), *Social Empiricism*, Cambridge, MA, The MIT Press.
Wray, K. B. (1997), "Rational Communities", *Perspectives on Science* 5, 232-254.

Nordeuropäische Beiträge aus den Human- und Gesellschaftswissenschaften

Scandinavian University Studies in the Humanities and Social Sciences

Herausgegeben von / Edited by
Hartmut Schröder

Band 1 Riitta Korhonen / Martin Kusch / Hartmut Schröder (toimittajat): Lopussa teksti. 1992.

Band 2 Marianne Nordman (Hrsg.): Fachsprachliche Miniaturen. Festschrift für Christer Laurén. 1992.

Band 3 Mariann Skog-Södersved: Wortschatz und Syntax des außenpolitischen Leitartikels. Quantitative Analysen der Tageszeitungen "Neues Deutschland", "Neue Zürcher Zeitung", "Die Presse" und "Süddeutsche Zeitung". 1993.

Band 4 Anna Mauranen: Cultural Differences in Academic Rhetoric. A Textlinguistic Study. 1993.

Band 5 Peder Skyum-Nielsen / Hartmut Schröder (eds.): Rhetoric and Stylistics Today. An International Anthology. 1994.

Band 6 Jouko Jokisalo: Vom Bockmist zur geschichtsmäßigen Kraft. Determinanten und Wirkung der Heilsversprechen des "deutschen Sozialismus" (1933-1939). 1994.

Band 7 Jarmo Pulkkinen: The Threat of Logical Mathematism. A Study on the Critique of Mathematical Logic in Germany at the Turn of the 20th Century. 1994.

Band 8 Eero J. Laine: Learning Second National Languages. A Research Report. 1995.

Band 9 Vesa Oittinen: Der Akt als Fundament des Bewußtseins? Zur Differenz zwischen G.I. Hartman und nachkantischer Transzendentalphilosophie. 1996.

Band 10 Christer Laurén / Marianne Nordman: Wissenschaftliche Technolekte. 1996.

Band 12 Harry Anttila: Prädikate und Satzmodelle in Zeitungstexten. Ein Beitrag zur historischen Syntax des Deutschen, gezeigt an der Leipziger Zeitung 1660-1914. 1997.

Band 13 Hannele Kohvakka: Ironie und Text. Zur Ergründung von Ironie auf der Ebene des sprachlichen Textes. 1997.

Band 14 Pertti Ahonen / Ari Salminen: Metamorphosis of the Administrative Welfare State. From Depoliticisation to Political Rationality. 1997.

Band 15 Gerhard Schmitt: Zyklus und Kompensation. Zur Denkfigur bei Nietzsche und Jung. 1998.

Band 16 Yves Gambier (éd.): Discours Professionnels en Français. 1998.

Band 17 Irmeli Helin: Vom Brodverein zur co op. Terminigeschichte der deutschen Genossenschaftssprache. Ein Beitrag zur Terminologieforschung. 1998.

Band 18 Pekka Kujamäki: Deutsche Stimmen der *Sieben Brüder*. Ideologie, Poetik und Funktionen literarischer Übersetzung. 1998.

Band 19 Jürgen Gercken: Kultur, Sprache und Text als Aspekte von Original und Übersetzung. Theoretische Grundlagen und Exemplifizierung eines Vergleichs kulturspezifischer Textinhalte. 1999.

Band 20 Christer Påhlsson: Decision-Making and Limited Resources. Studies in Interlanguage Vocabulary Space of Learners of English for Special Purposes. 1999.

Band 21 Urpo Kovala: Anchorages of Meaning. The Consequences of Contextualist Approaches to Literary Meaning Production. 2001.

Band 22 Eckhard Höfner / Hartmut Schröder / Roland Wittmann (Hrsg.): Valami más. Beiträge des Finnisch-Ungarischen Kultursemiotischen Symposiums *Zeichenhafte Aspekte der Veränderung* (25. – 28.11.1998 Berlin – Frankfurt (Oder) – Slubice). 2002

Band 23 Ewald Reuter / Marja-Leena Piitulainen (Hrsg.): Internationale Wirtschaftskommunikation auf Deutsch. Die deutsche Sprache im Handel zwischen den nordischen und den deutschsprachigen Ländern. 2003.

Band 24 Irmeli Helin (Hrsg.): Dialektübersetzung und Dialekte in Multimedia. 2004.

Band 25 Irmeli Helin: ... *so der Wetterbericht*. Evidentialität und Redewiedergabe in deutschen und finnischen Medientexten und Übersetzungen. 2004.

Band 26 Ingrid Simonnæs: Verstehensprobleme bei Fachtexten. Zu Begriffssystemen und Paraphrasen als Visualisierungs- bzw. Verbalisierungsinstrumente in der Kommunikation zwischen Fachmann und Laien. Eine Untersuchung anhand gerichtlicher Entscheidungen. 2005.

Band 27 Heikki J. Koskinen / Sami Pihlström / Risto Vilkko (eds.): Science – A Challenge to Philosophy? 2006.